The Turn of Moore's Law from Space to Time

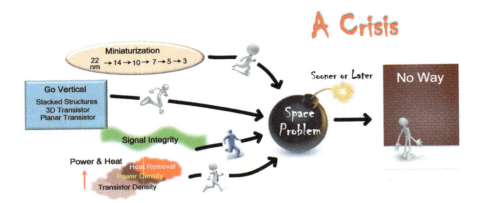
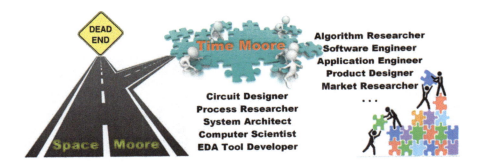

Liming Xiu

The Turn of Moore's Law from Space to Time

The Crisis, The Perspective and The Strategy

Liming Xiu
Plano, TX, USA

ISBN 978-981-16-9067-9   ISBN 978-981-16-9065-5 (eBook)
https://doi.org/10.1007/978-981-16-9065-5

© The Editor(s) (if applicable) and The Author(s), under exclusive license to Springer Nature Singapore Pte Ltd. 2022
This work is subject to copyright. All rights are solely and exclusively licensed by the Publisher, whether the whole or part of the material is concerned, specifically the rights of translation, reprinting, reuse of illustrations, recitation, broadcasting, reproduction on microfilms or in any other physical way, and transmission or information storage and retrieval, electronic adaptation, computer software, or by similar or dissimilar methodology now known or hereafter developed.
The use of general descriptive names, registered names, trademarks, service marks, etc. in this publication does not imply, even in the absence of a specific statement, that such names are exempt from the relevant protective laws and regulations and therefore free for general use.
The publisher, the authors and the editors are safe to assume that the advice and information in this book are believed to be true and accurate at the date of publication. Neither the publisher nor the authors or the editors give a warranty, expressed or implied, with respect to the material contained herein or for any errors or omissions that may have been made. The publisher remains neutral with regard to jurisdictional claims in published maps and institutional affiliations.

This Springer imprint is published by the registered company Springer Nature Singapore Pte Ltd.
The registered company address is: 152 Beach Road, #21-01/04 Gateway East, Singapore 189721, Singapore

To My Family:
Zhihong You, Katherine Xiu and Helen Xiu

Preface

Motivation for This Book

The first motivation for this book is the debate on *"The End of Moore's Law"*. The discussion around this topic is so wide-reaching that countless people from various fields, including many laymen, have spontaneously joined the exchange of views. The term Moore's Law is probably the most cited word in conversations whenever people chat about modern technology. It seems like everyone has an opinion on Moore's Law, or is proud of knowing something about it. As a practitioner in the profession of microelectronics for almost three decades, I am of course deeply interested in this topic of paramount significance.

My second motivation is pertaining to the book *"The Structure of Scientific Revolution"* by Thomas S. Kuhn. I have an intuitive feeling that microelectronics, although a practical engineering rather than a branch of pure science, must in its development inherit some of the major characteristics of science. Its progress could bear the same patterns and symptoms that the growth of science has endured, because microelectronics springs from physics, chemistry and mathematics. The terms *paradigm* and *paradigm shift*, which were introduced by Kuhn more than sixty years ago for studying the behavior of pure science, could find their applicability in microelectronics. Those terms have become quite popular in recent years. They are quoted in various technologically related or business-oriented occasions, sometimes rigorously and between times casually. I want to see how they were originally defined by Kuhn, under what circumstance and for what purpose, in the hope that insight acquired from this study could shine some light on this topic related to the life of Moore's Law.

Another drive for this book is my viewing of microelectronics, and the semiconductor industry at large, as something that is alive. I consider the development of this industry an evolving process, an omnipresent characteristic found among all living organisms in nature. The death of something only paves the way for new things; life never stops. The current psyche on Moore's Law can be loosely sketched as "making transistor smaller, packing ever-more transistors into a given space and then digging more computational power out of it". Its focus is on *space*. This more or less brute force approach has worked well for the past seven or so decades. When it is dying, or gradually fading away into background, something new might grow from its ashes. In my belief, the new evolving path must be around a being that we call *time*. Time is the only qualified rival in all aspects, especially in scope-of-influence, to space. *It's time to use time*, a potentially fruitful avenue for the future of Moore's Law.

Throughout my life, I have been fairly interested in history and philosophy. The stories on the burgeoning of our civilization, and the history of science, have always held my attention in my leisure time. By studying the history of the semiconductor industry, it is engrossing to me that the art of microelectronics is full of eventful dramas and colorful individuals. Those stories inspire me to fantasize that, maybe someday, I could be part of this history and leave my mark in the pantheon by making certain contributions to this field, in my own way, large or small. The sense of urgency and sense of responsibility coming out of this historical and philosophical grounding is another motive for this book. I am committed to making my mark in microelectronics, and will try my best, in an honest and decent way.

The last but not the least incentive for writing this book is probably my curiosity. As originally trained as a physicist and later becoming an electronic professional, I feel as if I have both a scientist's view and an engineer's view at the same time on microelectronics. The exercise in circuit design gives me a close-up view on how a sophisticated chip is developed from some very basic structures, like the experience you get from holding a magnifying glass to inspect the surface of the Earth. Meanwhile, the physicist's mindset allows me to explore the essence of the microelectronics as a whole, the kind of bird's-eye view you can only get by looking at the surface of the Earth from a spacecraft. Those two views are complementary rather than competitive, and the possession of both views at the same time arouses my curiosity.

The Electronic World, and The Moore's Law, as I See It

The Gaia hypothesis formulated in the 1970s likens the Earth to a self-regulating living organism. While not implying that the natural world is actually animate, it does entail complex and connected interactions between life and the environment. Similarly, the electronic world can be inspected using the Gaia principle. Our life in modern society is lived all around electronic devices. We oftentimes subconsciously isolate ourselves from the secularly real world by living in the virtual actuality of the electronic world. There, information is the life itself. The world of electronics is a self-regulating living organism wherein the digital bits, as the vital form of life, are generated, propagated and eventually consumed (equivalent to death). New information (bits) emerges from the ashes of old ones. Billions of years ago, chemicals randomly organized themselves into a self-replicating molecule. This sparking of life was the seed of every living thing we see today. That simplest life form, through the processes of mutation and natural selection, has been shaped into every living species on this planet. In this marvelous process, the mechanism of evolution has demonstrated its majestic power. In like manner, the growth of microelectronics from a basic cell of a transistor to current status can also be investigated through the lens of evolution.

As intricate as it seems, the electronic world is actually created from only two meaningful entities, namely the transistor and the electrical signal. If all the magics performed inside the electronic world were encapsulated as a magnificent show, there would be only two actors—the transistor and the signal. A transistor is a physically existing being that we can see and touch, assuming that we have a pair of very sharp eyes and an extremely small hand. An electrical signal is, however, something that we can only notice its existence through scientific instruments since it is beyond the reach of our five senses, like a ghost. If transistors were houses in a nicely planned community, electrical signals would be ghosts who visit the houses according to a predetermined schedule. Everywhere they go, the ghosts leave their traces. A Halloween show of such kind is what we could see from outside the chip.

By abstracting the complex electronic world into those imageries, I am able to gain a sense of comfortability while working on the various technical issues in my circuit design activities. More significantly, this type of comprehension helps ease the timidity when Moore's Law is on my radar of investigation. This cognition and personalization of the electronic world boosted my confidence when this formidable task of challenging the conventional wisdom on Moore's Law first emerged in my mind years ago. The use of the Gaia postulation, the viewing of microelectronics as an evolution process, and the metaphor of a show and its two actors, I realized, are the insights that I could use to scrutinize the microelectronics and Moore's Law.

Starting from the grown-junction transistor invented in the 1940s, the electronic system gradually grew from simple devices made of a few rudimentary transistors to monster systems with billions of sophisticated transistors on board. Perhaps someday in the future, human-like intelligence might even arise from those bits in this electronic Gaia world. In this short period of seven or so decades, compared to the billions of years of biological evolution, the structure of the basic cell (i.e., the transistor) has mutated multiple times and the surviving strategy is the same principle of "adapting to the environment" rooted in nature. In the biological world, the environment forces species to evolve along the direction of using resource in a fashion of ever-increasing efficiency, becoming faster, stronger and smarter generations after generations on a given amount of consumed resource. In microelectronics, environmental pressurization comes from the never-ending competition of using less power and smaller silicon area for processing the same amount, or more, information; echoing the *Principle of Least Action*. "*Nature is thrifty in all its actions*" is the notion that is at the core of physics and mathematics. This statement captures the soul of the Principle of Least Action. The world made of transistors shall not be an exception. I believe that if it ever wants itself to be any useful in guiding people's practice, Moore's Law must be in compliance with this basic law of Nature. In conventional wisdom, "making transistor smaller and more efficient, packing more of them into a given space and then getting more computational power out of it" are the solo drive expressed in Moore's Law. This is an attempt concentrating only on the space. When the Principle of Least

Action exhibits its power in Nature, it makes use of both space and time, with equal respect, as countless cases in physics and mathematics have demonstrated. Therefore, there is something missing in the current exercise of Moore's Law. This missed whatsit is the being that we name it as time, more precisely, the handling of the flow-of-time inside the electronic world.

A key subject in both evolutions, one occurring in nature and the other in the electronic world, is the management of flow-of-time. Lived time in the biological world is sustained by a circadian clock that is the master regulator of mammalian physiology. It is developed inside beings to keep life running in an orderly fashion. This orderly fashion, however, does not necessarily mean "fly-of-time" in a steady pace. As a matter of fact, most circadian clocks found in Nature exhibit a characteristic of nonuniform-flow-of-time. This nonuniform-flow-of-time has resulted from millions of years of evolution. It is efficient when the "activity/energy-consumed ratio" is used as the justification. Its high competence is proven by the very fact that it survives the relentless competition of natural selection during the enduring evolution. In the art of microelectronics, we are currently following a doctrine of uniform-flow-of-time. I believe that, by adopting an ideology of nonuniform-flow-of-time, the field of microelectronics can advance further towards the goal of higher information processing proficiency. This change will make the electronic world more resemble the biological world, which has endured the test of millions of years of natural selection. Learning from Nature, could nonuniform-flow-of-time be a path for the continuation of Moore's Law?

At this stage, we already have the hindsight of more than seven decades of microelectronics development. We have been always using electrical charge as the medium for embodying information. Putting it in another way, we are using *"volume of matter"* to convey message, since the amount of electrical charge is proportional to the volume of matter. In this ideology, the quantity of matter is used to differentiate connotation—*"Principle of Proportionality"* reflected in microelectronics. A different measure of electrical charge (voltage or current) corresponds to a unique meaning or message. This creed has worked well for us. However, matter occupies space. This leads to the inference that more information needs more matter and more space. Can we use time to directly convey message? Instead of using time merely as an indexing gadget in the v-t waveform (which is a visual tool for expressing all electrical signals), can we switch the role of v and t and use t for message and v for assistance? Using "switching rate of certain medium" to express our thought (i.e., message) is not unfamiliar to us, as many such examples found in life have demonstrated. Can this idea burgeon to a grand scale? A big bonus from this scheme is the unlimited resource—there is no end to time. In another front, as technology advancing, the transistor is switching faster and faster generation after generation. This is certainly a piece of good news when the resolution for signal processing is concerned. Eventually, the resolution in time is only limited by the *Heisenberg limit*, which leaves us with plenty of room to play for now. Using rate-of-switching as a means for expressing our thought, could this be another path for the continuation of Moore's Law?

The true connotation behind "The Turn of Moore's Law from Space to Time" is a directional shift-of-focus in microelectronics. It is a big thing, a really big thing. After all, the world is sitting on both space and time. The art of microelectronics cannot treat time as a second-class citizen forever. This is not fair.

My Regrets

A retrospect of my career reveals several regrets that I feel from time to time. The first one is that I missed the opportunity to earn a Ph.D. I was shortsighted at a young age, not able to see further into the future, a utilitarian young man who was eager to earn big money quickly. I am regretful about the actuality that I did not try my best to get into one of the three top schools in electrical engineering—MIT, Stanford or UC Berkeley. I missed the opportunity to learn from the masterminds in this field when I was young and, later, the chance to build a professional network around those influential people in this circle. My work on Moore's Law could have been started earlier if I had that help, and my career could have been looked like much more productive and my resume more impressive, I suppose. On the other hand, missing the Ph.D. opportunity might not necessarily be all negative. If I had indeed followed some big shots and worked on my Ph.D., my career could have been directed down to a different path, my mind could have been constrained, and I would have missed the opportunity to develop my own perspective on Moore's Law. Life is like a box of chocolate, as legend has it.

The second regret is related to my switching from science to engineering. I was originally trained in physics, a branch of science that focuses on the discovery-of-secret in Nature and the theorization of the phenomena found in Nature. Theoretically speaking, I was deprived of the hope for a Nobel Prize at the very moment when I switched to electrical engineering. In principle, that award is only given to science-related work. The work on Moore's Law, a topic having a wide range of influences and paramount financial impact, comforts me somehow in this regard. Switching from physics to engineering in the middle of the course also had another side effect. My craftsmanship as an engineer is not as well polished as those trained from the scratch in top engineering schools. I was able to make up for those disadvantages through my capability of developing deep insight into every problem that I worked on. This capability stems from my training as a physicist. Those insights have helped me read Moore's Law, or the microelectronics at large, from a unique perspective that others have not employed before.

The annoying regret that I often encounter is the lacking of resources to test my ideas. I wish that I had more control on capital so that many of my plans on solving various problems could be tested and be put into use for the good of the whole field. I want to use an example to illustrate this point. There are hundreds of semiconductor fabrication plants around the world. Every day, countless chips are shipped out of the door from those giant shops equipped with sophisticated and expensive equipment. At the very bottom, those chips are geometric patterns. Trillions of those patterns are configured each day to create chips that are subsequently used to consume huge amounts of energy for processing data, to serve people's appetite for information. It is always a pity for me to feel that something can be improved in this process, for the better use of such a large amount of money. What if those geometric patterns could be utilized just a little bit wisely by us, by paying a little bit more attention to the being that we call time? Those geometric patterns are formed in the regime of space, like a picture. Can we make the picture more vivid by adding some colors to it? This metaphor of color is the utilization of the dimension of time. Due to the enormous amount of those geometric elements, a small refinement on their usage (i.e., the better treatment on flow-of-time) can make a huge difference. It can surely help make the Earth a little bit greener. This is the biggest regret that I come across so far.

My Thankfulness

With those regrets hidden at the depth of my heart, I am still thankful for all the things that life has offered me so far. Mr. Deng, Xiaoping's reform that began circa the late 1970s in the land of China, which transformed China from a highly ideological society governed by Marxism and *"Class Struggle and Hatred Education"* to a pragmatical one aiming at the improvement of people's living standards, provided me with the opportunity of formal higher education. I was one of the luckiest few, among the millions of youngsters of my generation, to be admitted into Tsinghua University. Tsinghua University in the 1980s was known for its uplifting vitality, in contrast to the spiritual and philosophical void, seen nowadays. A generation of gifted and vigorous young men and women studied relentlessly day and night for the gain of knowledge. At that place, I was rigorously trained in the learning of physics and mathematics, among other good things.

Later, owing to Mr. Deng's "Open Door Policy", I was fortunate enough to secure a chance of studying in the United States of America through the help of my former professor, Dr. Li, Mingyang. I am so thankful to Dr. Li and his wife Ms. Aixinjuelo, Yulin, who helped me greatly in the first few years while I was gradually merging into this new country. After three years as a visiting scholar on accelerator physics, I was admitted to the Texas A&M University, college station, Texas. Texas A&M certainly has one of the top-quality electrical engineering schools in the world. While studying there, I switched from physics to engineering, which later led me to microelectronics and Moore's Law. Texas A&M is a word that always ignites a piece of warm feeling in my heart when I hear it.

As a professional in the semiconductor business, the first company that I worked for is Texas Instruments (TI), in Dallas, Texas. As one of the few pioneers in the semiconductor industry that survives up to this day, TI is a company that is full of history and seasoned engineers. I learned plenty of things in this great place. While working there, I became an all-rounder circuit designer who can do a little bit of everything, from mixed-signal circuit design to digital circuit design, to VLSI physical design and even to EDA tool development. This broad experience, augmented by the training in physics and mathematics, established the foundation for my comprehension on microelectronics at large. It also prepared me for the task of challenging the conventional wisdom on Moore's Law at a later stage. A consequential event to me while I worked for TI is the invention of Flying-Adder frequency synthesis architecture. This is a circuit technique aiming at the generation of clock frequency, which is a topic closely related to the flow-of-time in electronic devices. This work drew my attention forever to the subject of "time". Those two factors have played a significant part in my later investigation of Moore's Law. For that, I am deeply thankful for TI.

Novatek Microelectronics, Taiwan, is another company that I feel appreciative of. During my two years working there, I gained some first-hand sensibleness on Taiwan's semiconductor industry by roaming around the Hsinchu Science Industrial Park. It gave me the chance to watch in-person the diligent and hardworking people in Eastern Asia striving to make their homeland prosperous. The blend of a Western vantage point and an Eastern perspective on observing this industry has sharpened my discernment. An emotional-inspiring incident has surprised me in an interesting way. It is the first technical problem I encountered after joining the company. This problem of water-wave noise is the adventitious appearance of some annoying patterns on pixel-based display used for TV. People in there have been troubled by this problem for quite some time (several years as I was told). They presented this problem to me, more or less I guess, as a test for the capability of a newcomer. They did not know, or perhaps did not care, that this problem was actually not in my area of expertise. This is a system-level issue and I was a circuit-level designer. In any established company in the Western world, this problem would have been formally assigned to the appropriate people with the right skillset. Knowing no alternative, I assembled a full package of my skill and knowledge in physics, mathematics, circuit design and EDA tools and took the challenge. After a few days of examination, in a fleeting moment of insight, I was able to see my way to the root of the problem. I was impressed by myself. The shocking sensation growing out of this occurrence is that I was starting to realize how well I am already prepared for unexpected challenges in an unfamiliar environment. The rigorous training in physics and mathematics, the battle-hardened experience accumulated in TI and the vision established through extensive literature study have made me ready for something. This self-confidence certainly helps when, eventually, Moore's Law came onto my radar.

BOE Technology Group, Beijing, China, is the place where my thought on the already half-baked idea around Moore's Law gradually firmed up. BOE is a system company where chip design is not on its focus, which seems ill-suited for me. It turned out that I could still benefit from working there by observing the microelectronics business from another angle. I am thankful for the fact that the company did not put any constraints on what I do. This style reminds me of the Bell Labs of the early-twentieth-century, where creative individuals were respected and not bothered by management frequently. Although this company is relatively young in this business, its leadership team is ambitious. This is reflected on the company logo—BOE for

"Best On Earth". The team realizes that there is no shortcut to excellence. It keeps investing a significant chunk of revenue in R&D each year for the ultimate survival in the ruthless competition.

One particular individual I am indebted to is Mr. Wang, Dongsheng, former Chairman of BOE. Through a few exchanges with him, I was motivated to finally make a real move on this issue related to Moore's Law. Mr. Wang, as the spiritual leader of this company, has consistently placed great emphasis on the importance of R&D. As I understood, he is eager to see work in the grade of Nobel Prize be developed from this company. This work on Moore's Law is not in the regime of science and therefore his aspiration is still not fulfilled. But, at least in my opinion, the scope-of-influence of this shift-of-focus from space to time is at the same level of that grade. The few interactions with him have made me appreciate this motto a little bit more—sparks of creation often happen when exposed to a great diversity of ideas, especially in a way of spontaneous exchange among visionary people.

Finally, I am thankful for my family's support. During the period that I worked in Asia, my two daughters have shown exceptional self-discipline. After graduating from Yale with a degree in mathematics, my older daughter is now studying in the best law school, Yale law, en route to becoming a lawyer in New York. After earning a BA degree from Washington University in St. Louis, my younger daughter is now an assistant director of graphic design in a well-known firm in New York, pursuing her dream in art. I am very proud of them and believe that they can surpass their first-generation-immigrant parents in all aspects of life and make contributions to this great country for its prosperity. The most special person that I am deeply indebted to is my wife, You, Zhihong. She is a beautiful woman both outside and inside. She is extremely intelligent, a first-tier student in a top engineering school of China—Tsinghua University, and an excellent and productive analog circuit designer in TI. Moreover, she is elegant and graceful, having the typical characteristic of a "Beijing Girl". The trait I admire the most about her, however, is that she never compels me to become a billionaire, a status that many of our Tsinghua alumni have achieved. This leaves me with plenty of room for my daydreaming, including this one on Moore's Law. I am lucky in this regard.

A Long-Range Strategy

In the fields of science or engineering, there are cases where the problem in hand is of an extremely intellectually demanding nature and its solution requires a feat of great ingenuity. However, rather than inspiring new research, its resolution kills off much of the work in the field, leaving others with little to do in the way of follow-through. In a retrospective of science history, the good advances are not those that solve problems and thereby close off an area of research, but rather those that open up a whole new range of problems to explore. The topic discussed in this book is the latter kind. The shift-of-focus from space to time is the prologue of another big drama in microelectronics. It would forge openings into other areas that are waiting to be explored.

In the future, the prime advance in this new territory may well be made by others, rather than me. It would have to overcome lots of resistance before the perspective and strategy proposed in this book are widely embraced and adopted. Such resistance, as a guarded response, can sometimes serve as a helpful precaution, but it can oftentimes hold back progress in a field. That is why a new thing, be it a theory or a concept or a method, is seldom just an increment to what is already known. Its assimilation requires the reconstruction of prior theories or concepts, the reevaluation of prior facts and the modification or perhaps even the abandonment of known methods. This is an intrinsically revolutionary process that is, as history has taught us many times, never accomplished without a serious fight. I suppose that is the way with all new ideologies, especially this dramatically different one, are brought to the fore. For now, I intend on trying to get the ball rolling in this direction.

Working on this book is a source of satisfaction to me—personally, spiritually and psychically, as well as a source of frustration sometimes. To me, this is a passion rather than a drudgery. With the benefit of hindsight, the most important trait about me that has helped me finish this book is my concentration. There are many distractions in life, and each one of them can easily destroy your focus if you do not have a tough mind. I am grateful that my mental toughness has helped me endure all the ordeals and difficulties. I have not disappointed myself. One person, endowed with a bit of talent, drive and luck, can make a difference.

Given more time, I can surely make this book better, especially on the engineering detail of some of the technical issues discussed in Chapter six. But I can't wait any longer.

Plano, TX, USA
2022

Liming Xiu

Contents

1 **Preamble** 1
   References 3

2 **Microelectronics as a Normal Engineering Practice for Supporting an Industry** 5
   2.1 Major Milestones in Semiconductor Industry 5
   2.2 Innovative Business Models for a Dynamic Industry 8
   2.3 The Role of Creative Minds: Leadership Versus Command-And-Control 10
   2.4 The Road to Normal Engineering: From Science to Technology 12
   2.5 After the Formation of a Paradigm: Puzzle-Solving Engineering 15
   2.6 A Missing Attribute of the Puzzle 18
   References 20

3 **The Space Crisis in Current Paradigm** 23
   3.1 Perception of Space, Time, Change and Motion 23
   3.2 A Magnificent Show of Only Two Actors: Transistor and Signal 26
   3.3 Space and Time as Real Estate in Microelectronics Business ... 28
   3.4 Clock: The Most Important Signal and Its Three Functions 30
   3.5 Pursuit of Computation Efficiency from the Perspective of Space: Moore's Law 31
   3.6 The Recognition of a Crisis 34
   References 35

4 **Response to the Space Crisis** 37
   4.1 Time and Clock: Recourse to Philosophy and Debate over Fundamental 37
   4.2 Three Types of Time: Mechanical, Secular and Electronic 39

4.3	Microelectronics as an Evolution Process and the Emergence of Time Within	41
4.4	Connecting Space and Time by a Property of Clock Signal: Frequency	44
4.5	Clock Frequency: The Method of Indirect Period Multiplication (PLL) and Its Ineptitude to Solve Two Long-Standing Problems	46
4.6	Clock Frequency: A New Concept and a New Approach of Direct Period Synthesis for Meeting the Challenge	49
4.7	The Response: Time-Oriented Paradigm	57
4.8	Left and Right: Space and Time as Two Arms for Lifting the Weight of Signal Processing	61
4.9	Top and Bottom: From System to Circuit and from Circuit to System	62
4.10	The Good News: It Is Free, or Almost Free	64
References		65

5 Time Moore Strategy: Bridging the Old and New Paradigms 69

5.1	The Essence of Moore's Law and the Principle of Least Action	69
5.2	The Importance of Clock Technology: From Application Point of View	71
5.3	The Deleterious Consequence of the Two Long-Standing Problems	72
5.4	Whenever and Wherever: It's Time to Use Time	75
5.5	Time Moore Strategy as the Bridge	79
References		82

6 Old World and New Insight: Solving Problem with a Gestalt Switch 85

6.1	Recapitulation: Data Processing Investigated from Two Perspectives		85
6.2	A Novel Frequency-Tunable Clock Source: TAF-DPS DCXO		89
	6.2.1	Review on Tunable Frequency Source	89
	6.2.2	Architecture of TAF-DPS Based Frequency Tuning	91
	6.2.3	TAF-DPS DCXO Demonstration on FPGA	92
	6.2.4	In Comparison to Commercial Products	95
	6.2.5	Application Example #1: Correcting Frequency Error On-The-Fly	96
	6.2.6	Application Example #2: Counteracting Aging-Induced Frequency Drift	98
	6.2.7	Application Example #3: On-Chip Emulation of Si570 Function	99
	6.2.8	Application Example #4: A DCXO Module for Clock Synchronization in MPEG2 Transport System	101

	6.2.9	Application Example #5: A Method of GPS Disciplined Clock for Improving Frequency Accuracy and Steering Frequency	102
6.3		On-Chip Syntonistor for Assisting Time Synchronization in Network	105
	6.3.1	Notion of Time	107
	6.3.2	Frequency, Phase and Time Synchronization	109
	6.3.3	Establishing a Common Notion of Time	110
	6.3.4	Applications of Packet-Switched Synchronization	111
	6.3.5	Clock Synchronization in Network Standards	112
	6.3.6	Key Components in Clock Synchronization Algorithms	113
	6.3.7	Motivation for On-Chip Syntonistor: Eliminate Frequency Offset and Drift from a Hardware Perspective	115
	6.3.8	Suitability of Frequency Sources as On-Chip Syntonistor	118
	6.3.9	TAF-DPS as On-Chip Syntonistor for Assisting Time Synchronization	118
	6.3.10	A Method of Using TAF-DPS Syntonistor for Tuning Physical Clock	121
6.4		Spread Spectrum Clock Generation: Always-On Boundary Spread SSCG	131
	6.4.1	The Architecture of Always-On Boundary Spread (AOBS)	134
	6.4.2	Experimental Validation of AOBS SSCG	142
	6.4.3	Applying AOBS on Real System: Always-On	146
6.5		On-Chip Chirp Signal Generator	149
6.6		Frequency Lock on Time-Average-Frequency: All Digital TAF-FLL	155
	6.6.1	TAF-FLL Architectures	157
	6.6.2	TAF-FLL Experimental Verification	162
6.7		Random Number Generator Using Interplay of Frequency Sources as Entropy	171
	6.7.1	Frequency-Mixing TAF-DPS TRNG	173
	6.7.2	Frequency-Tracking TAF-DPS TRNG	180
	6.7.3	Some Open Issues for Further Investigation	182
6.8		Inscribing Temporal Encryption on Spatial MPV Imprints for PUF	183
	6.8.1	Brief Review on the Characteristics of TAF-DPS Operation	187
	6.8.2	TeS-PUF Architecture and Its Circuit	192
	6.8.3	Proof-Of-Concept on FPGA	198
	6.8.4	Directions for Future Research	202
6.9		TAF-DPS as a PWM Signal Generator	204

6.10	A Case of Signal Processing in Time Dimension: Deterministic Stochastic Computing	207
	6.10.1 The Scheme of Using TAF-DPS for Stochastic Operation	211
	6.10.2 Conversion from Binary Number to Desired Pulse Train	212
	6.10.3 Deterministic Bitstream and Stochastic Operation	214
6.11	Reducing FIFO Memory Size and Smoothing Data Flow	220
6.12	Clock Data Recovery by Explicitly Using Time-Average-Frequency	225
6.13	TAF-DPS for Dynamic Frequency Scaling	232
6.14	Flexible Clocking for Resource-Constrained Environment in Edge and IoT	234
6.15	TAF-DPS Generator in Large SoC for Producing Clock Frequencies	239
6.16	TAF-DPS for the Challenge of Global Clock Distribution	243
6.17	Frequency Issue in Clock Distribution Through Resonant Clock	248
6.18	TAF-DPS for Network-On-Chip GALS Strategy	253
6.19	One-Wire Communication Using Edge-For-Clock-Duty-Cycle-For-Data	258
6.20	Spectrum Pattern as Message for Communication	262
6.21	Conversion Between Time Span and Digital Value for Sensing and Actuating	264
6.22	Connection-Adaptive Dynamic-Frequency Clocking for Sparse CNN	270
6.23	Latency Awareness Computer: Frequency Scalable ISA and Microarchitecture	274
6.24	The Missing Piece in FPGA and Reconfigurable SoC: Programmable Frequency Source	279
6.25	TAF-DPS Clock Source as a Tool in Design for Manufacture	282
6.26	Digital-To-Frequency Converter and Frequency as a Variable in Programming	284
References		287
7	**Forefront of the New Paradigm: Using Time to Encode Information**	**299**
7.1	The State of Extraordinary Research and Development Activities	299
7.2	Computation Variable and Data Representation in Electronic System	301
7.3	Multi-value Rate-of-Switching as a Method for Signal Processing	303
7.4	The First Three Challenges	308
References		310

8 Epilogue: It's Time for the Big Ship of Moore's Law to Make a Turn ... 313
- 8.1 The Continuation of Evolution but Along a New Path ... 313
- 8.2 The Full Picture of Time Moore: Clock Moore and Rate Moore ... 315
- 8.3 No Science and Engineering Can Go Forever Unchallenged ... 318
- 8.4 Attitude Toward Paradigm Shift: Change-of-Mindset is Extremely Difficult ... 320
- Reference ... 323

Abbreviations

AOBS	Always-On Boundary Spread
ASIC	Application Specific Integrated Circuit
BJT	Bipolar Junction Transistor
CDR	Clock Data Recovery
CMOS	Complementary Metal-Oxide-Semiconductor
CNN	Convolutional Neural Network
DCO	Digital-Controlled Oscillator
DCXO	Digital-Controlled Crystal Oscillator
DDS	Direct Digital Synthesis
DFM	Design For Manufacturing
DFS	Dynamic Frequency Scaling
DPS	Direct Period Synthesis
DRC	Design Rule Checking
DSP	Digital Signal Processing
DTC	Digital-to-Time Converter
DVFS	Dynamic Voltage and Frequency Scaling
ECDD	Edge-for-Clock-Duty-Cycle-for-Data
EDA	Electronic Design Automation
EMI	Electromagnetic Interference
ESD	Electrostatic Discharge
FLL	Frequency-Locked Loop
FPFG	Field Programmable Frequency Generator
FPGA	Field Programmable Gate Array
GPS	Global Positioning System
IC	Integrated Circuit
IDM	Integrated Device Manufacturer
IoT	Internet of Things
IP	Intellectual Property
ISA	Instruction Set Architecture
ITRS	International Technology Roadmap for Semiconductors
MEMS	Micro-electromechanical Systems

MORO	Multiple Output Ring Oscillator
MOS	Metal-Oxide-Semiconductor transistor
MOSFET	Metal-Oxide-Semiconductor Field-Effect Transistor
MVRoS	Multi-Value-Rate-of-Switching
NEMS	Nanoelectromechanical Systems
PLD	Programmable Logic Device
PLL	Phase Locked Loop
PUF	Physical Unclonable Function
PWM	Pulse Width Modulation
RO	Ring Oscillator
RTL	Resistor-Transistor Logic
RX	Receiver
SoC	System-on-Chip
SSCG	Spread Spectrum Clock Generation
TAF	Time-Average-Frequency
TAF-CDR	Time-Average-Frequency Clock Data Recovery
TAF-DPS	Time-Average-Frequency Direct Period Synthesis
TAF-FLL	Time-Average-Frequency Frequency-Locked Loop
TDC	Time-to-Digital Converter
TRNG	True Random Number Generator
TX	Transmitter
VCO	Voltage-Controlled Oscillator
VCXO	Voltage-Controlled Crystal Oscillator

Chapter 1
Preamble

Microelectronics is the engineering practice of using electronic devices to collect, encode, process and eventually consume information. It is not a science that aims to discover the secrets of nature, but rather, a field of study that promotes the invention of apparatuses to facilitate a better life for mankind. At the core of this field is the integrated circuit, which is created from transistors. The industrialization of microelectronic engineering—which is generally referred to as semiconductor industry (sometimes also being called electronic industry)—has experienced tremendous growth in the past six or seven decades, and this growth has been largely guided by the so-called Moore's Law. Traditionally, on a given parcel of silicon real estate, Moore's Law has exploited it relentlessly for information-processing power and efficiency, mainly from the direction of space. As the key geometry of transistors approaches the size of several atoms, there is ever less opportunity to advance further in this dimension. Moreover, the heat generated from the space of densely-packed-transistors becomes harder to be transferred out. Furthermore, the signal integrity problem is getting harsher from generation to generation due to the reasons that signals are physically closer and travel relatively longer distance and, to make matter even worse, bear higher frequency. As a result, a space-related crisis gradually emerges and the end-of-life of Moore's law is slowly but surely becoming inescapable, as classically conceived. As such, the approaching towards this extremity presents an obstacle to the rapid technological advancement that has characterized this long and eventful period.

Unlike all other branches of natural science and practical engineering, the art of microelectronic design is unique in one aspect: it creates an electronic world of transistors wherein the sense-of-time has to be invented. This burden of handling flow-of-time is singular to the microelectronic engineering that no other science and engineering discipline needs to bear (except philosophy and psychology, and the pure science of physics that studies the very concept of time in great depth). Viewed from a different angle, however, this challenge of "fabricating time" presents a precious opportunity to electronic professional for creating something extraordinary. Such godlike responsibility of "creating time" opens a new possibility in microelectronics for us to continue the Moore's Law's pursuit of higher computing power and efficiency. Or, put it in another way, it allows us to circumvent this space-related crisis through a different route. This effort requires a much deeper investigation on the thing we called time and, more significantly, perhaps a change in the structure of the clock signal that reflects the flow of time. It leads to a new perception of time inside the electronic world: the nonuniform flow of time. Continuing to push Moore's Law forward from this new perspective of time, an approach hereinafter referred to as the Time-Moore Strategy, would necessitate a fundamental paradigm shift in regards to the practice of microelectronic engineering.

The objective of this book is to promote such a change-of-mindset in the perception of time and subsequently reevaluate the current practices of microelectronic engineering. The Time-Moore strategy is first introduced in the technical article of "Time Moore: Exploiting Moore's Law from The Perspective of Time", published in 2019 (Xiu 2019), which is the very first call for the turn of the Moore's Law from space to time. It is pointed out in that article that the Time-Moore strategy represents an answer to the current space crisis and it can help us in the continued advancement of modern technologies through the support from the most fundamental level. In that work, however, the discussion only narrowly focuses on the technical aspect of its implementation. In this book, the Time-Moore strategy is put into a much larger perspective and is elucidated against the background of entire microelectronic engineering.

History has repeatedly shown that crisis challenges conventional methodologies and stimulates progress, as elegantly and powerfully argued by Thomas S. Kuhn in "The Structure of Scientific Revolution" (Kuhn 2012). In light of this, the presentation style of the first half of this book is tilted slightly toward philosophical and methodological, by persuasion rather than proof, for the reason that change-of-mindset is extremely difficult and it requires effort much more than genuine technical argument. In this book's organization, the history of microelectronics is first briefly reviewed, to set up the necessary background for subsequent discussions. It is then argued that microelectronics has now become, using Kuhn's term, more or less a normal puzzle-solving engineering after the field transformed from initial phase of scientific discoveries to later stage of technological inventions. The imminent space crisis, which will be defined and justified in Chap. 3, is identified as the uncrushable obstacle if we keep going in this direction after seven or so decades of rapid growth. In Chap. 4, a deep investigation around the concept of time is first carried out, which smoothly leads to the new time-oriented paradigm. After that, an opportunity naturally emerges as a way to circumvent this obstacle. From the background set in Chaps. 2, 3 and 4 are more serious discussions on space and time, from a scientific perspective as well as from historical and philosophical ones. The discussion is however not the kind of general discussion on space and time, which is surely a much bigger task than this book can handle. The discussion is entirely focused on only serving the microelectronic engineering. Due to the historical and philosophical nature of this discussion, it is suggested that readers keep an open mind and use patience with those two chapters.

The second half of the book is more pragmatic. In Chap. 5, the Time Moore strategy is proposed as the bridge connecting the old space-dominant paradigm and a new time-oriented paradigm. In Chap. 6, the change on the perception of time is implemented as a workable tool, within the frame of the new paradigm, in dealing with various kinds of old and new problems. The innovation power of this new school of engineering is demonstrated through examples against the background of the emerging paradigm. Moreover, in Chap. 7, a "unorthodox" idea of using time directly as a means to encode information is described. The goal is, as always, to improve the information processing power and efficiency. It however starts the action from a much more fundamental level of defining a new computation variable. After that, the need for a new set of arithmetic logics and basic circuits is discussed.

In the end, a conclusion is made at Chap. 8 that Time Moore strategy can continue the Moore's Law's pursuit of higher computing power and efficiency for some years and decades to come. It's time for the big ship of Moore's Law to make a turn, steering into a widely open but uncharted water for exploring.

References

T. S. Kuhn, "The Structure of Scientific Revolution," the *Univ. of Chicago Press*, 2012.

Xiu, L., "Time Moore: Exploiting Moore's Law from Time Perspective", *IEEE Solid-State Circuits Magazine*, vol. 11, no.1, pp. 39–55, 2019.

Chapter 2
Microelectronics as a Normal Engineering Practice for Supporting an Industry

2.1 Major Milestones in Semiconductor Industry

The history of semiconductor is long and intricate. It is impossible, and also unnecessary, for all those great stories to be replayed in this book. The brief review given in this chapter is solely for the purpose of serving the main theme of this book. Given this limitation, only the most important facts will be presented and the justification can never be fully impartial. It is therefore apologized in advance if readers find this short review of semiconductor industry incomplete or inaccurate.

Roughly speaking, the field of electronics is born when vacuum tube was invented by John A. Fleming in 1904. Gradually, vacuum tube was replaced by smaller-size and less-power-hungry transistor, which was first patented by Bell Labs in 1947. The operation of transistor is based on semiconductor, hence the semiconductor industry hereinafter. Semiconductor is a generic term referring to any material that conducts electricity imperfectly. By introducing various chemicals into a substrate of such semiconductor materials, it is possible to make a transistor that can switch electric current on or off, which provides the fundamental building block for representing the logic values of one and zero. Microelectronics is a subfield of electronics; it relates to the study and manufacture of electronic designs and components in very small scale (micrometer or smaller). These devices are typically made from silicon, the most widely used semiconductor material. The design and manufacture of such small-scale electronic components, which are typically used inside certain electronic products such as TV set and cell phone, is often called microchip or integrated chip, or simply chip or IC.

The first point-contact transistor (a signal amplifying circuit) was demonstrated in Bell Labs in 1947 (Bardeen and Brattain 1949; Riordan and Hoddeson 1997). Then in 1948, William Shockley, worked in Bell Labs at that time, carried out the fundamental study of carrier injection, which leads to the understanding on the function of point-contact. This work laid the groundwork for the p–n junction theory, the dominant physical mechanism of the *transistor effect* (Shockley 1949; Shockley et al. 1951). It was then given the name bipolar junction transistor (BJT). Hence the era of transistor, and it spelled the doom of vacuum tube. However, the assembly of the early transistors was difficult to automate. A step forward is the creation of npn transistor with mesa ring-dot structure, which enabled the production of a large number of transistors simultaneously. The next key invention, by Jean Hoerni of Fairchild in 1959, was the planar manufacturing process, which allowed multiple transistors to be created and connected together simultaneously. It flattened the transistor and allowed it to be mass-produced (Hoerni 1959). In the same year, Jack Kilby at Texas Instruments and Robert Noyce at Fairchild independently developed the technology of integrated circuit (Kilb 1959; Noyce 1959). It demonstrated that multiple transistors, diodes, resistors and capacitors can be formed on the same piece of silicon substrate. This is a very important step for large scale industrialization. Also critical is the elimination of the flying wires that connect the components electrically together since the wire is the main cause of problem and thus the trouble for high yield. Interestingly, although not being appreciated appropriately, the key development that solved this problem and made the integrated circuit feasible for mass-production was the planar technology invented by Jean Hoerni.

When the early BJT transistor was used to build functional circuits, one must start with system requirements and then design around transistors, rather than simply substituting transistors for vacuum tubes one-for-one. Several design technologies for realizing digital functions have emerged during this period: Transistor-Diode Logic (DTL), Resistor-Transistor logic (RTL) and Transistor-Transistor logic (TTL). They are all constructed around BJT, which acts as the switching device. In later years, bipolar transistor's role as the key player in electronics was gradually replaced by the Metal–Oxide–Semiconductor transistor (MOS). The journey of MOS transistor started also in Bell Labs when, in 1946, William Shockley made a predication of *field effect*: a significant modulation of conductivity of thin layers of semiconductor could be produced by inducing a surface charge through electric field. In 1959, Dawon Kahng and Mohamed M. Atalla at Bell Labs created the first Metal–Oxide–Semiconductor Field-Effect Transistor (MOSFET) (Kahng and Atalla 1960). Operationally and structurally different from the BJT, the MOSFET was made by putting an insulating layer on the surface of the semiconductor and then placing a metallic gate electrode on top of that. It used crystalline silicon for the semiconductor and a thermally oxidized layer of silicon dioxide for the insulator. Additionally, the method of coupling two complementary MOSFETs (p- and n-channel) to make one switch was used to create CMOS (Complementary Metal–Oxide–Semiconductor). CMOS digital circuit dissipates very little power except only when it is actually switched. It then has become the most widely used type of transistor in integrated circuits.

Comparing to BJT, CMOS has the advantages of small size, low power drain, small heat generation, innately long life and high shock/vibration capabilities. However, the most important factor for CMOS IC (integrated circuit) eventually ending the supremacy of bipolar IC is its process scalability. In the 1970's, Robert Dennard and his colleagues in IBM recognized the potential of downsizing MOS devices. A scaling theory was formulated in 1974, which observes that MOS transistor would continue to function while all key figures of merit such as layout density, operating speed, and energy efficiency would improve provided that geometric dimensions,

2.1 Major Milestones in Semiconductor Industry

voltages, and doping concentrations were consistently scaled to maintain the same electric field (constant-field scaling). The three key scaling factors are the thickness of the insulator between gate and underlying silicon, the channel length, and the power-supply voltage. This so-called Dennard's scaling (Dennard et al. 1974; Frank et al. 2001) is the enabling factor behind the well-known Moore's Law (Moore 1965). After three decades of continuous scaling, Dennard's scaling stopped around the middle of 2000s. Extensive research then follows to find alternatives, in different directions such as new materials, new transistor structures, new circuit architectures, and new computing models. Among them, several novel transistor structures were proposed such as SOI (silicon on insulator) MOSFET, Mutigate transistors (double gate, FinFET, surrounding gate). Eventually, FinFET (Fin Field Effect) emerged as the winner for mainstream applications. Instead of having a planar inversion layer, FinFET transistor creates a three-sided silicon fin that the gate wraps around, creating an inversion layer with a much larger surface area (Liu 2012). Nowadays, the channel length of the most advanced processes has reached 3–5 nm range (Xiu 2019). Beyond that, quantum effect will come into play and we have to again search for more options, as discussed in (Xiu 2019). One possibly way is to go vertical, packing more transistors using the 3rd dimensions. Hence, the so-called 2.5D and 3D ICs (Lau 2013).

Besides the development around basic transistor structures, innovations at the higher level of circuit design and implementation also play important role in shaping the industry. One such technology is the Programmable Logic Device (PLD). Unlike logic gates that have fixed functions, PLD devices are essentially blank slates that can be programmed to perform any number of tasks. With PLD, a chip company can realize the economy of scale by manufacturing high volumes of ICs that other system companies can then customize to fit the chip into any number of different products. PLD later became Field Programmable Gate Array (FPGA) that reduces manufacture time and cost by preprinting all the transistors in wafer (in the stage called front-end-of-line or FEOL) and only adding wire later to achieve desired functions. FPGA allows the on-chip devices to be quickly programmed in the field based on an application's requirements. Hence, it dramatically reduces a product's time-to-market (Nenni et al. 2014).

Application specific integrated circuit (ASIC) is another important term in semiconductor history. Literally, it refers to a chip that is custom designed for a specific application rather than for a general-purpose application. More importantly however it creates a new type of companies that specialize in just designing IC, not involved in system design or chip manufacture businesses. Those so-called ASIC companies paved the way for the transition from IDM (Integrated Device Manufacturer) model to the now popular fabless-design-house vs. pure-foundry-manufacturer model. The ASIC model also cultivated the electronic design automation (EDA) industry. It further fostered the System-on-Chip (SoC) design methodology, which integrates all the components of a computer or other electronic system into a single chip. A SoC chip may contain, all on a single silicon substrate, digital, analog, mixed-signal, and radio-frequency functions. A SoC chip can be used for a wide variety of applications such as networking, video and image processing, graphics, motor control, ubiquitous computing, IoT, AR/VR, Big Data, AI and etc. Naturally, SoC design methodology promotes the emergence of semiconductor IP (Intellectual Property) industry, with ARM® as its symbolic business-model icon (Nenni et al. 2014).

In summary, from a technological perspective and with the benefit of hindsight, the invention of transistor in 1947, the invention of planar process technology in 1959 and the invention of monolithic integrated circuit in 1959 are the three most important cornerstones that establishes the very foundation for microelectronics. In next section, from a business perspective, we will briefly discuss how the semiconductor industry is created, evolved and matured.

2.2 Innovative Business Models for a Dynamic Industry

The semiconductor industry has been a driver of global economic growth and social change. The widespread application of semiconductors has transformed computing, communication, education, medicine/health, commerce, banking, entertainment, military, industry, government and etc. This dynamic industry constantly embraces rapid technological changes and then transmits them to the rest of the economy, leading to an ever-better standard of living. Steady progress in semiconductor performance and reduction in its cost has been one of the major drivers in the improvements of worldwide productivity and growth. It could be argued that if the atomic bomb were not invented, or if human did not reach the Moon, the life of majority of people would not be affected. However, without the invention of transistor and the consequential semiconductor industry, our life will be totally different.

From a business operation perspective, the seven or so decades of semiconductor industry can be roughly divided into three stages: (1) the genesis of the inventions of transistor and integrated circuit, lasted about three decades from 1950s; (2) the first transition from off-the-shelf component to ASIC, started circa 1980s; and (3) the second transition from IDM to the fabless model, from late 1990s to present. During these stages, this dynamic industry has interminably experienced crises of technical or managerial nature (Nenni et al. 2014; Lojek 2007; Brown and Linden 2009). People in this industry respond to those crises with various attitudes and strategies, resulting in companies' and even countries' up-and-down. As a whole, the industry's collective well-being survives the challenges and it grows to be the key source of productivity for the modern economy.

The business, or financial, side of story on semiconductor industry is solely controlled by the powerful law of supply and demand. The whole story can be summarized as chasing two goals: best-technology and economy-of-scale. From technology aspect, the non-stop pursuit for ever-advanced technology is to reduce the cost per transistor. For economy, the aim is to "fill the fab". Their combined effects are the ever-better-and-cheaper products for consumer and the competitive advantage, or simply survival, of the individual companies. This game is played in three fields: design, manufacture (including assembly and test, material and equipment) and market. For a particular company, it can involve in one or more of those

2.2 Innovative Business Models for a Dynamic Industry

fields. The ultimate driving force is the market, which experienced a major shift from performance-minded corporate purchasers to price-sensitive consumers in the 1990s.

The market force shapes the industry. In the first stage of genesis, the main electronic products are mainframe computer and then the minicomputer and personal computer. The major customers were corporate purchasers whose key concerns are performance and reliability. The good marketing strategy then was single design of high-volume and long-life-time. Relatively speaking, this type of market is stable and the semiconductor industry can simply follow the proven IDM model. In other words, those companies who owned the manufacturing assets researched, developed, designed, fabricated, and marketed their own products. When consumer became the main focus, the market was however fragmented and good opportunities for products with high volume are more and more rare. This new market also has the tendency of large demand swing, which leads to more severe business cycle. This situation made the goal of "fill the fab" harder to be fulfilled. The industry then responded with the ASIC strategy, which separates the design and manufacture functions so that the segmented market can be better served. With the help of venture capitalist culture, some ASIC-style start-ups grown up and eventually matured (the second stage). This group of pure design companies and the ecosystem built around them, combined with pressure from the ever-increasing cost of building manufacturing fab, set the background for the emergence of the fabless-design-house and manufacture-foundry model (the third stage).

Semiconductor industry has a long value chain from fabrication (silicon wafer, material, manufacturing equipment), design (logic design, physical design, verification, analog design, system design, IP (intellectual property) design, EDA tools), to assembly and test, to market research and product creation. Each firm in this value chain needs to create its unique value. To be successful, however, it also faces the issue of capturing the value. The challenges here are sometime engineering in nature, sometime business-related. Those challenges require change in the way of conducting business. Some are incremental such as the extension of existing optical lithography from generation to generation. Others are as radical as the change of company leaders from technology-oriented-entrepreneurs to professional financial managers, reflecting the profound change of philosophy from valuation creation to valuation capture. This drama of constant challenges-and-response makes this industry dynamic. This dynamism is also evidenced in the strategies used by the people of this industry: positioning (change to different markets, create new markets), offshoring (establish design and manufacturing centers in low-cost countries, global brain circulation), restructuring (organizational innovation such as merger and acquisition, spin-off, downsizing), cooperation (cooperation in industry standards or collaboration through alliances for participating in industry-critical basic research beyond each company's own medium-term development programs). For example, TSMC's "Grand Alliance" is a business model innovation that gathers all the significant players together to support customers, not just EDA and IP but also equipment and materials suppliers, especially for high-end lithography.

This dynamism is further shown in global scale. The rise of new lower-cost global competitors and the growing importance of price-sensitive consumer markets have not only put pressures on the Return-On-Investment (ROI) for most companies but also have brought the battle to the level of national security. Because of the semiconductor industry's strategic economic importance, countries in the world have taken a keen interest in studying how global competitive advantage can be won in this swiftly moving and vital industry. As a result, governments often adopt protective policy of supporting domestic development of the industry and regulating export of its technologies and products.

From the discussion of Sects. 2.1 and 2.2, it is clear that innovations in both technical and business aspects are crucial to this industry. In next section, we will look at the people of great minds who have played decisive roles, in a variety of arenas and in different degree, in making those innovations happen.

2.3 The Role of Creative Minds: Leadership Versus Command-And-Control

The births of all modern industries are fundamentally enabled by science, and then further propelled by technology and accumulated knowledge-skill-experience. During the development, a key factor that cannot be overstated is the individual's creativity. Wright Brothers is instrumental to the birth of airline industry. So is Henry Ford to automobile industry. For modern information society, the three biggest names arguably are Alan Turing, John von Neumann and Claude Shannon. Alan Turing showed us how to describe the solution to computable problem using a method of programming. John von Neumann architected a computer for running the program. Claude Shannon taught us how to create a world-of-information using the computer. To complete this chain-of-significance, years later, Gordon Moore described how semiconductor scaling makes the computer grow exponentially more capable over time.

The common characteristic of those creative individuals is the leadership, which is the willingness, courage, and capability to surmount obstacles. Any creative endeavor will undoubtedly face obstacles because such endeavors threaten the established or entrenched interest, or status quo. It is believed that creative individuals, not business managers or government initiatives, are the most important factor that shapes the semiconductor industry. In the early history of transistors and integrated circuits, it is a sad fact that there are only a few cases where well-planned and carefully-managed projects resulted in success (Lojek 2007). The inventions of transistor, diffusion technology, integrated circuit and the planar process were not the result of coordinated and supervised effort, but the fruits of individuals' relentless effort. In many cases, as semiconductor history repeatedly showed, the company establishment was frequently one of the biggest obstacles, if not the biggest, which needed to be

2.3 The Role of Creative Minds: Leadership Versus Command-And-Control

overcome in the introduction of new ideas. The mental toughness and perseverance of creative individuals are critical to the success of innovative process.

Unfortunately, highly creative individuals are frequently regarded as trouble maker. The reason is that creative individuals often exhibit atypical thought processes and mental content. They are less constrained by conventional expectations, less concerned with making the right impression on others. They often do not respect common practices. Their methods, style, authoritarian control, and temperament are frequently at odds with conventional norms. Taking Bob Widlar for example, he almost single-handedly created the business of linear integrated circuits, from creating new circuits, to new products, to new applications, and eventually to new market. His masterpiece $\mu A709$ is the most widely used operational amplifier in electronics for many decades (Widlar 1964; Widlar et al. 1965). His achievements demonstrated that, at least in the early years of semiconductor industry, an individual creative engineer could have a major impact on the business of a large company. On the other hand, his genius is only matched by his personality. Bob Widlar was a fiercely independent individual (Lojek 2007). He did almost everything in a stunning way, which was natural to him but completely weird to the so-called "normal people."

The greatest innovations are usually created by the brightest minds. Ambition is what drives the good scientist and artist to try great things, and try again when they failed. For great engineers with strong personality like Shockley or Widlar, technical work was glamorous and perhaps seductive. They were born engineers. They enjoyed the satisfaction in focusing on a problem, educating themselves and mastering the knowledge. From solving technical problems, they gained a sense of pride and accomplishment. The inventions of planar manufacturing process (Jean Hoerni), integrated circuit (Jack Kilby), FPGA (Ross Freeman) are some other good examples of not-planned-official-projects but individuals' efforts.

On the business side, individuals of creative mind show their leadership in the form of futurist's vision. The "Traitorous Eight" of Fairchild Semiconductor virtually founded the Silicon Valley through their restless pursuit of creating ever-better electronic products, all in their unique paths when their opinions were not aligned with the authority (Lojek 2007; Moore 1998). Gordon Moore and Robert Noyce, founders of Intel, changed the tide of semiconductor industry by leaving the memory market and creating a new market of microprocessor. This visionary strategy not only made Intel the most powerful company in semiconductor industry but also enabled US to gain back the semiconductor leadership from Japan during the 1970s~1980s (Brown and Linden 2009; Moore 1996, 1997). Morris Chang of TSMC devised a new model of foundry that has grown into a business of billions of dollars. This is also an achievement due mostly to an individual's great vision. The most influential figure in recent years is Steve Jobs of Apple, who revolutionized the way of mobile communication, computing and entertainment through his great vision, creating a new market with the potential of hundreds of billions of dollars.

Those examples provide some proof that creativity lies in the mind of individual, not in organization. Organizations, especially those large ones, often are the enemy of progress and change. It is usually a convenient hideaway for the mediocre and the weak. Luckily, when semiconductor industry was in its infant stage, the environment was much friendly to creativity. For example, Bell Labs' culture respected creative individuals. Its leaders understood that great engineers often have little social skills and they cannot comply with politically correct and hypocritical rules. Such "incapability" is balanced by other qualities such as smartness, never satisfied with status quo, aggressiveness and willingness to take a risk, driven to excel and ultimately able to solve difficult problems. Unfortunately, in current business environment, the managerial position is frequently a "political" position related to power and financial compensation. In this situation, the best strategy to keep the managerial position is to maintain the status quo with no disturbance. Engineers and scientists of creative mind often do not understand the concerns of such commercialization personnel. Both sides have different objectives and rewards stimuli. Technical people find value in novelty and push for the frontier of knowledge while commercialization people need products to sell and they often consider the value of novelty as marginally-valuable, if not entirely useless. To them, accepting even a small level of risk associated with novelty or new creation could be risky and it is always safer to reject such new idea (Lojek 2007; Brown and Linden 2009). This philosophy is managerial command-and-control. It is not the leadership that is required to move the technology forward. Leadership comes from creative minds, which are not necessarily managers. Leaders have vision. To break through the current obstruction and march forward, we undoubtedly need a force that shows its power as leadership instead of just command-and-control. Such force is especially precious today when this industry has clearly reached its maturity and is now facing an imminent crisis.

2.4 The Road to Normal Engineering: From Science to Technology

Precisely defining what science is can be a difficult task. Science can however be loosely defined as the reasoned investigation or study of world phenomena. It aims at discovering enduring principles among elements of the phenomenal world by employing certain rigorously proven formal methods. The practice of science focuses heavily on theory: the creation, the analysis and the generalization of theories. The result from this practice is the cumulative addition to the knowledge of *laws of nature*, constantly building upon previous benchmarks to scale new peaks. On the other hand, technology is a form of applied science. While the goal of science is the pursuit of knowledge for its own sake, the goal of technology is to create system to meet the need of people and improve human life. Science has a quest of explaining things, while technology is leaning more to developing things for use.

The realm of technology is geared towards real world application and execution; and the action of taking science out of laboratory and placing it in the real-world context is called engineering. In the activities of engineering, the term innovation is often used in referring to new ideas, creative thoughts or new imaginations in the form

2.4 The Road to Normal Engineering: From Science to Technology

of device or method. In essence, innovation is the creation of better solutions that meet new requirements and existing or unarticulated needs. Innovation takes place through the provision of more effective processes, services, techniques, products, or business models. In this process of innovation, any new idea generated and the means of its embodiment or accomplishment is called invention. Innovation is not in the domain of science but technology. This point can be illustrated by the case of Isaac Newton. Newton's work is generally regarded as pure science. But his idea unlocks technology. His systematic investigation of motions and forces led to the invention of new machines. As examples, the spinning jenny and the stream engine were the first in a series of innovations that transformed the economic landscape of England and then the world.

Unlike universities and government-sponsored research institutes whose primary goal is the pursuit of understanding of the nature, industries' Research & Development (R&D) programs is the systematic use of the knowledge or understanding gained from pure scientific activities. This effort is directed toward the production of useful materials, devices, systems, or methods, including the design and development of prototypes and processes. While science is focused on discovery, analysis and explanation (theory), the majority of industry R&D activities does not fall into this category. It is instead in the domain of technology because industries are more interested in creating product, or the synthesis of design. However, industry R&D must stand on the solid ground of science.

For the case of semiconductor industry, its life starts from fundamental research in physics and chemistry. The term *semiconducting* was used for the first time by Alessandro Volta in 1782. The first documented observation of a semiconductor effect is made by Michael Faraday in 1833. In 1878 Edwin H. Hall discovered that charge carriers in solids are deflected in magnetic field, the Hall Effect. This phenomenon was later used to study the properties of semiconductors. The discovery of electron in 1897 by Joseph J. Thomson inspired several theories of electron-based conduction in metals: in 1930 Bernhard Gudden reported that the observed properties of semiconductors are due exclusively to the presence of impurities and that chemically pure semiconductor does not exist; in 1931, based on the idea of empty and filled energy bands, Alan Wilson developed the band theory of solids; in the same year Heisenberg developed the concept of hole; in 1938, Walter Schottky and Neville F. Mott developed the models of potential-barrier and current-flow through a metal–semiconductor junction; In 1938 Boris Davydov presented a theory of a copper-oxide rectifier including the presence of a p–n junction in the oxide, excess carriers and recombination (Busch 1989; Laeri et al. 2003; Hoddeson et al. 1992).

Based on those fundamental researches of discovery and theorization, the work associated with semiconductor industry gradually moved into the phase of technology for creating useful things to improve human life. In 1945, William Shockley proposed a conceptual device of signal amplifier operating on the semiconductor field-effect principle. In 1947 John Bardeen and Walter Brattain built the first point-contact transistor and demonstrated that this device exhibited some power gain. Shortly after that, Shockley formally developed his p–n junction theory (Ross 1998; Riordan et al. 1999). The mechanism of forming p–n junction was hence better understood and the resultant devices were heavily used in military communication and radar equipment. Then came the grown junction transistor in 1952, which is much stable than the point-contact transistor and its electrical characteristics is much easier to be controlled. From that time on, technology marched forward rapidly thanks to strong industrial interest. The first commercially available grown junction-based silicon device was manufactured in 1954; the first diffused silicon transistor appeared in 1955. In 1958, the first integrated circuit was demonstrated. The formation of interconnects by means of deposition of aluminum on a layer of SiO_2, which is a key part of integrated circuit, was achieved in 1959. A transistor with epitaxial layer was reported in 1960. In the same year, Jean Hoerni proposed the landmark planar transistor which paved the way for industrial mass-production in low cost and high yield fashion. In 1963 a work on silicon MOSFET was published (Hofstein and Heiman 1963). In the same year the first CMOS circuit was proposed (Wanlass et al. 1963). Meanwhile, the fundamental characteristics of the band structure had been understood step by step towards the end of the 1960s, which led to the industrial breakthrough of cost-effective manufacturing process of high-purity silicon. The extraordinary properties of the Si-SiO_2 pair finally made silicon the selected semiconductor material. This series of technology developments opened the silicon era, in the same sense as the Stone, Bronze, and Iron Age where a specific material that predominantly characterizes the advancements made during that time is chosen for the name of that era.

Figure 2.1 illustrates this development path of semiconductor R&D from science to technology. Although the development is a gradual process, the characteristics of science and technology in different periods can be identified. The science nature of early semiconductor R&D is evidenced by the number of Nobel Prize awarded. There are about ten or so Nobel Prizes in physics awarded to semiconductor-related fundamental research, most of them are due to awardees' work in the infancy of the semiconductor research (Łukasiak and Jakubowski 2010).

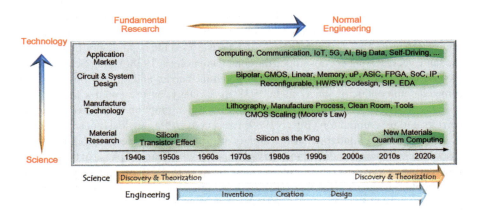

Fig. 2.1 Semiconductor R&D: from science to technology

The creation of point-contact transistor is more a scientific discovery than a technological invention since its working mechanism was not understood at that time. For the case of junction transistor, the theory was developed before the device was made. In other words, Shockley predicted the junction transistor by first articulating a scientific theory. This theory on electron & hole puts the infant industry on a solid physics foundation. From this point onwards, the semiconductor industry can be viewed as shifting from science to technology, and the majority of development thereafter has been carried out in the form of engineering. Between 1960 and 2000s, as another example, the industry steadily marched by simply following the path of CMOS scaling (the good time of traditional Moore's Law). This type of work was technology-oriented and can be clearly classified as normal engineering. However, as Dennard's scaling failed circa the middle of 2000's, science-oriented work started to gain more share in R&D funding. It reflects the change-of-trend from normal engineering back to basic research of searching for new physical phenomena and new materials, to continue the technology advance.

2.5 After the Formation of a Paradigm: Puzzle-Solving Engineering

In "The Structure of Scientific Revolution" (Kuhn 2012), Kuhn spent great effort in defining, or describing from various aspects, the terms *paradigm* and *paradigm shift* (or paradigm change). In his view, a field of science gradually progresses by a series of individual discoveries and theorizations. At some point of time, the resultant facts, concepts, laws and theories, when gathered together, constitute a body of knowledge to such an extent that the achievement is sufficiently unprecedented to attract an enduring group of adherents. Simultaneously, it is sufficiently open-ended to leave many sorts of problems for the attracted group of practitioners to resolve. At this stage, a paradigm is formed. A paradigm provides a base for the relatively unproblematic character of professional communication and for the relative unanimity of professional judgment. The commitment and the consensus that a paradigm produces are prerequisites for science, and also for engineering. People whose work is based on a shared paradigm agree on the same facts, honor the same theories, and are committed to the same rules and standards for their practice. In essence, paradigm is a map for direction. Further, it provides guidance that is essential for future mapmaking.

After a paradigm is formed, according to Kuhn, the activities become normal science. Normal science does not aim at novelty, conceptual or phenomenal. Novelty for its own sake is not a desideratum in the normal science as it is in many other creative fields, such as art and music. Instead, again according to Kuhn, normal science is a puzzle-solving activity. A paradigm becomes tools for engineering, which leans even more to the nature of puzzle-solving. Using this justification, it is reasonable to state that the first paradigm for microelectronics was roughly established around the time when W. Shockley proposed his p–n junction theory near the

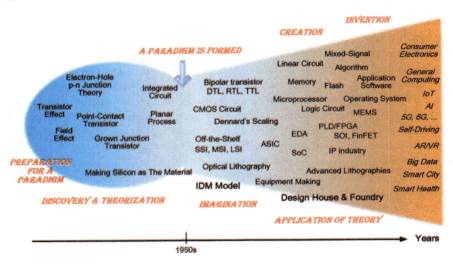

Fig. 2.2 Microelectronics: the formation of first paradigm

end of 1940s. Figure 2.2 depicts the two phases of pre-paradigm and post-paradigm in the development of the semiconductor industry.

After the first paradigm of microelectronics was formed with the mechanism of transistor being understood, microelectronics entered the era of the puzzle-solving engineering. As transistor-based microelectronics gradually penetrated into more and more areas for facilitating applications. And as new applications became more sophisticated, the scope of microelectronics expanded and the number of pieces in the puzzle increased, as seen in Fig. 2.2. Then specialization became necessary to handle the various pieces of the puzzle. Specialization is a fact of human civilization, and it is a means of science and engineering. The development process of science and engineering is like Darwinian. As time goes by, specializations are like speciation events in which one species splits into two, or one species continues but with variants on the side following their own trajectories. In this newly formed paradigm, new subdisciplines emerged and developed, each with its own symbolic generalization, model, and exemplar, just as modern science gradually specializes in general (It can be reasonably argued that all branches of science and engineering start from Astronomy. During this long process, progressive specialization is the driver.). As a result, it becomes increasingly difficult for the practitioners of one field to understand what the people in other fields are doing. In other words, each piece of the puzzle becomes increasingly harder to be recognized by outsiders.

In this puzzle of semiconductor R&D, there are two primary types of engineering: process development and chip design. Semiconductor equipment for manufacture process emerged as a separate specialty in the 1970s. Those equipment companies are R&D intensive and are responsible for overcoming the industry's most fundamental challenges. On the other hand, as time flees, chip design becomes increasingly more diversified. The variety of markets forces chip design adopt different design

2.5 After the Formation of a Paradigm: Puzzle-Solving Engineering

strategies for different applications, and hence the specialization in chip design gradually becomes more colorful. This can be clearly seen in Fig. 2.2. Today's scientist or engineer of semiconductor industry has difficulty in reading the research paper of a colleague in another field that might be just slightly away from his own. Oftentimes, he or she doesn't even try. In academia, universities are organized into "departmental silos". Young professors build reputations and are awarded tenure by becoming world experts in their narrow subfields.

The magnitude of this puzzle is also evidenced by the large number of technical journals published and conferences held every year, each is dedicated to a particular subfield (there are hundreds of them, perhaps thousands). When all microelectronics-related journals are scrutinized, five types of problems are addressed: (1) formulation, articulation and proliferation of theory (theory on well-defined specific subject, not theory at fundamental level), (2) matching of experimental observation with theory, (3) description of procedure for creating device, circuit and system, (4) method and apparatus of performing a task, and (5) review or tutorial of a field. All those works are normal engineering and can be labeled as puzzle-solving.

The puzzle-solving nature of this type of normal engineering can be elucidated by examples. In the infancy of microelectronics, for example, there were three primary limitations regarding transistors: poor reproducibility, poor reliability and poor understanding of circuitry built with transistors. This circuit design limitation was due to not-yet understood phenomena in gain, noise, frequency and power. The resolution of those issues did not involve novelty but required step-by-step detailed knowledge accumulation on the behaviors of transistor. Another good exemplary case is Bob Widlar's effort on creating the masterpiece μA709 operational amplifiers. In this design of textbook-excellence, he brilliantly traded the restrictions imposed by the limited types of components, poor tolerances and limited range of integrated component values against the use of a large number of active devices with free choice of device geometry and the close matching of active and passive devices over wide range of temperatures. This is not an act of discovery or theorization but an excellent demonstration of great engineering skill. It much resembles the analytic skill needed by solving puzzle.

The engineering effort around the widely-used circuit component, Phase Locked Loop (PLL), demonstrates the characteristic of puzzle-solving in an even greater degree. PLL is a component for producing electric pulse train. This pulse train, the so-called clock signal, can be used as a time marker for indexing events occurred inside electronic world. It can also be used as RF carrier upon which information is encoded and transmitted. This piece of puzzle in designing PLL is complicated by several facets such as loop stability analysis, loop response speed study (bandwidth tradeoff), device noise analysis, noise transfer and suppression, frequency operating range, concerns of power consumption and process migration, cost analysis and etc. Those issues are so intricately connected that there are over twenty thousand technical papers found when keyword PLL is searched only in a single database of IEEE Xplore. Intricate as it is, this topic however is not science but normal engineering. It is a matured craftsmanship with over a hundred of textbooks dedicated to its study.

Computer science is a major discipline brewed from semiconductor industry. It takes shape only after the first paradigm of microelectronics is formed. Computer is a coldly mechanical and deterministic system. It applies rigid deductive logic to solve problems given to it, making decisions by exhaustively enumerating options and grinding out exact right answer no matter how long and hard it has to think. Behind its cold appearance are the living algorithms, tirelessly creating magnificent wonders every day in our life. Algorithm is at the heart of computer science. However, the activity of creating algorithms can hardly be categorized as science. An algorithm is just a finite sequence of steps used to solve a problem, or a precise recipe that specifies the exact sequence of steps required to perform a task. This kind of mechanical feel is a typical characteristic of normal engineering. Modern computer faces more challenging tasks, such as conversing with people, auto-piloting in harsh environment, or winning the game of Go. Therefore, today's algorithms are presented with problems where the rules are not clear, some of the required information is missing, or finding exactly the right answer would require considering an astronomical number of possibilities. Those challenges have moved the modern algorithms away from the extreme reliance on exhaustive calculation to being comfortable with chance, trading off computing time with accuracy and using approximation. However, complicated as it is, the task of algorithm development still bears the nature of engineering. But this piece of puzzle becomes increasingly larger and convoluted.

In summary, after the formation of the first paradigm, microelectronics becomes a practice of rapid specialization, resulting in ever-increased number of subdisciplines. Each of them is one piece of the puzzle which, when pieced together, represents an increasingly intricate picture as time goes by. As a whole, microelectronic engineering resembles more and more the puzzle-solving activity. Within the frame of this paradigm, it gradually moves away from pure science of fact-finding and theorization to normal engineering of creation and innovation.

2.6 A Missing Attribute of the Puzzle

The purpose of the lengthy discussion given in previous sections is to print a picture that today's microelectronics is made of a puzzle of many pieces. There is only one piece in the beginning: the invention of transistor. All the subsequent developments follow the path of continued miniaturization of the transistor and, at circuit and system levels, the continued use of the ever-large-number of available transistors to create more and more functions. The essence in those activities is to make more transistors out of a given silicon space, or to pack more transistors into a given space for doing more work. This process is more or less similar to biological evolution in that things gradually becomes more complicated and specialization promotes the emergence of ever-more subdisciplines. All those subdisciplines however are established within the frame of this paradigm described above, which has a set of rules and standards that are agreed upon by all its practitioners. Among them, one of the most fundamental issues is about how to handle time.

2.6 A Missing Attribute of the Puzzle

Table 2.1 The perception of time in two different worlds

World	Subject of study	Perception of space & time	Measuring space	Measuring time
Earthly world (Including all other engineering & science disciplines)	Human activities	Newtonian space & time Einstein's four-dimension spacetime	Meterstick Kilometer Light-year	Celestial motion Waterflow Clockwork Crystal
Electronic world (the discipline of circuit & system design)	Electrical signals	Newtonian system is used in circuit & system design	Meterstick of much smaller scale	Electrical pulse train of certain frequency

Unlike all other fields of science and engineering, the art of microelectronics faces a unique problem of creating "a sense of time" inside the world of electronic. In almost all the fields of science and engineering, people deal with problems of earthly world. In their practices, they employ common sense when the issue of time is concerned (except in philosophy and physics). In other words, in their professional works, they do not need to think too much on the thing that we call "time", other than simply using the one they are already familiar with in their daily life. This is however not the case for engineering practice in microelectronics. In microelectronics, although not obvious in the first sight, people have actually created an entirely new world made from transistors. In this world, the sense of time needs to be created from scratch. Table 2.1 compares the situations regarding sense of time in those two different worlds. In electronic world, an electrical pulse train is used to help establish the sense of time. This pulse train is referred as clock signal. The speed of this clock pulse train is gauged by a parameter called frequency. More in-depth discussion on the concept of time will be left to Chap. 4.

In the currently established paradigm of microelectronics, the law on time follows the philosophy of Newtonian's absolute time. Time marches uniformly, homogenous in all the directions. Following this guideline, the clock pulse train is made from uniformly shaped pulses. For a given application, the clock's frequency is fixed. In the entire past history of electronic engineering, after the principle of using this simple clock signal for representing time-marching is adopted, the issue regarding time is considered done, or problem solved. After that, the edifice of microelectronics as a whole focuses its effort on just utilizing the space, without paying much attention to time any more. However, science and common sense teach us that matter exists in both space and time. To create more efficient computing machine, transistors (and circuit and system) need to be scrutinized under the scope of time as well. Is there something different that we can do, inside this world of electronic which is entirely different from the Newtonian-mechanics-dominant world that we live in, on handling this thing called time? The significance of this time-handling task increases with the ever-large number of transistors at our deposal, also with the gradually diminishing potential in the direction of space. Microelectronics needs to utilize time more sophisticatedly if further advances need to be achieved. Figure 2.3 illustrates

Fig. 2.3 The missing attribute in the first paradigm of microelectronics (left), and the new paradigm (right)

the essence of this point. The current microelectronics edifice is a puzzle made of black-and-white since it focuses only on space. A more vivid picture needs color, a more efficient handling of the time must be merged into the picture to stimulate interests among the puzzle-solvers.

References

J. Bardeen, W. Brattain, *Phys. Rev.* 75, 1208 (1949).
Clair Brown and Greg Linden, "Chips and Change: How Crisis Reshapes the Semiconductor Industry," *The MIT Press*, 2009.
G. Busch, "Early history of the physics and chemistry of semiconductors - from doubts to fact in a hundred years", *Eur. J. Phys.*, vol. 10, no. 4, pp. 254–263, 1989
R. H. Dennard, F. H. Gaensslen, V. L. Rideout, E. Bassous, and A. R. LeBlanc, "Design of ion-implanted MOSFETs with very small physical dimensions," *IEEE J. Solid- State Circuits*, vol. 9, no. 5, pp. 256–268, Oct. 1974.
D. J. Frank, R. H. Dennard, E. Nowak, P. M. Solomon, Y. Taur, and H. S. P. Wong, "Device scaling limits of Si MOS FETs and their application dependencies," *Proc. IEEE, vol. 89*, no. 3, pp. 259–288, 2001.
L. Hoddeson, E. Braun, J. Teichmann, and S. Weart, Out of the Crystal Maze: Chapters in the History of Solid State Physics. New York: *Oxford University Press*, 1992.
J. A. Hoerni, "Method of Manufacturing Semiconductor Devices," *U. S. Patent 3,025,589* (Filed May 1, 1959. Issued March 20, 1962).
S. R. Hofstein and F. P. Heiman, "Silicon insulated-gate field-effect transistor", *Proc. IEEE*, vol. 51, no. 9, pp. 1190–1202, 1963.
Jack S. Kilb, "Miniaturized Electronic Circuits Application," *U.S. Patent 3,138,743*, filed in 1959, issued in 1964.
D. Kahng and M. M. Atalla, "Silicon-silicon dioxide field induced surface devices", *in Solid State Res. Conf.*, Pittsburgh, USA, 1960.
T. S. Kuhn, "The Structure of Scientific Revolution," the *Univ. of Chicago Press*, 2012.
F. Laeri, F. Schüth, U. Simon, and M. Wark, Host-Guest-Systems Based on Nanoporous Crystals, Weinheim: Wiley, 2003, pp. 435–436
John H. Lau, "3D IC Packaging and 3D IC Integration," available on http://www.ewh.ieee.org/r3/orlando/2013/Nov/3D Packaging&3D Interconnect-John Lau.pdf.
Tsu-Jae K. Liu, "FinFET: History, Fundamentals and Future," short course, *2012 Symposium on VLSI Technology*, 2012.
B. Lojek, History of Semiconductor Engineering, Berlin-Heidelberg: Springer, 2007

References

L. Łukasiak and A. Jakubowski, "History of Semiconductors", *Journal of Telecommunications and Information Technology*, pp. 3–9, Jan. 2010.

G. E. Moore, "Cramming More Components Onto Integrated Circuits," *Electronics*, vol. 38, no. 8, pp. 114–117, 1965.

G. E. Moore, "The Role of Fairchild in Silicon Technology in the Early Days of Silicon Valley," *Proc. IEEE*, Vol. 86, (1998), pp. 53–62.

G. E. Moore, "Intel – Memories and the Microprocessor", *Daedelus 125*, No. 2. 1996.

G. E. Moore, "The birth of Microprocessor," *Scientific American*, September 22, 1997.

D. Nenni and P. Mclellan, Fabless: The Transformation of the Semiconductor Industry, *CreateSpace Independent Publishing Platform*, 2014.

Robert Noyce, "Semiconductor Device and Lead Structure," *U.S. Patent 2,981,877*, filed in 1959, issued in 1961.

M. Riordan and L. Hoddeson, "The origins of the p-n junction", *IEEE Spectrum*, vol. 34, no. 6, p. 46, 1997.

M. Riordan, L. Hoddeson, and C. Herring, "The invention of the transistor", *Rev. Mod. Phys.*, vol. 71, no. 2, pp. S336–S345, 1999.

I. M. Ross, "The invention of the transistor", *Proc. IEEE*, vol. 86, no. 1, pp. 7–27, 1998.

W. Shockley, "The Theory of p-n Junctions in Semiconductors and p-n Junction Transistor," *Bell System Technical Journal*, Vol. 28 (1949), p. 435.

W. Shockley, M. Sparks, G. K. Teal, "Physical Principles Involved in Transistor Action," Phys. Rev. Vol. 83 (1951), p. 151.

F. M. Wanlass and C. T. Sah, "Nanowatt logic using field-effect metal-oxide semiconductor triodes", in *Proc. Techn. Dig. IEEE 1963, Int. Solid-State Circ. Conf.*, Philadelphia, USA, 1963, pp. 32–33.

R. J. Widlar, *US Patent 3,364,434*, Filed April 19, 1965.

R. J. Widlar, A monolithic high gain DC amplifier, *Proc N.E.C.* 1964, pp.169–174.

Xiu, L., "Time Moore: Exploiting Moore's Law from Time Perspective", *IEEE Solid-State Circuits Magazine*, vol. 11, no.1, pp. 39–55, 2019.

Chapter 3
The Space Crisis in Current Paradigm

3.1 Perception of Space, Time, Change and Motion

The stage for performing the magic of microelectronics is set with the backdrop of *space*, *time*, *change* and *motion*. The conceptual framework built on those fundamental elements deserves some serious discussion before further investigation on microelectronics paradigm can take place. The concepts of space and time seems trivial at first sight and yet are so profound that a precise and complete definition on them is elusive, if not impossible. Newton is the first person known in trying to describe the space and time in a scientific fashion. His view on space and time are much like our intuitive ideas. Space is simply there: a three-dimensional nothing that goes on forever. It is an empty and unchangeable backdrop against which all events in the universe play themselves out. Meanwhile, Newton imagined time as an intrinsic property of the Universe, flowing linearly and immutably. (In classical Newtonian physics, mechanics is perfectly reversible. In other words, time is bidirectional, t and −t are indistinguishable in equations.) Further, Newton considered space and time to be absolute concepts, and they are independent of each other.

We human have no trouble imagining space and time as the way Newton did. It's the natural way to think of them. Our brains are wired to comply with this convenient view. An object, such as a person, an electron or a planet, can sit at a certain location in space, or it can move from one location to the next. If we choose a particular reference point, all other locations can be identified by just three coordinates. Starting from your reference point, the three numbers tell you how far you have to go forward or backward, left or right, and up or down to reach the other location. Human perception of time also seems obvious: we feel time passing even though we have no actual sense that can physically perceives time. We experience events in time and then assign them an order as they pass from the instantaneous "now" into our memory. Time is the absolute, infallible metronome of the cosmos, tagging each and every event with a unique time stamp. It's one-dimensional: if you choose a reference moment, you need only one number to tell you at what time an event takes place.

Einstein blew the Newton's space-time idea, and our intuitive sense, out of the throne with his revolutionary theory. He argued that space and time are linked: three-dimensional space and one-dimensional time are actually interwoven in four-dimensional Spacetime. His theory is called theory of relativity on the belief that both space and time are not absolute, but relative. Within the frame of relativity theory, what's the distance between two points in space? The answer depends on whom you ask. For someone traveling at half the speed of light, the distance between two points in space is much smaller than someone at rest. The same is true with the time interval between two events. Time is supple and it depends on the amount of motion and the local force of gravity. According to relativity theory the only thing that is absolute is the four-dimensional separation between two events, occurring at two locations, in the Spacetime.

The definition for the concept of *change* is to make the form, nature, content and future course of something different from what it is at current state, or from what it would be if left alone. In the earthly world, motion is the change in the position of an object over time. The motion of a body is observed by attaching a frame of reference to an observer and measuring the change in position of the body relative to that frame. Motion is mathematically described in terms of distance, speed and acceleration, which are all built on the foundation of space and time. In Newtonian mechanics space and time, although independent of each other, can be indirectly linked by the motion of an object. In Einstein's world, the four-dimensional Spacetime is centralized on the motion of light. The speed of light is postulated as a constant, the highest achievable speed of any object. The motion of light straightly couples the space and the time.

In astronomical scale, it takes millions of years for light ray to travel from far distance to earth. Thus, distant light is old light. Thus, interestingly, telescope is a time machine since it allows us to look back into the past. According to astrophysicist, the Universe is expanding and its expansion speed might be fast than the speed of light. As a result, there are regions where the light ray emitted from object there has no chance to reach earth and we will never be able to see that part of the Universe. Amazingly, we are physically bounded by time, not space. Even more contrast to our common sense is the following picture that, on Einstein's relativity, mass exerts influence by inducing curvature on the four-dimensional Spacetime. Gravity is the effect of curvature of Spacetime on the motion of other bodies. Time is also influenced by the presence of massive bodies. Closer to massive object such as black hole, clock starts to tick slower and slower.

In our common sense, the direction of time flow is naturally forward. However, at molecule level, atoms know nothing of time and reactions can flow back and forth. The arrow of time hence seems to be an emergent property of large collections of atoms (i.e., the formation of the Universe) and it always points to the direction of entropy increase. With this arrow inscribed on the thing that we called time, the imaginary clock of nature ticks away the moments that make up a dull day, as well as all the seconds since the birth of the Universe.

The same series of events might be an instant to one observer but a glacial progression to another. Timekeeping thus needs an objective way to keep track of events.

3.1 Perception of Space, Time, Change and Motion

For time used in our daily life, we use wrist watch, pendulum clock or cycles of sky. For longer scale, timekeeping uses gravity: tree ring, geological processes, and astronomical cycles that imprint periods much longer than one year (such as the interplay of gravity between Sun, Earth and Moon). For even longer timescales people use radioactivity as clock, the half-life decay. Clock embedded in the rock often uses carbon of half-life 5730 years. Uranium of 4.5 billion years half-life is used to calibrate astronomical events.

Another important aspect of timekeeping is accuracy, which depends on the predictability of the periodical cycles. Vibration is a fundamental property of atoms. Vibration or oscillation of small objects, which usually has high resonant frequency, provides good means for periodicity. Their rates of vibration or oscillation are naturally better in predictability than those obtained from the aforementioned methods. In 1967, "one second" was officially defined as 9192631700 oscillations of cesium atom. From the sky, the celestial object called pulsar (a rapidly rotating neutron star) emits regular pulses of radio wave (and other electromagnetic radiation) at rates of up to one thousand pulses per second, and it is very stable. This is hence also a good time source, may be even better than atomic clock.

Inside electronic world, when transistor and signal are used to make circuit and system, engineer uses the framework of Newtonian space and time. This is mainly due to the reason that, after circuit layout is settled down, the locations of all the active and passive components are fixed when the physical die is used as the frame of reference. An electrical signal travels between transistors located in different locations. The signal's change of state over a period of time is the motion, which can be considered as transmission-of-intelligence (i.e., information and communication). This type of signal is an electromagnetic wave. Similar to the traveling motion of light, this motion also links space and time. Although its traveling speed in media (such as copper wire) is comparable to the speed of light, the Einstein's relative effect is not directly involved in circuit designer's calculation since the traveling distances are extremely short. Instead, the signal's traveling time is quantified by a term called propagation delay, a number generated from circuit simulator and then used in designated abstract model. Propagation delay is the TOF (Time of Flight) for a signal to travel, inside the structure of a chip, from one place to another. To regular the activities and to properly marker the events inside circuit and system, a special signal called clock is created. It is a train of electrical pulses where every pulse's length-in-time is fixed and all of them are ideally equal. The value of this length-in-time is determined mainly by the propagation delay of signal traveling through a typical circuit. This value is referred as clock period and its inversion is called clock frequency.

In summary, the concepts of space, time, change and motion establish the stage for the show of microelectronics to take place. The significance of those concepts is summarized in Table 3.1. It is however worth mentioning that the laws of quantum mechanics need to be applied when we move into the regime of transistor's internal space. In that domain, the world is controlled by quantum laws, neither Newton's nor Einstein's.

Table 3.1 The link between space and time in different worlds

World	Space and time relation	Object of study	Changeable variable	Characteristic motion	Linkage for space and time	Governing law
Earthly world	Space and time are separated, independent of each other	Earthly object	Space: 3-D distance Time: 1-D of past, present, future	Change within a 3-D frame over a period of time	Velocity of moving objects	Newton's laws
The universe	Space and time are one entity, a 4-D spacetime	Celestial body	Spacetime: 4-D distance	Change within a 4-D spacetime	Speed of light	Einstein's relativity
Electronic world	Space and time are separated, independent of each other	Electrical signal	Space: 2-D distance Time: 1-D of past, present, future	Change within a 2-D frame over a period of time	Propagation delay of traveling signals	Kirchhoff's laws, quantum mechanics

3.2 A Magnificent Show of Only Two Actors: Transistor and Signal

The edifice of microelectronics is built on transistor which is in turn made from resistor, capacitor, inductor and memristor. The collective effort of those transistors is directed to the handling of electrical signals. Although there are hundreds, or even thousands, types of chips available on market today, they all share the same and only purpose-of-life, namely, to process information. The information processed by all those chips is embodied in electrical signals, each one of those signals is the effect of a group of electrons' collective movement in the dimension of time. This comprehension is illustrated in the drawing on the left of Fig. 3.1.

For electrical signal, its strength is quantified by a parameter indicating level-of-magnitude, which is materialized through either electrical voltage or current. This parameter (i.e., the strength of the electrical voltage or current) must also be qualified

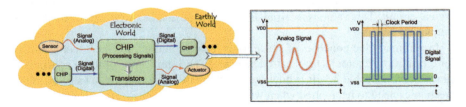

Fig. 3.1 All chips are created to process electrical signals: in either analog or digital fashion

3.2 A Magnificent Show of Only Two Actors: Transistor and Signal

by a time marker, which is embodied through the special signal called clock. Hence, for each one of all those signals, two qualifiers are created to characterize it so that information can be faithfully expressed. There are two approaches in the construction of this information-bearing electrical signal: the analog way where every point in the scale of magnitude is meaningful, and the digital way in which the magnitude is only distinguished by the two levels of high and low. For silicon chips and its upper-level structure of electronic systems, large or small, simple or complex, they all can be abstracted and modeled, without exception, as electrical machine for processing those signals.

Inside electronic world, each electrical signal, which is the collective effect from the movement of a group of electrons, must originate from some transistors and terminate at other transistors (Here, the term transistor is used in its general sense. It can include diode, resistor, capacitor, inductor and memristor). Hence the cast in this drama of microelectronics are transistor and signal. There are no other actors than those twos. For this magnificent show, the stage is set with the two fundamental entities of space and time as the backdrop. The name of the show performed on the stage is called "change and motion". After the chip or system is physically fabricated, transistors become stationary (not alterable any more). Signals then start to play the leading role in performing this magic. The essence of this magnificent show is illustrated in Fig. 3.2.

For our observer lived in real world, what we can see though our eye and perhaps touch through our hand, is the physical structure of the transistors; what we can feel, through some special instruments such as oscilloscope, is the behavior of the electrical signal. Therefore, from the perspective of an audience of this show, the two actors have their distinguished characteristics: the physically touchable transistor and

Fig. 3.2 The drama of microelectronics is played by transistor and signal on the stage set by space and time, the show is change and motion

the electrically observable signal. The signal, like a ghost, moves among transistors which can be viewed as houses. Everywhere he goes, he leaves a trace.

3.3 Space and Time as Real Estate in Microelectronics Business

Whatever majestic power a chip seems to possess, interaction among signals is the only thing that matters in achieving its function and it is the only observable activities inside the chip. As said, electrical signal is the collective behavior of a group of electrons. Under the hood, a variety of states emerging from the movement of this group of electrons are used to create something meaningful. The change-of-state is a physical phenomenon occurred inside chip that bears discernible characteristics in its behavior and thus can be utilized to encode information. Change-of-state by definition must occur in two different points in the one-dimensional time, resulting in a motion of electrical nature. When a chip or electronic system is powered on, the family of electrons continuously creates states; and the states constantly change their status over time. This continual flux-of-state can be utilized to express sophisticated connotation, and thus information springs up. Going one step deeper, viewing from the lens of physical structure, the very existence of signal and state is only possible when there are electrons. Electrons originate from real material such as transistors and metal wires. Transistors and wires are made of matter. Matter consists of atoms, which have sizes and hence occupy space. Following this chain of logic, space is the first piece of real estate for microelectronics business (i.e., the semiconductor chip industry).

When the Universe is inspected at atom level, there is no view of time. The Universe at its most basic level of fundamental particles is perfectly reversible. Irreversibility was only a statistical effect affecting aggregated particles. The arrow of time emerges only when a large number of atoms congregate together and then collectively create material object (Coveney and Highfield 1991). Transistors, and its super structures named chip or system, are made from aggregated particles, as everything else in this world. Therefore, the thing that we call time will play a role when we use transistors to create circuit, chip and system. The role that the time plays is to create order so that the electrical signal can be temporally qualified and the continual flux-of-state discussed previously can be sequentially played out. Without time, there is no order and all the activities in the world become meaningless noises. Hence, time must be the second piece of real estate in creating chip. By this argument, a statement can be made that space and time are the real estate necessitated to carry out the business of microelectronics. This point is illustrated in Fig. 3.3.

Against the background of four-dimensional space and time, as shown in Fig. 3.3, the foundational physical devices that control the electrons' movement are abstracted and modeled as resistor, capacitor, inductor and memristor. The collective effect of electrons' movement is modeled as electrical current (or voltage when the current

3.3 Space and Time as Real Estate in Microelectronics Business

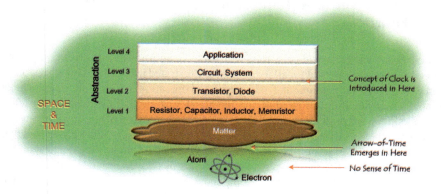

Fig. 3.3 Space and time are the real estates in the business of microelectronics

passes through a resistor). In next level of abstraction, transistor and diode models are created from those basic devices. Transistor can act as signal amplifier (for analog circuit) or electrical switch (for digital circuit). Further above, in circuit and system level where transistor is treated as black box with pins for communication, circuit designers have two set of tools available to them for creating the circuit and system: voltage (or current) magnitude and clock frequency. Electrical current is created from matter (movement of electrons) and its level (or magnitude or strength) is used to encode information. In light of this, in microelectronics engineering, level is proportional to the number of electrons involved in the process, and voltage and current are two artificial metrics for representing the magnitude of this level.

In this abstract world made of current and voltage, Kirchhoff's Laws take control. Meanwhile, at this abstraction level, the crucial concept of clock is created, as illustrated in Fig. 3.3. In its founding principle, clock signal is created to artificially indicate the passage of time. In circuit operation, clock frequency is used to regulate the pace of all actions inside the circuit. Hence, in the pedestal of microelectronics lies the two fundamental cornerstones of our universe: space and time. When creating electronic devices for information processing, both the space and time are real estates that we can use. They are equally valuable and we need to use both of them wisely. The rationale of using just space is straightforward and incontrovertible, but not intelligent enough.

3.4 Clock: The Most Important Signal and Its Three Functions

Clock signal is an electrical pulse train with a fixed length-in-time on all its composing pulses (i.e., a stable period for all its cycles), as illustrated in Fig. 3.4. This signal plays three distinct roles: fundamental, sequential and operational. The first one is its most fundamental function of materializing the arrow of time. All elementary particles, such as silicon atoms or electrons, do not possess the capability of sensing the flow-of-time when they act as individuals. Only when a large number of them group together as one entity, the sense of time emerges. The mechanism can be elucidated with the help of an important concept in physics called entropy (Coveney and Highfield 1991). As an example, the phenomenon of "component aging" observed in chip products and correspondingly used as a term in semiconductor industry concerning reliability is a continuous process of entropy-increase. This process has a deterministic direction, which is inherently an indication of the flow-of-time.

For the art of circuit design, we need a method to reveal the direction of time flow since there is no wrist watch or wall clock inside electronic world. Therefore, here comes the clock signal. The arrow-of-time is explicitly materialized by this clock pulse train since each pulse is sequentially generated and each pulse is inherently assigned with a number, which increases monotonically. In practice, a threshold voltage is assigned (either implicitly or explicitly) to qualify the clock pulse train. Whenever the voltage level of a pulse in this train of pulses crosses the threshold, a moment-in-time "*now*" is generated. Each "now" in this series of moment-in-time is labeled with a marker (either implicitly or explicitly), illustrated as t_1, t_2, t_3 ... in the Fig. 3.4.

The second function of clock signal is to index events occurred inside chip. Inside silicon chip, there are two types of observable entities as discussed in Sect. 3.2: physically the transistors and electrically the signals. The numbers of both the transistors and the signals can be very large, in the range of millions or billions for a large chip. Oftentimes, a specific situation or state collectively created by a group of electrical signals is abstracted as an event. This abstraction of reality is invented to facilitate the process of circuit design, especially for large complicated digital design. For events to be properly utilized, each of them needs to be tagged with a stamp of sequential order. The numbers ①, ②, ③, ..., labeled on the cycles of the clock signal serve this purpose. They are not the same as t_1, t_2, t_3 ..., which only have symbolic meaning. Those number are used to identify the clock cycles for doing certain real operations.

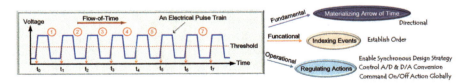

Fig. 3.4 The clock signal and its three functions

Only after events are properly indexed, order can be established within the chip. This action of tagging events is the clock's functional role.

Clock signal's third role is to regulate all the operations occurred inside the chip. When analog signal needs to be converted to digital signal, or vice versa, the conversion has to take place at some purposeful, and usually predetermined, moments. This process is controlled by the clock signal. On the other front, for modern digital design, synchronous design principle is the mainstream methodology. In this approach, clock signal regulates the structure of the synchronous digital circuit (i.e., the circuit structure is shaped by the clock frequency). Moreover, in some cases, clock signal also functions as the global switch for certain types of circuit design styles. For example, in dynamic domino circuit and DRAM (dynamic random access memory) circuit designs, clock signal's edge serves as the switch for "pre-charge and discharge".

Without the three functions provided by clock signal, a silicon chip is just a collection of atoms. It is possible that some random (or meaningless) electrical signals could spontaneously appear from this collection of atoms. Life however emerges only after the clock signal is defined. Depending on a chip's complexity, there could be hundreds, thousands, millions, or billions signals inside a silicon chip. Among those signals, without any doubt, the most important one is the clock signal (the second most important one perhaps is the global reset or enable signal). From a craftsman's perspective, clock signal is also the most difficult one to deal with. This is due to the reasons that: (1) clock signal electrically has the greatest fanout (loading); (2) it physically spreads to the largest area; and (3) it operates at the highest speed of the chip and it consumes the largest chunk of overall power budget of the chip. For any chip design project, clock planning is an issue of top priority. The importance of clock signal can never be overstated, as discussed in topic 25 of (Xiu 2007), Chap. 1 of (Xiu 2012) and Chap. 1 of (Xiu 2015).

3.5 Pursuit of Computation Efficiency from the Perspective of Space: Moore's Law

Moore's Law is a symbol of technology innovation. The term is deeply rooted in semiconductor industry. It is the observation made by Gordon Moore in 1965 that the number of components per integrated circuit doubles every year (Moore 1965). In 1975 the forecast was revised to doubling every two years (Moore 1975). The period is often quoted as 18 months when the combined effect of more transistors and transistors being faster is considered. Although Moore's Law is an observation or projection and not a physical or natural law, its prediction is proven to be rather accurate for over five decades. It has been used in the semiconductor industry to guide long-term planning and to set targets for research and development (International Technology Roadmap for Semiconductors (ITRS) and Schaller 2004). Advancements in world economic growth in the past decades are strongly linked to the Moore's law. It describes a driving force of technological change, social change, productivity increase, and economic growth (Bondyopadhyay 1998; Rupp and S. Selberherr 2010; Flamm 2017). The unremitting effort around Moore's Law can be captured in three phrases: *More Moore*, *More Than Moore* and *Beyond Moore*, each is accompanied by a profound insight in its way of pursuing the goal of achieving the highest computation efficiency. The computation efficiency discussed here is referred to the computation power obtained from unit resource. The resource is something that is monetarily measurable. In the case of chip design, the content of resource is consisted of two key elements: silicon area and power consumption. Those three strategies of More Moore, More Than Moore and Beyond Moore all aim at squeezing more computation power from a given silicon area by using as less power as possible.

More Moore is the strategy of continually engaging the transistor scaling down. It evolves from constant-field scaling to constant-voltage scaling to equivalent scaling. Miniaturization is its distinguished characteristic. The issues involved are: lithography, power supply and threshold voltage, short-channel effect, gate oxide, high-field effect, dopant number fluctuation, and interconnect delay (Taur et al. 1997; Frank et al. 2001). In the early stage of Moore's Law, transistor scaling down also improves speed and reduces energy consumption. This is known as Dennard Scaling (Dennard et al. 1974; Bohr 2007). While Moore's Law states that more transistors could be packed into the same area from generation to generation, Dennard Scaling ensures that each individual transistor in a new generation would be faster and draw less power. This triple benefits of smaller, faster and cooler led to the rise of affordable personal computers in the 1980s. However, the breakdown of Dennard Scaling in the middle of 2000's stopped this continuous trend of twenty years. The heat-removal problem has prevented the clock-based scaling (i.e., smaller and faster). Instead, compute capability is enhanced by adding more CPU cores (multi-cores architecture) and improving single-threaded CPU performance. Moore's Law hence continued once more from 2005 through 2014: transistors' speed gain might not be greater than that of their predecessors but they were more power efficient and less expensive to build; chips might have more transistors on-board but not all of them are able to be turned on simultaneously (dark silicon) (Gargini 2017; Theis and Wong 2017).

Meanwhile, semiconductor manufacturers have come up with innovations like strained silicon, hi-k metal gate, FinFET, and FD-SOI. But none of them has re-enabled the continuous geometrical scaling that Dennard Scaling has offered. Then comes the More Than Moore strategy. It takes the challenge from the other direction: rather than making the chip better and letting the application follow, it starts with application. From smartphones and supercomputers to data centers in the cloud, it works downward to see what chips are needed to support them. The idea of More Than Moore is not to focus solely on the computing power of a single chip but also to observe the efficiency of the whole system from a higher perspective. It encourages functional diversification, which refers to the integration of those functionalities that do not necessarily scale according to Moore's Law but provide additional value to the end application in different ways. It changes from a single technology transition to the integration of various technologies.

Moore's Law was initially proposed and verified in the development of the logic and memory circuits. More Than Moore examines the opportunity of integrating myriad functions at system level, which typically includes non-digital functionalities such as analog, RF, sensor, actuator, embedded DRAM, MEMS, high voltage circuit, power control, passive components, and so on. From new type of transistor structures, process compatibility of various types of circuits, to advanced packaging technologies, More Than Moore is to improve the overall integration efficiency and make a system capable of supporting more functions, and at the same time reduce overall system cost. In essence, it evolves from the "cheaper, better, faster" of More Moore to "better and more comprehensive".

Currently we are pushing the limit of silicon-based CMOS. The fundamental physical limit on the size of an atom will cause a hard stop. New technologies like multiple patterning, immersion lithography, and 3D Tri-gate transistors can probably support chips with process of few nm. The question now is what will happen after that, when quantum effect comes into play and continued scaling is no longer possible? This is the domain of Beyond Moore. One option is to go 3D, which is an architectural approach: stick with silicon but configure it in a new way (DeBenedictis et al. 2017). Instead of just simply etching flat circuits onto the surface of a silicon wafer, we can stack many thin layers of silicon with circuitry etched into each of them. In principle, this should make it possible to pack more computational power into the same space. In practice, however, it works only with memory chips that do not have serious heat problem.

On the other hand, there are several prospects on the radar to replace CMOS transistor. Many of those alternative devices operate on state variables other than charge and some of them may offer functionalities beyond those of a binary device. They could be useful for more complex operations. Examples of such novel devices include novel materials FET (III–V, Ge, carbon nanotubes and graphene), SpinFET, Spin-Torque, Spin-Wave, tunneling transistor (TFET), piezo-electric transistor (PET), molecular switch, NEMS, thermal transistor and etc. Beyond Moore can also include ideas of biologically inspired ways to compute (neuromorphic computing, which aims to model processing elements on neurons in the brain), approximate computing, superconducting computing and etc. (Cavin et al. 2012; Nikonov and Young 2013; Bernstein et al. 2010; Theis and Solomon 2010; Shalf and Leland 2015).

The strategies of *More Moore*, *More Than Moore* and *Beyond Moore* all share one common belief: more computation power only comes from using more processing units. This creed on brute-force of using ever-more processing units (transistor, diode, resistor, capacitor, inductor) to gain ever-stronger computation power is reflected, in one way or another, on all the practices so far in the entire semiconductor history. Matter occupies space. The number of processing units is simply proportional to the silicon space allocated to the task. Therefore, it is reasonable to say that Moore's Law is the practice of squeezing computation power from space.

3.6 The Recognition of a Crisis

The force of Moore's Law enables us now to produce monster chips with multi-billion transistors on-board. However, when the sizes of wire and transistor get too small, electrical current (collection of electrons) begin to stray from their designated paths and short circuit the chip. One solution to alleviate this problem is to go vertical, from planar transistor to 3D transistor. But the trend of shrinking-transistor-channel-length still makes continuous minimization a futile endeavor. We are hitting the brick wall of physics: transistor is running up against its atomic size limit (Zhirnov et al. 2003; Powell 2008).

On another front, as we pack more transistors into a given area, or stacked more transistors into a given volume (to include the utilization of vertical direction), power density increases quickly and the problem of heat removal becomes increasingly difficult to be dealt with. Moreover, as more transistors and signals are squeezed into an ever-smaller space, the quality of signals suffers badly from the phenomenon of signal-interference since the influence of inter-signal electromagnetic effect become stronger. As signal traveling relatively longer distance in the ever-larger chip, as signal toggling at higher frequency, as the resistance and capacitance of interconnect wire increasing, as supply voltage lowering and its variation widening, as process variation deepening, this signal integrity problem keeps getting worse for every new generation of technology.

All the aforementioned problems are due to the very reason that we try to make the basic processing unit (i.e., the transistor) smaller and pack more of them into a given space as we move along this path of pursuing ever-higher computation power. Since 1990s, the semiconductor industry has released the research road map ITRS every two years to coordinate the industry-wide effort of marching forward. In 2017, the update on ITRS is stopped. This is a clear sign that the end of size-minimization is near and we now face a crisis.

With the benefit of hindsight, the emergence of a crisis often starts with a period of extraordinary, rather than normal, research and development activities. As Kuhn pointed out in 1962 (Sect. 8 of (Kuhn 2012)), the symptoms of the pre-crisis extraordinary period can include "proliferation of competing articulations, the willingness to try anything, the expression of explicit discontent, the recourse to philosophy and to debate over fundamentals". Those symptoms are clearly what we are seeing right now in the semiconductor industry. The argument of "Moore's Law is dying" has been floating around for more than a decade. This is an explicit expression of anxiety for an uncertain future or discontent of status quo. "The willingness to try anything" is evidenced in many aspects of current semiconductor R & D, such as the search for CMOS transistor's replacements in Beyond Moore effort (quantum logic, superconducting logic, photonics/plasmonics, spintronics and valleytronics), the billion dollar investment on advanced lithography tools (e.g. High-NA EUV), the courage of spending hundreds of millions of dollars to go through several rounds of testchip design for a not-yet-secured future product (initial design on speculative model, updated design on silicon-influenced model, final design on silicon-based

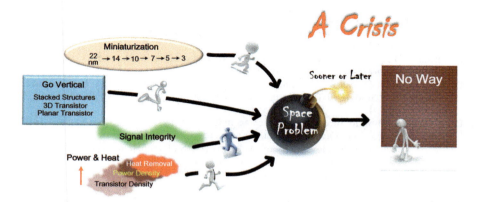

Fig. 3.5 The crisis caused by current practice of Moore's law

model), and the determination of tedious and labor-intensive design-iteration effort on dealing with technology-imposed non-idealities (speculative device model update, R & C parasitic extraction, layout-dependent effects, strict layout density rule, DRC update, strict ESD & latch-up rules and new effects) (Yeric 2019 and Loke 2019). Moreover, the "proliferation of competing articulations" is obviously embodied in the three strategies of different focuses around the Moore's Law and in the great diversity found within each of them.

Therefore, now, it can be reasonably said that the current research and development effort in semiconductor industry is no longer in the path of a steady progress through puzzle-solving activities towards a pre-established goal (i.e., the miniaturization of transistor). Rather, it progresses away from what once worked well but can no longer handle its new problems as efficient. We are running into a crisis. And, as discussed in previous texts and illustrated in Fig. 3.5, it is a space-related crisis.

References

K. Bernstein, R. K. Cavin, W. Porod, A. Seabaugh, and J. Welser, B, "Device and Architecture Outlook for Beyond CMOS Switches," *Proc. IEEE*, vol. 98, no. 12, pp. 2169–2184, Dec. 2010

M. Bohr, "A 30 Year Retrospective on Dennard's MOSFET Scaling Paper," *IEEE Solid-State Circuits Society Newsletter*, vol. 12, pp. 11-13, 2007.

P. K. Bondyopadhyay, "Moore's Law Governs The Silicon Revolution," *Proc. IEEE*, vol. 86, no. 1, pp. 78–81, 1998.

R. K. Cavin III, P. Lugli, V. V. Zhirnov, "Science and Engineering Beyond Moore's Law," *Proc. IEEE*, vol. 100, pp. 1720–1749, 2012.

P. Coveney and R. Highfield, "The Arrow of Time: A Voyage through Science to Solve Time's Greatest Mysteries," *Fawcett; 1st American ed edition*, 1991.

E. P. DeBenedictis, M. Badaroglu, A. Chen, T. M. Conte, and P. Gargini, "Sustaining Moore's Law with 3D Chips," *Computer*, vol. 50, no. 8, pp. 69–73, 2017.

R. H. Dennard, F. H. Gaensslen, V. L. Rideout, E. Bassous, and A. R. LeBlanc, "Design of ion-implanted MOSFETs with very small physical dimensions," *IEEE J. Solid- State Circuits*, vol. 9, no. 5, pp. 256–268, Oct. 1974

K. Flamm, "Has Moore's Law Been Repealed? An Economist's Perspective," *Computing in Science & Engineering*, vol. 19, no. 2, pp. 29–40, 2017.

D. J. Frank, R. H. Dennard, E. Nowak, P. M. Solomon, Y. Taur, and H. S. P. Wong, "Device scaling limits of Si MOS FETs and their application dependencies," *Proc. IEEE*, vol. 89, no. 3, pp. 259–288, 2001.

P. A. Gargini, "How to successfully overcome inflection points, or long live Moore's law," *Computing in Science & Engineering*, vol. 19, no. 2, pp. 51–62, 2017.

International Technology Roadmap for Semiconductors (ITRS), [Online], available: www.itrs.net

T. S. Kuhn, "The Structure of Scientific Revolution," the *Univ. of Chicago Press*, 2012.

Alvin L.S. Loke, C. K. Lee, and B. Mike Leary, "Nanoscale CMOS Implications on Analog/Mixed-Signal Design," *CICC*, Austin, TX, 2019.

G. E. Moore, "Cramming More Components Onto Integrated Circuits," *Electronics*, vol. 38, no. 8, pp. 114-117, 1965.

G. E. Moore, "Progress In Digital Integrated Electronics," *Proc. Technical Digest Int'l Electron Devices Meeting*, pp. 11–13, 1975.

D. E. Nikonov and I. A. Young, "Overview of Be-yond-CMOS Devices and a Uniform Methodology for Their Benchmarking," *Proc. IEEE*, vol. 101, no. 12, 2013, pp. 2498–2533.

J. R. Powell, "The Quantum Limit to Moore's Law," *Proc. IEEE*, vol. 96, no. 8, pp. 1247–1248, 2008.

K. Rupp and S. Selberherr, "The Economic Limit to Moore's Law," *Proc. IEEE*, vol. 98, no. 3, pp. 351–353, 2010.

R. R. Schaller, "Technological Innovation in the Semiconductor Industry: A Case Study of the International Technology Roadmap for Semiconductors (ITRS)," Ph.D. dissertation, 2004, George Mason University.

J. M. Shalf and R. Leland, "Computing Beyond Moore's Law," *Computer*, vol. 48, no. 12, pp. 14–23, 2015.

Y. Taur, D. A. Buchanan, W. Chen, D. J. Frank, K. E. Ismail, S. H. Lo, G. A. Sai-Halasz, R. G. Viswanathan, H. Wann, S. J. Wind, H. S. Wong "CMOS Scaling into the nm Regime," *Proc. IEEE*, vol. 85, no. 4, 1997, pp. 486–504

T. N. Theis and P. M. Solomon, B, "In Quest Of The 'Next Switch': Prospects For Greatly Reduced Power Dissipation In A Successor To The Silicon Field-Effect Transistor," *Proc. IEEE*, vol. 98, no. 12, pp. 2005–2014, Dec. 2010

T. N. Theis and H. S. P. Wong, "The End of Moore's Law: A New Beginning for Information Technology," *Computing in Science & Engineering*, vol. 19, no. 2, pp. 41–50, 2017.

L. Xiu, "VLSI Circuit Design Methodology Demystified: A Conceptual Taxonomy," *John Wiley IEEE press*, Piscataway, NJ, 2007.

L. Xiu, "Nanometer Frequency Synthesis Beyond the Phase-Locked Loop," *John Wiley IEEE-press*, Piscataway, NJ, 2012.

L. Xiu, "From Frequency to Time-Average-Frequency: A Paradigm Shift in the Design of Electronic system," *John Wiley IEEE-press*, Piscataway, NJ, 2015.

G. Yeric, "IC Design After Moore's Law," Session ES2–1, *CICC*, Austin, TX, 2019.

V. V. Zhirnov, R. K. Cavin, III, J. A. Hutchby, and G. I. Bourianoff, "Limits to Binary Logic Switch Scaling—A Gedanken Model," *Proc. IEEE*, vol. 91, no. 11, pp. 1934–1939, Nov. 2003.

Chapter 4
Response to the Space Crisis

4.1 Time and Clock: Recourse to Philosophy and Debate over Fundamental

People have a tendency to see what they expect, even when it is not there. It often takes a long time for a crisis, something contrary to the established order, to be seen for what it is. In the case of microelectronics engineering, we intend to believe that we can always get more processing power from packing more transistors into given space. The space crisis argued in previous discussion is difficult to be recognized. But once recognized, it leaves us with no option but takes serious action. Now, in this author's belief, we are forced to turn our attention to the other piece of resource in the real estate available to us: the thing we called time. In the past course of microelectronics, time is only used in a very straightforward fashion. For any given design, a clock pulse train of fixed period (frequency) is simply employed as event-indexer and operation-regulator. However, to dig more potential from time for designing better circuit and system, we need to understand time better, we need to scrutinize time in a much greater depth.

"What, then, is time?" inquired Saint Augustine around 400 AD in his *Confessions*, notably pointing out a paradox in our conception of time. Nothing was more intuitive yet more complicated than time itself: "If no one asks of me, I know; if I wish to explain to him who asks, I know not." The irreparable time that flees, the impossibility of remaking the past and also of knowing the future, are some of the most thought-provoking problems in our civilizations. Since ancient times, philosophers had debated about the relation of the discreteness in the form of still instants to the continuity in the form of movement. How can a flying arrow both move and occupy a fixed length in space? Zeno's paradox showed just how hard it is to answer that question. In modern times, these ancient questions are still debated, perhaps not by reference to flying arrow but to cinematographic machine. Instead of debating about how to read the Bible, thinkers across a wide variety of disciplines debated about how to read the complex unfolding of nature through time. The discussions can happen from the most private to the most public places, can take the form of from the most scientific to the most philosophical, can carry out at the level of from the most profound to the most informal.

In history, the deepest discussions on time had taken place among scientists (especially, physicists in modern science) and philosophers. For Isaac Newton, time was God-given and as absolute and perfect as the deity himself. He had described time in both theological and scientific terms. For him, absolute time was defined by recourse to an absolute observer, an omniscient consciousness, which he attributed directly to God. The notion of clockwork universe was created by Newton, who believed that the solar system, once wound up and set in motion by God, had simply continued ticking along with occasional help from the Almighty. The "sensorium of God" guarantees the existence of absolute time, "endures forever and is everywhere present; and, by existing always and everywhere, he constitutes duration and space."

A century later after Newton, a secularized interpretation of time gained prominence. The philosopher Immanuel Kant argued that time, alongside space, could not be studied directly. Kant taught us that absolute Newtonian space and the principle of uniform causality are a priori principles of thought, necessary conditions on how human beings comprehend the world in which they live. In Kant's view, our sense of physical time and space arises from a certain primary proclivity of our minds to organize experience in these terms. Although in his view space and time are no longer a feature of the God-given universe itself, time nonetheless remains a universal and single concept.

Einstein's special relativity work dispenses with these prior notions by the fact that it is based on clock, which is the measurement of the time. In his theory, time is represented by variables of t_1 and t_2, which could be expanded in an infinite series represented by t_n. It needs neither God nor consciousness to sustain it; it could be described perfectly by simple recourse to a measurement device of clock. The theory of relativity breaks with classical physics in three main respects: first, it redefines the concepts of time and space by claiming that they are no longer universal; second, it shows that time and space are related; and third, the theory does not set its footing on the concept of the ether, a substance that allegedly fills empty space and that scientist had hoped would provide a stable background to both the Universe and their theories of classical mechanics. Different from Newton and Kant, Einstein mathematizes the time. In his four-dimensional view of the Universe, the past, the present, and the future are already laid out and predetermined; our sense of time is merely an illusion.

Einstein's work pushes the time-related debate to the limit of classical mechanics. Quantum mechanics goes still one step further by directly challenging the causality. Its differences from relativity theory and classical physics are reflected in various respects. First, it revolutionizes the concept of measurement by claiming that the act of measurement itself changes the experimental system. Second, it introduces limits to what could be known with measurement. The measurement of a particle's position introduces a certain degree of uncertainty in the measurement of its momentum. These two principles have a dramatic consequence. Quantum mechanics forces people to reevaluate the idea of physical causality, introducing an essential indeterministic quality into the Universe. Cause and effect are mere appearance, and indeterminacy is at the root of reality.

There are two different ways of noting time, one human and subjective, and the other clocklike and objective. Subjective understanding of time is frequently related to meaningful moments, places, and events, which are popularly described in literary and poetic ways. With the spread of wristwatches and electronic time displaying devices, objective measurement of time becomes an integrated part of our daily life. The difference hence becomes increasingly worthwhile noting. The pure mathematical treatment of time by relativity theory, and later the role of measurement and the indeterministic nature of universe brought up by quantum mechanics, forced thinkers to turn to the subjective side of the time: the unquantifiable aspects of time. The philosopher Henri Bergson insisted that time should not be understood exclusively through the lens of science. It had to be understood philosophically. Bergson's central contention is that time is not measurable by any objective standard. For him, time includes aspects of the Universe that could never be entirely captured by instruments (such as clock or recording devices) or by mathematical formulas. Determining time, according to Bergson, is a complex operation. "To know what time it is" was not simply about reading a number given by an instrument (e.g. a clock). It is an assessment of the overall meaning of that moment. Time, he argued, is not something out there, separated from those who perceives it. It does not exist independently from us. It involves us at every level (Canales 2015).

By measuring time with a device, in Bergson's view, scientist destroys some of it. He espouses a theory of time that explains what clocks does not: memories, premonitions, expectations and anticipations. Yes, clock is bought to "know what time it is", admitted Bergson. But "knowing what time it is" presupposes that the correspondence between the clock and "an event that is happening" is meaningful for the person involved so that it commands his attention. That certain correspondence could be significant for us; it could explain our basic sense of simultaneity (a crucial piece of special relativity theory). Clock, by itself, could explain neither simultaneity nor time, Bergson argued. He insisted that when scientists measure time, they remove from it what is the most important aspect of time: its flow and its relation to duration (Bergson and Jacobson 1999).

4.2 Three Types of Time: Mechanical, Secular and Electronic

Since the beginning of recorded human thought, the distinction between past, present, and future is one of the most puzzling issues in our exploration of the nature and the development of our self-awareness. When time is concerned, there is a clear separation between subjective and objective factors. It could be determined physically, physiologically, and psychologically. As a result, this time-related subject has been studied in many aspects, from physical astronomy to moral philosophy. We can probably all agree that the "lived time" that we experience (a.k.a. the secular time) is different from the measured "clock time" (a.k.a. the mechanical time). These two experiences of time, the lived time and the clock time, mark the difference between the living and the mechanical. Further, they are symptomatic of the broader divisions of irrationality and rationality where the first is associated with experience and the second with science. It is therefore logical to ascribe the study of the subjective realm to psychology and philosophy, the study of objective events to physics.

The clock time and the lived time, one mechanical and one living, serve two different worlds of the Universe and our lives. "The time of the Universe" studied by scientists like Newton and Einstein and "the time of our lives" associated with philosophers such as Kant and Bergson have been historically in the conflicting paths, splitting the intelligent society into two cultures and pitting scientists against humanists, expert knowledge against lay wisdom (or intuition). This leads to the longstanding rivalries between science and philosophy, physics and metaphysics, objectivity and subjectivity, with heavy influence on pragmatism, logical positivism, phenomenology, and quantum mechanics.

The mechanical time is sometimes also referred to as the concrete time wherein real systems develop, and the lived time as the abstract time that enters into our speculation on artificial systems. The intuition of time is the aspect of our temporal sense that is short of repetitiveness, cadence, and homogeneity. It is the intuition of the moving character of reality. It could not be used to quantify the time. Lived time is supported by the circadian clock that is the master regulator of mammalian physiology. It regulates the daily oscillations of crucial biological processes and behaviors. In most living things, circadian clock makes it possible for organisms to coordinate their biology and behavior with daily environmental changes in the day-night cycle. It controls a number of bodily functions, from feeding to sleeping. For plants, it regulates a diversity of processes including photosynthesis, metabolic activity, flowering and responses to pathogens. On the other hand, mechanical time is machine-like. It is based on nature phenomena. The most notable example of mechanical time is the system based on the rotation of the earth against the stars. People refer to this method as the sidereal clock, which probably is the first vehicle from which our sense-of-change and flow-of-time develop.

The groups embracing lived time and mechanical time have battled each other in various stages. Philosophers accuse scientists of confusing reality with measurement by mathematizing time. They argue that the time in scientific theories' equations is not our everyday common notion of time; physicists simply deal with mathematical time whereas common sense and philosophers are concerned with real time. Some key philosophers view Einsteinian time in large part as fictional or imaginary since, in their opinion, a system of time measurement based on the constancy of the speed of light should not be confused with time itself. The twin brother paradox brought up by Langevin in the interpretation of special relativity promoted Einstein to ask if the delay in the time marked by a clock would also affect biological processes, not just physical ones. Quantum physics heats the debate further by creating a link between psychic time and physical time.

Time is an issue of scientific, technological, philosophical, historical, and everyday knowledge all together. The debate on time in relating to the Universe and our lives reflects how science migrates from the concrete to the abstract and how mathematical representation is transformed into transcendental reality. From those thought-provoking debates, people recognize the irreversible time of evolution toward equilibrium, the rhythmic time of structures whose pulse is nourished by the world they are part of, the bifurcating time of evolution generated by instability and amplification of fluctuations, and microscopic time which manifests the indetermination of microscopic physical evolutions. The contradictions in different school-of-thoughts divide the world into science and art, public and personal, objective and subjective, abstract and concrete. It represents two competing strands of modern times: vitalism is contrasted against mechanization, creation against ratiocination, and personality against uniformity. It is a show of sign that, deep down in our heart, nobody could draw a firm distinction between the world of sensations and the world of the mind (Canales 2015).

When practitioners in microelectronics creates the electronic world from transistors, it brings up the fresh issue of establishing order inside this new world (i.e., inventing some form of flow-of-time), hence the emergence of a new type of time (a.k.a. the electronics time). This third kind of time raises challenges and also presents opportunities. The most important concern regarding this type of time is the controversy that whether it is a time attached to a living thing with consciousness (does an electronic machine have self-awareness?) or simply a non-living mechanical clock just for measurement and control (i.e., everything inside this electronic world, including the creation of clock signal, is controlled by the almighty circuit designer of the system, the "God" of this electronic world).

4.3 Microelectronics as an Evolution Process and the Emergence of Time Within

Billions of years ago, according to theory of evolution, chemicals randomly organized themselves into a self-replicating molecule. This sparking of life was the seed of every living thing we see today. That simplest life form, through the processes of mutation and natural selection, has been shaped into every living species on this planet. During this astonishing process, the mechanism of evolution has demonstrated its magnificent power. In a layman's term, evolution is the noticeable change in the heritable characteristics of biological populations over successive generations. These characteristics are the expressions of genes that are passed on from parent to offspring during reproduction. Different characteristics tend to exist within any given population as a result of mutation and genetic recombination. This biological process of evolution helps species to adapt to their environments over years and it results in the colorful diversity of life-forms on the earth.

The growth of microelectronics from nowhere to current status can also be investigated through the lens of evolution. Staring from the grown-junction transistor invented in the 1940s, electronic system gradually grows from small devices made of few simply-structured transistors to monster systems with billions of sophisticated transistors on board. In this short period of seven or so decades of electronic evolution, compared to the billions of years of biological evolution, the structure of the basic cell (i.e., the transistor) has mutated more than multiple times and the surviving strategy is the same principle of "adapting to environment". In biological world, the environment forces species to evolve along the direction of using resource in a fashion of ever-increasing-efficiency, becoming faster and stronger and smarter, generations after generations on a given amount of resource consumed. In microelectronics, environmental pressure comes from the never-ending-competition demand of using less power and smaller silicon area for processing the same amount, or more, information. The resemblance between the two evolution processes in the biological- and the electronic-world can be appreciated from Fig. 4.1 and Table 4.1.

Both the biological world and the electronic world are sitting on the background of earthly world (i.e., the Universe) wherein time is reflected by sidereal clock.

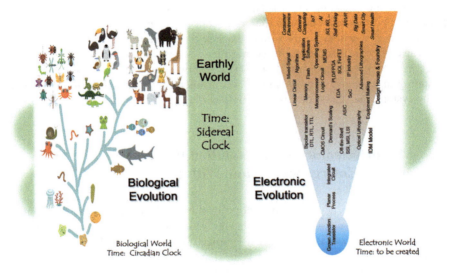

Fig. 4.1 The resemblance between biological evolution and electronic evolution

Table 4.1 The comparison of biological and electronic evolutions

	Biological evolution	Electronic evolution
Motivation	Multiplication of life	Generation and exchange of information
Goal	↑ Number-of-life/energy	↑ Number-of-bit/energy
Format	Becoming ever-diversified	Becoming ever-specialized
Mechanism	Gene mutation	Technological innovation
Governing law	Competition for survival	Competition for survival (Quantified as Moore's Law)

4.3 Microelectronics as an Evolution Process ...

For biological world, circadian clock is developed inside lives to keep life running on an orderly fashion. Inside electronic world, there is a similar demand when the task of sequencing is concerned. The electronic world, as a lifeless entity with no self-awareness (at least at current stage), does not however have the capability of developing something like the circadian clock. This is due to the reason that, at microscopic level, when all the things in nature are investigated, there is no evidence of directionality in their activities. This is most notably confirmed by the Brownian motion that shows perfect reversibility at molecular level. As a matter of fact, in all the laws of physics, scientists so far have found that there is seemly no any distinction between the past and the future.

In our study of the nature, certain aspects of life seem to flourish in ways that could only be explained as statistical exceptions. In physics, a temporal consciousness has been developed from the statistical and molecular theories of the thermodynamics and the law of entropy. For thermodynamics, the first law suits reversible phenomena, the second law explains irreversible processes. Although in our macroscopic dimension we can perceive time flowing in a certain direction, however, according to statistical and molecular theories of thermodynamics, our sense of arrow-of-time and of temporal irreversibility is actually based on the reversible effects at the microscopic or molecular level.

Inside electronic world, referring back to Fig. 3.3, the arrow-of-time emerges when the silicon material is formed from the aggregation of a large number of silicon atoms. This level of sophistication regarding time however is only useable for identifying some simple phenomena, such as material aging. It is not sufficient enough for the quantification of time. For the purpose of sequencing various tasks, something repetitive like circadian clock or sidereal clock is required. Thus, the electrical pulse train illustrated in Fig. 3.4 (i.e., the clock signal) is created to serve this purpose. As a device for time-keeping, this clock pulse train is unique in that neither is it self-developed from within a living thing (like the circadian clock) nor is it something existed in the nature (e.g., the sidereal clock). It is created artificially by human (i.e., the electronic circuit designer). Unlike the mechanical time inborn within the Universe where we have no control and different from the secular time yielded in our mind where we cannot peek into, the time in the electronic world is created by us and we can command it. For making sense of "this time", we have the right and responsibility to create whatever mechanism that can serve us the best. In other words, for our convenience, we are entitled to create the clock signal in whatever way we desire.

4.4 Connecting Space and Time by a Property of Clock Signal: Frequency

In this Universe, all objects are constantly moving and it takes time for an object to move from one physical location to another. In Newtonian and Einsteinian mechanics, space and time are only related through the motion of objects. Mechanically, time is reflected through motion, or the change of state. Among all the movable things in the Universe, according to Einstein's relativity theory, light ray is a very special one. Space and time are mathematically integrated by a property of light: the speed of light. This value is presupposed as a constant in relativity theory and it is the speed limit of the Universe (According to the Big Bang Theory, nothing can move faster than light unless it is the space–time itself. During the inflationary period, the Universe's expansion broke this cosmic speed limit.). Abided by this constancy, space and time influence each other through mathematical manipulation.

On another front, if we shift our attention from the magnificent Universe to the invisible electronic world, electrical signal is the information carrier inside this world fabricated by human. Once a chip is manufactured, the physical locations of all the information processing units (i.e., the transistors and etc.) are fixed. Electrical signals travel among the units located in various places. Hence, in this world, time is reflected through the motion of signals. A special signal, clock signal, is created as a reference frame for measuring and controlling other signals. By the same token, although not in exactly the same manner, we can use a property of this clock pulse train, its frequency, to connect the space and time inside this electronic world. Clock frequency is considered as the speed up-limit of all other signals, similar to the fact that speed of light is the speed limit of the Universe. The parallelism of the two mechanisms is illustrated in Fig. 4.2.

Both the light ray and the clock signal are traveling waves that link the space and time in their respective worlds. They all belong to the class of "transverse wave" where the direction of oscillation is perpendicular to the direction of wave traveling

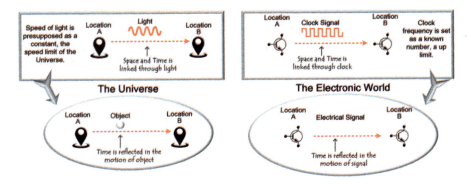

Fig. 4.2 Space and time are linked through motion: in Universe (right) and in electronic world (left)

4.4 Connecting Space and Time by a Property of Clock ...

(the other type is "longitudinal wave" where vibration moves parallelly in the direction of wave traveling). It is however important to point out that the analogy shown in Fig. 4.2 is presented here just for illustration purpose. Microcosmically, clock signal is the superposition of electromagnetic waves traveling among transistors and wires in a controlled way. In the eye of a physicist, light ray and clock signal essentially are all electromagnetic wave. The light ray has a frequency about 500 trillion Hz. The frequency of clock signal, after the superposition process, is much lower, typically in the range of a few million to several billion Hz.

Functionally, the light ray in Universe and the clock signal in electronics are both used for measurement. To measure with precision, something has to be compared against "something else" of unique quality. Ideally, this *something else* should be unchanging, so that subsequent measured values could remain comparable. Using mechanical, electrical, optical, chemical and biological methods, whatever process is employed for measuring, all the measurements should lead to concordant results. Preferably, the reference used for measurement should not be damageable by human or natural actions, so as to withstand the vicissitudes of time and history. In addition, it should be mobile, easily accessible, and reproducible, so that the standard, or a copy of it, could be used over and over again.

Hypothetically, all the involved parties should agree on the same standard. However, as it turns out, some of those ideals discussed above conflict with each other. A good example is the case of measuring space (i.e., the measurement of length), which consequently influences the measurement of time (at least before the adoption of the atomic definition of the second in 1967). Since its inception in 1800, scientists could not fully trust the very definition of the meter bar (the standard for measuring length). Then, by midcentury they could not trust the stability of the solar system and of the sidereal clock based on it. After that, a quarter of a century later they could no longer trust the Earth-Sun distance (which was considered a fixed length and used in astrometry intensively). Soon, they would admit that they could not even determine that the Earth moves through space (i.e., the failure on the search for ether).

During the period of late nineteenth century and early twentieth century, electromagnetism has rapidly entered into human's daily life in the form of various applications: telegraph, telephone, radio communications, cinematographic cameras and film, and etc. One distinguished quality unites all those applications is that they are all used for the storage and transmission of events-occurred-in-time over long physical distance. As electromagnetic communication networks (both telegraphic and wireless) increasingly crisscrossed the globe at that period, scientists became increasingly sure about one thing: the behavior of light on the surface of the Earth. Light signals can be optical (torch and semaphore), electrical (telegraph), and electromagnetic (wireless). Its usage was widespread as new technologies developed at that time. Eventually, Einstein proposed that the speed of light is a universal constant. Furthermore, he claimed that it is the highest speed of the Universe. From this presuppose, the space and time are no longer independent of each other but linked forever. Analogously, clock signal plays the similar role inside the electronic world. It sets the base for all the measurements since, from signals' point of view, there is no other straight-edge (i.e., reference object) for measurement. From the perspective of time, clock signal is the foundational reference framework for all the activities occurred inside the electronic world (die is the reference framework regarding space). Different from the light ray that is the creation of Mother Nature, clock signal is however artificially made through human engineering. It is created by the almighty circuit designer.

4.5 Clock Frequency: The Method of Indirect Period Multiplication (PLL) and Its Ineptitude to Solve Two Long-Standing Problems

Figure 4.3 shows the mechanism of generating the so-called electronic time from a clock signal, adding to the family of mechanical and secular times discussed in Sect. 4.2. From a source of mechanical vibration or electrical oscillation, a repetitive pattern is produced. A conversion unit, if the repetitive pattern is originally created in the form of mechanical vibration (such as in the case of crystal), is employed to convert the repetitive motion into an electrical pulse train that uses voltage's low- and high-level alternately to reflect the repetitiveness in electrical form. This electrical repetitiveness is a means that can be used to represent the flow of time, the electronic time mentioned previously. A threshold voltage, which is often implicitly incorporated in the driven circuit (such as the clock pin in a flip-flop which opens the "gate" when the voltage applied reaches a certain level), is set to convert the electrical repetitiveness into a series of moments of "now", $t_0, t_1, t_2, t_3, \ldots$ as shown in the figure.

This electrical pulse train is derived from a vibration or oscillation originated in nature, wherein each alternation from one state to the other state (or vice versa) is an integrated whole which cannot be divided into any smaller parts. As a result, a pulse in this electrical pulse train cannot be segregated into anything smaller. The time-span of the pulse therefore is an undividable unit and is determined by the originating vibration or oscillation. Oftentimes, the rate of the repetitive motion of the vibration or oscillation is shaped by the physical dimension of the oscillating device. Unless the physical structure is changed, this rate is fixed. In other words, the clock frequency associated with this pulse train is fixed once the device responsible for the vibration or oscillation is chosen. Fortunately, in the art of circuit design,

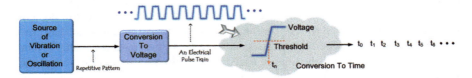

Fig. 4.3 The mechanism of generating electronic time from a clock signal

4.5 Clock Frequency: The Method of Indirect Period ...

a circuit called frequency divider can lower the rate (i.e., frequency) of a signal by an action of dividing-down, through a mechanism of event-counting plus event-triggering. Using this divider circuit, an elegant electronic component called Phase Locked Loop (PLL) is invented that can boost the frequency of its output pulse train to some multiples of its input's rate. Hence, from a primeval single-value-frequency oscillating source, PLL and divider provide us with some additional selectable values for clock frequency.

In the 1940s, the first widespread use of PLL was in the synchronization of the horizontal and vertical sweep oscillators to the transmitted sync pulses in television receivers. Those circuits were called synchrolock and/or synchroguide. Since then, the PLL principle has been extended to many other applications. In modern microelectronics wherein integrated circuit design is the core of the art, PLL is the most widely used, if not the only, tool for generating clock signal. Most electronic devices would not be possible if PLL technology is not invented.

Figure 4.4 depicts the structures of frequency divider and the typical PLL circuit. PLL is a feedback control loop wherein an electrical oscillator is embedded (called voltage-controlled oscillator, or VCO). The VCO's output frequency is controlled by a voltage applied on its input, which is fed from a loop control unit. The control unit receives input from a frequency and phase comparator, which compares the instantaneous phases and frequencies of an oscillating source and the VCO output. Optionally, the VCO output can be divided by a frequency divider before being fed into the comparator. When the loop reaches equilibrium (a state commonly called "lock"), the frequency of the VCO output f_o is related to the input source's frequency f_i as $f_o = N \cdot f_i$, where N is a number set in the divider. By this way, PLL is able to generate other higher rates (frequencies) from the rate of the original source (often called reference source, or reference for short).

Although the value in the frequency divider can be in principle set to a rational number with both integer and fractional parts, the core circuit of frequency divider can only be implemented in integer fashion. This is due to the reason that the operation of division is cycle based, or event based, which cannot be divided into any sort of sub-events. In practice, fractional-number frequency division is sometimes seen in use. However, the effect of fractional-number frequency division is achieved by averaging between two or more integer numbers over a period of time (the so called fractional-N PLL). The fractional-N PLL, compared to its integer-N counterpart, comes with a price tag of higher cost (in terms of silicon area and power consumption) and degraded performance (in terms of jitter or phase noise). For those reasons, inherently, PLL

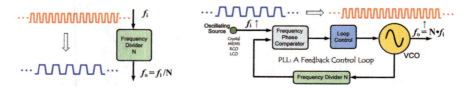

Fig. 4.4 Using frequency divider (left) and Phase Locked Loop (right) to generate other frequencies

is only good for producing integer multiples of the reference frequency. In other words, for a given design, the number of frequencies that can be generated from PLL is limited.

From Fig. 4.4, it is seen that PLL is a feedback control loop. Therefore, when the value of N is changed, it takes certain amount of time (typically several tens or hundreds of input cycles) before the PLL can reach a new equilibrium. This leads to another unpleasant fact that we cannot change its output frequency quickly. This is due to the "compare-then-correct" feedback mechanism innately built inside the PLL. Hence, slow response is another painful problem in the domain of PLL design.

PLL is an amazing system that blends digital- and analog-circuit design techniques beautifully in one package. It unites the frequency flexibility of voltage control oscillator with the frequency stability of the reference oscillator, to produce other desirable frequencies. Interestingly, it is also worth to point out that PLL is a great example of such a group of electronic devices that, in a philosophical sense, link the future with past through the mechanism of feedback. PLL is one of the most widely studied subjects in the field of circuit design. During the years, many technical papers and books have been developed around this topic. Several of commonly referenced books on PLL are listed in reference (Gardner 2005; Best 2007; Egan 1999; Razavi 2008).

However, as powerful and beautiful as it is, PLL is unable to fulfill two requirements that are badly needed in applications: generating arbitrarily demanded frequency and changing frequency in a rapid fashion. This unfortunate reality is illustrated in Fig. 4.5. In the right-hand side of that figure, it is seen that we need two dimensions of voltage and time to fully describe an electrical signal, either analog or digital. In order to better serve the purpose of embodying information, we demand these two features or capabilities of "arbitrary generation of any value" and "quick change of the current value" on both the voltage and time. In voltage domain, after several decades of great effort, it is fair to say that the problems have been solved. For any particular project, sophisticated circuit techniques are available to arbitrarily generate any voltage level we want and to change voltage potential quickly from one level to another. In time domain, however, the problems still remain unsolved in spite of the great PLL technology. The primordial reason behind this reality is that time, unlike voltage level that structurally ties to electrons, cannot be materialized by itself. Time has to show its face through the help of other media, such as the

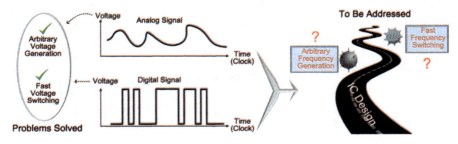

Fig. 4.5 The two long-standing problems in clock frequency generation

mechanical movement in the case of mechanical watch or the water flow in water clock.

In microelectronics, time is reflected through voltage as illustrated in Fig. 4.3. In other word, compared to the case in voltage domain, the solution to those two problems in time domain is much harder to reach. Putting this in a more formal phrase, we demand a clock signal generator having the capabilities of

- arbitrary-frequency-generation, at least in ppm range
- instantaneous-frequency-switching, within one or two cycles.
- furthermore, for a given design, the above two features have to be achieved concomitantly.

4.6 Clock Frequency: A New Concept and a New Approach of Direct Period Synthesis for Meeting the Challenge

To resolve the two long-standing problems discussed in previous section and depicted in Fig. 4.5, a different approach from PLL has been proposed: direct period synthesis (DPS). The idea is illustrated at the bottom part of Fig. 4.6. From a base unit (a small time-span), an electrical waveform is created by concatenating multiple copies of the base into a pulse with both low and high parts. Further, from many pulses of such kind, a pulse train is made by sticking them together in series. In the top part of Fig. 4.6, the PLL working principle is included for side-by-side comparison. The output pulse from PLL is generated from the VCO. As seen, a whole pulse is created all once by a low-to-high (or high-to-low) transition in one step, from the oscillation of the VCO. The length-in-time of the pulse (i.e., the period) is then multiplied by the frequency divider. The resulting lengthened pulse is compared with the input pulse. In essence, this operation is a period multiplication and it can be symbolized as "multiply plus compare". It is a closed loop mechanism. In contrast, the DPS directly synthesizes each individual pulse by putting together the whole from smaller pieces.

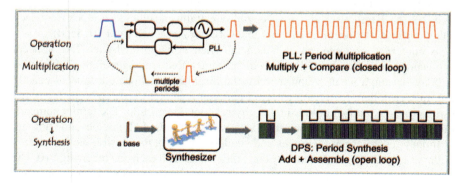

Fig. 4.6 Two different principles of generating pulse train: PLL (top) and waveform synthesizer (bottom)

Its operation can be symbolized as "add plus assemble" and it is an open loop action. A direct benefit resulted from this philosophy of "add plus assemble" is the rapid switching speed. Unlike the case of PLL where frequency change takes multiple iterations of compare-then-correct, the pulse's length (the period, or frequency) can be altered in next cycle by the DPS after a command is received. Another benefit is the individual control of the pulse's length since each one is synthesized from the base individually. This feature is helpful to the resolution of the problem of arbitrary-frequency-generation. Before getting into the detail of this issue, the very concept of clock frequency needs to be investigated in a greater depth.

In common parlance, the method customarily accepted by people for calculating frequency is to use the inversion of the time-span between the two moments making up a particular pulse (the so-called instantaneous frequency). This practice is inherited from a belief adopted in the early years when electromagnetic wave was discovered in nineteenth century: the inversion of the wavelength. By this action, it implies that all the composing cycles in an electromagnetic wave are identical. Following the same postulation, all the pulses in a clock pulse train must have the same length-in-time. This local-oriented view is straightforward and easy for understanding. However, it could unfortunately lead to a deleterious effect on the comprehension of the essence of the clock frequency. To deal with the difficult challenge of arbitrary-frequency-generation, it is worthwhile to investigate the very concept of clock frequency at a higher level, to look beyond the locality of each individual pulse. In a higher view, frequency is defined in a larger frame of one second. It is the number of clock pulses existed in the time frame of one second when clock signal itself is concerned. It is the number of operations executed within the time window of one second when functional operation is concerned. This view is actually in coherence with the official definition of frequency applied on electromagnetic wave.

This broader view on clock frequency leads to a radical rethinking on the construction of clock pulse train: the constraint of "all cycles must have same length-in-time" could be removed. It is NOT an essential element in the definition of clock frequency but just a convenience for implementation. This revivification of clock frequency concept points us to a new direction in attacking the problem of arbitrary-frequency-generation. In 2008 a novel concept, Time-Average-Frequency (TAF), is introduced (Xiu 2008b). It removes the "equal-length" constraint and elucidates clock frequency solely on activities occurred in the framework of one second (Xiu 2015c, d).

A numerical example can be beneficial when explaining why the TAF concept is powerful in dealing with the issue of "arbitrary-frequency-generation". During a holiday season, a wealthy and merciful family plans to allocate a fund of exactly ten million dollars for giving to small children in its city for goodwill. By initial estimation, the number of kids in this city is about one million. The family demands fairness in distributing the fund that, ideally, all the kids shall get the same amount of money regardless of their ages and genders. To fulfill this desire, the ten million dollars can be distributed to exactly one million kids with each one receiving $10. The problem however is that the actual number of kids is not known for sure at the time of planning. It could vary slightly around one million. Realizing this uncontrollable

4.6 Clock Frequency: A New Concept and a New Approach ...

Table 4.2 Distribution plans using U1 = $10 and U2 = $11 → number of choices = 90,910

Index of plan	Number of kids: A getting U1 = $10	Number of kids: B getting U2 = $11	Number of kids Sharing the Fund	Total of fund A · U1 + B · U2
1	1,000,000	0	1,000,000	Ten millions
2	999,989	10	999,999	Ten millions
3	999,978	20	999,998	Ten millions
4	999,967	30	999,997	Ten millions
5	999,956	40	999,996	Ten millions
...	...	...	...	Ten millions
90,908	23	909,070	909,093	Ten millions
90,909	12	909,080	909,092	Ten millions
90,910	1	909,090	909,091	Ten millions

factor, the family translates the desire of fairness into two constraints: (1) spend exactly ten million dollars, no more no less. (2) to be as fair as possible to each kid.

In light of those constraints, the organization helping the family on this project develops a plan of using two types of slices for distribution: one is $10 and the other $11. Table 4.2 lists the number of choices that the whole pie (the ten million dollars) can be sliced into pieces. Under this plan, there are 90,910 combinations that can accommodate a group of kids whose number can be anywhere from 909,091 to 1,000,000. Some kids will get $11 while other $10. For comparison, Table 4.3 is the case with two types of slices of $10 and $10.1. Table 4.4 is the case of $10 and $15. It is interesting to note that fairness and number-of-choices move in opposite trends. In the case of U1 = $10 and U2 = $15, some kids can get $5 more which is troublesomely unfair. But this plan gives a large number of options of 333,334. On the other extreme, for U1 = $10 and U2 = $10.1, all the kids get almost same

Table 4.3 Distribution plans using U1 = $10 and U2 = $10.1 → number of choices = 9901

Index of plan	Number of kids: A getting U1 = $10	Number of kids: B getting U2 = $10.1	Number of kids sharing the Fund	Total of fund A · U1 + B · U2
1	1,000,000	0	1,000,000	Ten millions
2	999,899	100	999,999	Ten millions
3	999,798	200	999,998	Ten millions
4	999,697	300	999,997	Ten millions
5	999,596	400	999,996	Ten millions
...	...	...	...	Ten millions
9899	302	989,800	990,102	Ten millions
9900	201	989,900	990,101	Ten millions
9901	100	990,000	990,100	Ten millions

Table 4.4 Distribution plans using U10 = $1 and U2 = $15 → number of choices = 333,334

Index of plan	Number of kids: A getting U1 = $10	Number of kids: B getting U2 = $15	Number of kids sharing the fund	Total of fund A · U1 + B · U2
1	1,000,000	0	1,000,000	Ten millions
2	999,997	2	999,999	Ten millions
3	999,994	4	999,998	Ten millions
4	999,991	6	999,997	Ten millions
5	999,988	8	999,996	Ten millions
...	...	...	...	Ten millions
333,332	7	666,662	666,669	Ten millions
333,333	4	666,664	666,668	Ten millions
333,334	1	666,666	666,667	Ten millions

amount of money. This plan is much closer to the ideal model of fairness. It however can only accommodate 9901 combinations.

This fund-distribution example faithfully illustrates the philosophy and principle adopted by TAF methodology for producing more frequencies, to address the problem of arbitrary-frequency-generation. The comparison between conventional frequency and TAF in making clock pulse train can be illustrated by an analogy to pie slicing. The analogy is depicted in Fig. 4.7. The two approaches face the same challenges: (1) slicing a whole pie into pieces as impartially as possible and (2) accommodate as many people as possible. Further, the size of the cutter cannot be arbitrarily chosen but is constrained by some physical limitations (for example, the smallest monetary unit is the cent, the "cutter" cannot be made smaller than that). Under those constraints, using two types of cutters can provide us with much more choices as evidenced from the numbers in Tables 4.2, 4.3 and 4.4.

Using the TAF concept, which removes the constraint that all the cycles must have same length-in-time, two or more types of cycles can be used in the creation of clock pulse train, resulting in much more frequencies. The spirit embedded in this

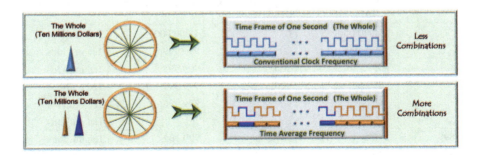

Fig. 4.7 Using Time-Average-Frequency for producing more frequencies

4.6 Clock Frequency: A New Concept and a New Approach ...

methodology is exactly the same as that of pie slicing. The different type of cycles is similar to the use of different cutters. In this regard, the period synthesis principle depicted in Fig. 4.6 is formulated with the idea in mind of using different types of cutters.

TAF based clock signal seems risky in the first sight since the use of different types of cycles can be easily confused with a crucial and somewhat puzzling concept of clock jitter. TAF concept, from all the aspects, is fundamentally different from the jitter. Detailed discussion on this issue can be found in Sect. 3.6 of Xiu (2015a). In essence, jitter is the inborn instability associated with nature oscillation and vibration, or the unintended effect induced by some uncontrollable factors found in the operating environment or in the process of design and manufacture. Jitter is not known to designer and user beforehand. In contrast, TAF clock signal is created by designer. The exact locations of the clock signal's all transiting edges are planned ahead by the clock designer and known to the clock user.

In the example of fund-distribution, all the kids must plan their spending under the budget of $10, not $10.1, $11 or $15 since no kid can be sure that he or she can get more than $10. As long as all the kids keep their spending plans below $10, no one will be disappointed. Similarly, when TAF clock signal is used to drive circuit, the setup-constraint must use the shortest one of all cycles. In circuit design, if the designer guaranties that all the circuit can finish their tasks within the time span of the shortest cycle, then everything will be fine. Regarding hold-constraint (i.e., the hold check in digital circuit timing closure), fortunately, it has nothing to do with TAF (a pleasant actuality). Hold-constraint is not affected by TAF since hold-check only concerns one clock edge. Please refer to question 89 and 90 in Xiu (2007a), Sect. 1.3 and 4.22 of Xiu (2012), Sect. 3.2 and 3.6 of Xiu (2015a) for more discussions.

Based on this novel Time-Average-Frequency concept proposed in Xiu (2008b), a new type of clock generation technology, Time-Average-Frequency Direct Period Synthesis (TAF-DPS), is developed in the following years (Xiu 2012, 2015a). From circuit operation perspective, it differs from PLL by directly synthesizing each clock pulse from a base unit, using the principle illustrated in Fig. 4.6. By adopting TAF philosophy, the clock pulses have the option of adopting different types of lengths (Xiu 2015c, d). For the resulting pulse train, the plan of making more frequencies available become feasible. Its comparison with conventional clock technology is extensively discussed in Xiu (2016).

Figure 4.8 depicts the TAF-DPS architecture, highlighting its work principle. Starting from a base time unit Δ (e.g., 10 ps), TAF-DPS is able to create two type of cycles $T_A = I \cdot \Delta$ and $T_B = (I+1) \cdot \Delta$, where I is an integer. These two types of cycles then can be cascaded in an interleaved fashion to make a pulse train. The possibility of T_A (and thus T_B) occurrence is determined by a fraction r, $0 \leq r < 1$. As a result, the output period can be calculated using an equation of $T_{TAF} = 1/f_{TAF} = F \cdot \Delta = (I+r) \cdot \Delta$, where F (=I + r) is named as frequency (more precisely, period) control word. The comparison of waveforms between conventional frequency and TAF is presented at the bottom left of the figure. The comparison of period distributions is at the right. As seen, the TAF-DPS output utilizes two types of cycles (more is also allowed) to match the theoretical frequency value, using the principle elucidated in the fund

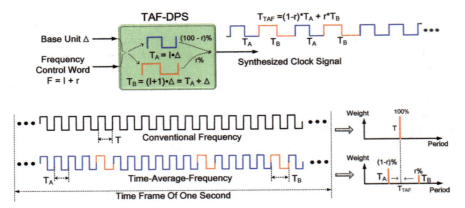

Fig. 4.8 TAF-DPS work principle (top), the comparison of TAF and conventional clock signals (bottom)

distribution example (Tables 4.2, 4.3 and 4.4). From $T_{TAF} = 1/f_{TAF} = (I + r) \cdot \Delta$, it is understandable that, for a specific design, almost any period (frequency) can be generated by given enough resource (silicon area and power budget). The frequency resolution (or granularity) is determined by r (the LSB of the register holding the value of r). In experiment, frequency granularity in the range of deep sub-ppb has been achieved. For this reason, TAF-DPS is regarded as a clock generator capable of "arbitrary-frequency-generation".

Figure 4.9 shows more details of TAF-DPS circuit implementation. The base time unit Δ is generated from a plurality of K phase-evenly-spaced signals of a known frequency f_i. The value of Δ is the time span between any two adjacent such signals. The said plurality of signals are fed into the direct period synthesizer that synthesizes each output pulse's waveform (i.e., the low and high portions). In contrast to

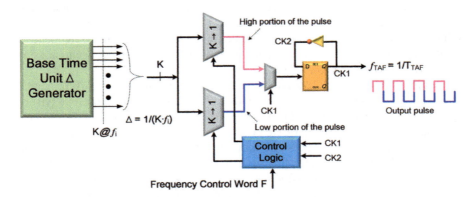

Fig. 4.9 Circuit implementation of TAF-DPS

4.6 Clock Frequency: A New Concept and a New Approach ...

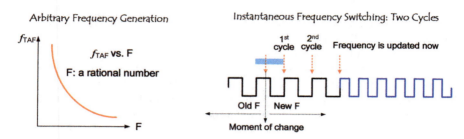

Fig. 4.10 The two features of arbitrary-frequency-generation and instantaneous-frequency-switching

PLL which is an indirect approach (period multiplication through compare-then-correct), each TAF-DPS output cycle's structure is directly constructed and individually controlled. Thus, its period (and frequency) can be instantly changed. Further, its frequency switching speed is fast and quantifiable. This leads to the so-called "instantaneous-frequency-switching".

The two features of arbitrary-frequency-generation and instantaneous-frequency-switching, accomplished by TAF-DPS, are graphically illustrated in Fig. 4.10. The relationship between frequency control word F and output frequency f_{TAF} takes the shape of $1/x$ since TAF-DPS's output period is linearly proportional to the value of control word F. Along this curve, in a practical sense, almost all frequency points can be reached as long as enough bits are used in the fraction r. The frequency resolution/granularity can be calculated as $df/f = -dF/F$, where dF is the LSB used for representing the r. As an example, $r = 2^{-27}$ and $I = 4$ leads to a frequency granularity df/f of 1.89 ppb from calculation. In the corresponding lab measurement, the real measurement number obtained is 1.99 ppb. Such a fine frequency granularity can hardly be realized by other techniques. It is also worth to note that the frequency transfer function f_{TAF} versus F is precisely predictable (traceable with 100% fidelity by equation $T_{TAF} = 1/f_{TAF} = F \cdot \Delta$). This fact is extremely valuable when TAF-DPS is used in applications.

The first and most-known TAF-DPS implementation is the Flying-Adder frequency synthesis architecture, which was invented in the late 1990s (Mair and Xiu 2000; Xiu and You 2002). In the following years, more improvements on TAF-DPS circuit were developed (Xiu and You 2003, 2005; Xiu 2009, 2017; Xiu et al. 2012, 2013, 2019; Xiu and Chen 2017). In the meantime, the technology has been used in a variety of commercial products for decades (http://www.ti.com/lit/ug/sprugx8c/sprugx8c.pdf). Some application examples can be found in Xiu et al. (2004), Xiu (2007b, 2008a, c, 2015b), Ma et al. (2020), and Yang and Haider (2012). On another front, the spectrum of TAF-DPS output is complex due to the use of different types of cycles. The spectrum is no longer a simple picture that contains only lines representing its fundamental frequency and higher harmonics but bears much more intricate patterns. Those patterns are determined by and can be derived from the frequency control word F. On this topic, theoretical studies have been carried out in the direction of rigorously deriving its spectrum under various scenarios (Sotiriadis 2010a, b, c, 2016; Talwalkar 2012, 2013; Xiu et al. 2010, 2011).

In short, after almost two decades' work, a solution to the two long-standing problems has emerged and it is becoming ever mature over time. This new technology is still facing challenges. The most difficult task is however not technical-related. Instead, the biggest hurdle is the battle against the old-school mindset. As said, clock signal is used to reflect the flow of time inside electronic world. TAF based clock signal disturbs the sense of uniform-flow-of-time congenitally existing in the conventional clock signal since, instead of identical pieces, it stirs the uniform flow with intermittent irregularities (pieces of different sizes) (Xiu 2015c, d). Intuitively, this irregularity is annoying. Historically, however, this phenomenon is not uncommon in our study of the usage of time. Researches in physiology have found plenty of evidences that physiological time does not flow uniformly like physical time. Even for the physical time that has been described by physicist as uniform movement, philosopher and mathematician sometimes still have different opinions. The great French mathematician Henri Poincaré once said that scientists "did not" measure time, "but cut it up into pieces that they declare to be identical so that their equations be as simple as possible." In his view, cutting the flow of time into identical pieces is just a convenience for operation. Following this reasoning, for serving our purpose better, nothing can prevent a physicist, a mathematician or a circuit designer from treating the time flow as a non-uniform one.

In our study of Mother Nature, mathematics, including its treatment of time, is mere a tool for assisting us to do our tasks. The tool itself must not be qualified as a true reflection of how the world actually is. The most notable example of differentiating tool from reality is the case when Einstein chose multidimensional geometry, instead of Euclidian geometry, to carry out his arguments on relativity theory. Regarding time and clock, the philosopher Henri Bergson once said: "when our eyes follow on the face of a clock, the movement of the needle that corresponds to the oscillations of the pendulum, I do not measure duration, as one would think; I simply count simultaneities, which is quite different." This statement shows a deep feeling that something different, something novel, something important, something outside the watch or clock itself needed to be included in our understanding of time. Only that could explain why we attribute to time-keeping device such a great power: why we buy them, why we use them, and why we invent them in the first place. In microelectronics engineering, this is the spirit that we shall embrace. The clock signal that we create is for serving our purpose of processing information. We do not create clock signal simply for the clock signal itself. The conventional approach of clock generation (e.g., the PLL) is simply for the reasons of easy comprehension and implementational convenience. It makes the work of clock circuit design easier ("so that their equations be as simple as possible", as Poincaré put it). There is nothing wrong for doing it in some other ways, such as the TAF way, if it can yield more favorable result in the end.

What, then, can TAF clock do for us?

4.7 The Response: Time-Oriented Paradigm

According to Thomas Kuhn, normal science is a highly determined activity wherein there are mostly three classes of problems: determination of significant fact, matching of facts with theory, and articulation of theory. Like it or not, the overwhelming majority of the problems undertaken by science and technology practitioners fall into one of those three categories. In practice, scientists and engineers conduct their businesses by committing to same rules and standards. The existence of this strong network of commitments, including conceptual, theoretical, instrumental, and methodological, is a principal source of the metaphor that relates normal science to puzzle-solving. And, it is the sign that a paradigm has been formed or existed.

In the development of a science, within qualitative paradigm there exists quantitative laws. The first accepted paradigm in a branch of science with its associated laws in action can usually account quite successfully for most of the observations and experiments that are easily accessible to most practitioners in that field. Further progress will call for the construction of elaborate apparatuses, the establishment of an esoteric vocabulary and a new set of skills, and a refinement of concepts that increasingly lessens their resemblance to the ordinary common-sense prototypes. The paradigm provides rules that tell the practitioner of a mature specialty what his science is like so that he can concentrate with assurance on the perplexing problems that these rules and existing knowledge define for him. On the other hands, this development of professionalization however could lead to an immense restriction of the scientists' vision and then to a considerable resistance to paradigm change. The science hence would become increasingly rigid. This is a picture painted by Thomas Kuhn several decades ago.

Normal science is characterized by a paradigm, which legitimates puzzles and problems on which the community works. Those puzzle-solving problems almost exhaust the literature of normal science, both empirical and theoretical by its nature. They do not however quite exhaust the entire literature of science. The study of science history reveals that normal science first goes with a paradigm and is a dedication to solving puzzles. It is then followed by serious anomalies, which lead to a crisis. The anomalies emerge only on special occasions prepared by the advance of normal research. Anomalies appear only against the background provided by the paradigm. All is well until the methods legitimated by the paradigm cannot cope with a cluster of anomalies. Those are extraordinary problems, and it may well be their resolution that makes the scientific enterprise as a whole so worthwhile and advances science in great leap. The more precise and far-reaching that paradigm is, the more sensitive an indicator it provides of anomaly and hence of an occasion for demanding change. From crisis comes breakthrough. Crisis persists until a new achievement redirects research and serves as a new paradigm. That is the so-called paradigm shift. With the benefit of hindsight, based on the history of science, especially the recent history of last two hundred or so years, it is believed that work under a paradigm can be conducted in no other way. To desert the paradigm is to cease

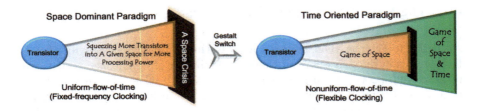

Fig. 4.11 The space-dominant paradigm is hitting a wall (left) and the time-oriented paradigm expands the scope for exploration (right)

practicing the science it defines. In Kuhn's view, anomalies and crises are the pivots about which scientific revolutions turn.

Armed with this understanding on paradigm and paradigm shift, now let's turn our attention to microelectronics. As discussed in Chap. 3, a space-dominant paradigm had been gradually formed in microelectronics after the invention of transistor in the late 1940s. After seven or eight decades of rapid growth, we are currently running into a space-related crisis. Due to several fundamental limitations, it is no longer possible for us to continue on the course of squeezing more transistors into a given space for more processing power. This is a case of anomaly, which cannot be coped with by "business as usual". It is time for visionary thinkers to explore ventures of other possibilities, and to frame new paradigm when needed. One of the palpable options is to delve deeper into the domain of time for more capacity, because, as discussed previously, time is the only alternative when real estate for developing microelectronics is concerned. This change of direction is more or less a gestalt switch as illustrated in Fig. 4.11. In its face value, the gestalt switch in this case is from space-dominant paradigm to time-oriented paradigm. Under the surface, it requires a change-of-mindset from only accepting the rigorous view of uniform-flow-of-time to the adoption of a more flexible attitude of nonuniform-flow-of-time. This can subsequently lead to a shift of design philosophy from fixed-frequency clocking to flexible clocking. At the very bottom, the technical underpinning for this shift is the resolution of the two long-standing problems: arbitrary-frequency-generation and instantaneous-frequency-switching.

This paradigm shift requires a crucial change of commitment (e.g., adopting the concept of Time-Average-Frequency) and the consequential employment of a new set of tools (e.g., the TAF-DPS technology) for handling problems. In science and engineering, so long as the tools that a paradigm supplies continue to prove capable of solving the problems it defines, development of the field moves fast and it can penetrate into unexplored frontiers through confident employment of those tools. Retooling is only an extravagance to be reserved for the occasion that truly demands it. The severity of the current space crisis is an indication that the occasion for retooling might have arrived. History shows that, for any science or engineering, the rigidity of normal science or normal engineering will not go forever unchallenged. This judgment must be true for today's microelectronics as well. This campaign

4.7 The Response: Time-Oriented Paradigm

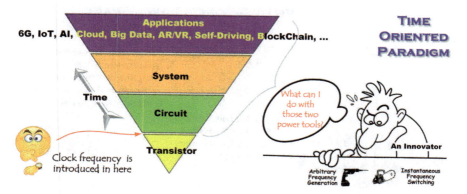

Fig. 4.12 The retooling in microelectronics for preparing the time-oriented paradigm

of shifting to a new paradigm can start from the abandonment of the conventional clock frequency concept that insists on the equal-length for all the clock pulses, a commitment that is derived from and popular in the space-dominant paradigm. It is the most distinguished characteristic of this old paradigm.

Figure 4.12 illustrates the master plan of retooling for the time-oriented paradigm. The fact of retooling is reflected in the resolution of the two long-standing problems of arbitrary-frequency-generation and instantaneous-frequency-switching. Against the background of conventional clock generation technologies (e.g., PLL), the Time-Average-Frequency based direct period synthesis method stands out as a new set of power tools. With these newly available tools, the landscape of microelectronics is expected to transform at all levels from circuit to system to application. This is simply due to the reason that time-handling is at the very bottom of signal processing.

The current microelectronics landscape shaped by Moore's law can be characterized by a phrase of "Space Moore (i.e., to use space more)". Future landscape will be "Time Moore (i.e., to use time more)". This is a significant turn in the direction of Moore's Law, as illustrated in Fig. 4.13. The transition will be a multidiscipline campaign. It requires effort from many professionals, from process researcher, circuit designer, system architect, computer scientist to algorithm developer, software engineer, application engineer, product engineer, market researcher and etc. All those practitioners need to accept and share a new commitment: the nonuniform-flow-of-time inside electronic world. They need to acknowledge the Time-Average-Frequency principle as a legitimate means for devising time-keeping devices (Xiu 2015c, d). This is a nontrivial task and it might take years or decades to see its full impact. When this transition of landscape is accomplished, people would have changed its view of the field, its methods, and its goals. After the transition, it is believed that scientists and engineers may view the microelectronics differently, have a different feeling on how it works, be puzzled by new difficulties, notice unfamiliar social impacts, and interact with it in new ways. After all, it is a new paradigm.

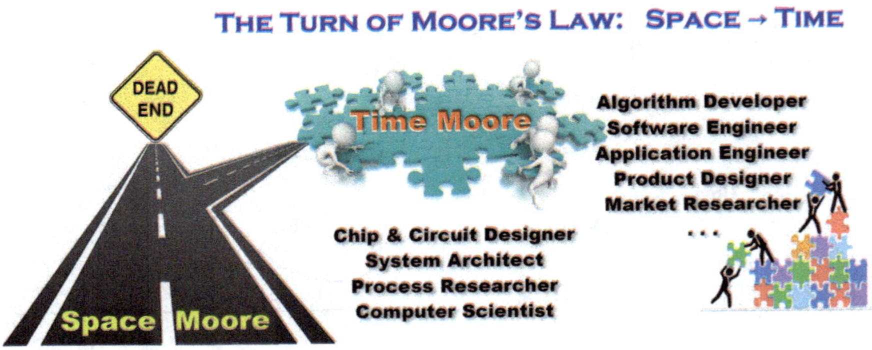

Fig. 4.13 The turn of Moore's Law is an effort of multidiscipline

Change-of-mindset is the most difficult thing in the practice of science and engineering. Retrospect to the history of technology might help ease some of the pain. One good example is the time-handling in the development of cinematographic projection. Early cameras and projectors were driven by hand. Later, automatic clockwork mechanisms were installed in commercial cinematographic cameras to record and display images at fixed intervals. Those changes have affected how people thought about film. In the famous horse-in-gallop photography by Eadweard Muybridge in 1878, the intervals between the frames were neither fixed nor properly determined. The cameras went off every time the horse broke with his stride a series of strings strewn across the running path. The resulting photographs were therefore taken according to the horse's speed, not according to fixed-intervals clock time. Should we always opt to keep the intervals fixed in order to better display movement? Should cinematography become a time-recording technology as much as an imaging technology? There are surely many photographers who want to stress the need to record and display images with clocklike precision since, deep in their hearts, time must flow uniformly. Certain topics however, like the horse in gallop, seems to call for expanding or contracting the intervals between frames or regulating the recording speed according to the speed of the filmed object. Otherwise, interesting visual effects could be lost. Which technique is better? What do these competing methods reveal about the nature of time? As the film industry grew, it turned out, an increasing number of filmmakers abandoned the strict timekeeping standard advocated by clocklike-precision photographers and used strictly by astronomers and physicists. They changed their minds for the good of their businesses.

As another example, variable-interval films became essential for cell biology. In this field, some key exponents of the "new cytology" (cell-based biological research) have argued how these films reveal a different aspect of time, biological rather than physical. Biological films are usually projected at the ordinary rate of 16 frames per second or less. But when filmed, they could be done at variable speeds that vary according to the activity of the culture, some at rates of once every 10, 15, 20 or 30 s and during periods varying from 24 to 72 h. The need to introduce variations in recording speed, in its evident contrast with other regularly cadenced films, comes as a relief of the difference between "biological" and "physical" time and between living and dead matters.

The notion of lived time, as the one in opposition to mechanical time, has obtained gradually stronger legitimacy in a number of scientific disciplines during recent years. In light of this, it is argued by the author that the time has arrived for microelectronics to have its own "lived time" since electronic devices might very well have their own soul sometime soon. This type of "lived time" can be materialized, or at least assisted, by the approach of flexible clocking. Its aim is to process information at a rate demanded by the operating environment, not too slow, not too fast, but just right. The essence is to use resource and power wisely since there will be too much information waiting to be processed in the years and decades to come.

4.8 Left and Right: Space and Time as Two Arms for Lifting the Weight of Signal Processing

The transition from old space-dominant paradigm to new time-oriented paradigm can be comprehended through an analogy to weight lifting, as illustrated in Fig. 4.14. The task of signal processing has been studied, thoroughly from all possible aspects, within the frame of space-dominant paradigm during the past seven or so decades. It has achieved tremendous success by just exploiting the capacity from space. As illustrated, it simply lifts the whole weight of signal processing using just one arm: the space. As we are heading towards the three practically uncrushable space-related obstructions, namely the signal integrity problem (signals are physically much closer to each other, signals are traveling relatively longer distance, signals' frequencies are higher), the power density problem (more and more processing units operate within ever-smaller space and with higher speed, leading to the issue of heat removal) and the quantum limit (continuing miniaturization pushes transistor to atom size), the weight inevitably becomes ever-heavier. At this crucial moment, a helpful hand is definitely welcomed. Hence emerges the time-oriented paradigm. In this weight-lifting metaphor, more-efficient-time-utilization is the other arm that can help offload some of the weight. After all, to become a fully functional body, two arms are needed.

The transition from space-dominant paradigm to time-oriented one will not be accomplished through a cumulative process. It cannot be achieved by an articulation

Fig. 4.14 Space and time as two arms for lifting the weight of information processing

or extension of the old paradigm. Rather, it will be a reconstruction of the field from a new fundamental, a reconstruction that changes the field's most foundational concept as well as many of its methods and applications. During the transition period, there will be a large but never complete overlap between the problems that can be solved by the old and the new paradigms. Many of the weight-lifting tasks (i.e., technical problems regarding signal processing) can probably be accomplished by using just one arm. But there will also be a decisive difference in the modes of the solutions. Put it in another way, some of the problems cannot be handled as efficient if the other arm is not there.

More broadly speaking, the gestalt switch discussed around Fig. 4.11 can be appreciated from two aspects. First, the new time-oriented paradigm is able to preserve a large part of the concrete problem-solving ability that has accrued to microelectronics through its predecessor (i.e., the space-dominant paradigm). Second, the new paradigm will resolve some outstanding problems that, generally recognized, cannot be settled in other way. Back to the weight-lifting metaphor, the help from the extra arm provides a feasible path to circumvent the barrier caused by the current space crisis. This discussion on handling easy and difficult problems in old and new paradigms will be carried out in great details in Chap. 6.

4.9 Top and Bottom: From System to Circuit and from Circuit to System

A good design of electronic system requires effort from the directions of both top-down and bottom-up. This design philosophy has been proven effective countless times in real battles, spectacularly through some of the most famous killer applications. To pursue high efficiency and low cost, the task of doing signal processing has to be investigated from system level down to circuit level, as well as from circuit level up to system. This guideline-for-practice is depicted illustratively in Fig. 4.15. System architects present to circuit designers what they want, in the form of specific requirements on chip's functionality and performance. Those demands are expected

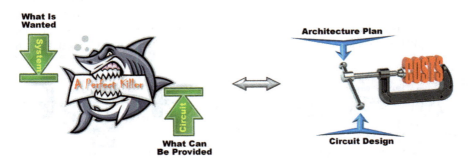

Fig. 4.15 Good design requires effort from both directions: top-down and bottom-up

4.9 Top and Bottom: From System to Circuit and from Circuit ...

to be fulfilled from lower-level circuit blocks. On the other hand, circuit designers have the responsibility to inform architects what they are able to do, because there might be some requirements that cannot be satisfied due to various constraints and difficulties in circuit level work. Only by a full synergism and effective communication from both directions, a perfect killer application can be possible. The ultimate goal is to create a signal processing device that can process the maximum amount of information (number of bits) in the shortest time frame, and at the same time, using the minimum amount of resource (area and power).

In this tug-of-war game between the two groups of professionals, misunderstanding can easily occur and it does happen quite often in real work. Sometimes, system architects may ask too much that their counterpart, namely the circuit designers, just cannot fulfill their will at a reasonable cost, or simply impossible to do so due to some technical difficulty. In other times, circuit designers might fail to inform the architects certain new techniques that they have just invented and developed. Worse yet, in some cases, circuit designers are simply unable to convince the architects on the advantage of their new solutions over the old ones, to stimulate the architects for taking the risk and using the new solutions. As a result, the architects cannot take advantage of those new developments and, consequently, miss the opportunity for making the device overall more competitive. This type of unfortunate circumstances is often caused by the lack of effective communication between the two groups.

In the transition from space-dominant paradigm to time-oriented paradigm, one crucial message, which can be regarded as those crucial two-ways communication described above, is related to the two long-standing problems discussed in Sect. 4.5. This message is of great significance and it can never be over-emphasized. The message can be elucidated as the following. One way or another, system architects need to be aware of the fact that there is already a solid solution to the two long-standing problems of arbitrary-frequency-generation and instantaneous-frequency-switching. Those problems, when viewed individually in an isolated environment, lie in the level of circuit design. However, by the resolution of those lower-level problems and from adopting a new design doctrine in a higher level, innovative ideas in system level can be spontaneously, naturally, and widely incited. Referring back to Fig. 4.12, a new generation of system architects is expected to grow up with this new doctrine and be skillful at using the newly available power tools. When this generation of freshly-minded system architects realize that they can take advantage of those two new features of arbitrary-frequency-generation and instantaneous-frequency-switching in their architecture planning, many new possibilities will follow and the field can be benefited as a whole.

In Sect. 3.5, we have argued that Moore's Law is a practice that focuses on digging more processing power from space by packing more transistors into a given volume. By providing a solution to those two long-standing problems, a new direction for exploration emerges. From the perspective of time, when electronic system is devised, higher processing efficiency can become possible by using flexible clocking style. This argument is depicted in Fig. 4.16. This has to be a coordinated effort from both

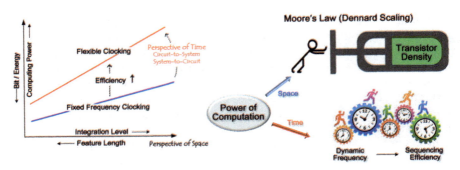

Fig. 4.16 Adopting the flexible clocking style for higher processing efficiency (i.e., a higher level of complexity is emerged inside an entity)

system architect and circuit designer, from circuit-to-system and from system-to-circuit. It is truly believed that there will be a tremendous opportunity for innovations once this design doctrine gains ground. In Chap. 6, we will demonstrate this point through some examples.

4.10 The Good News: It Is Free, or Almost Free

From the history of recorded human thought, it is evident that great ideas are priceless in advancing our civilization, both civically and technologically. The power of ideas is the driving force that keeps our achievement marching forward, to reach new heights one after another. Interestingly, it is a delightful fact that the truly great ideas ever occurred to our civilization are not necessarily tied to money. More often than not, they are free of charge to anyone who want to learn them. The jewel of thought belongs to the entire human society. It is the treasure of whole mankind. In microelectronics, the greatest examples of this type are Alan Turing's method of computing, John Neumann's computer architecture, Claude Shannon's information theory and, arguably, Gordon Moore's Moore law. Alan Turing shows us how to describe the solution to computable problem using a method of programming. John von Neumann architects a computer for running the program. Claude Shannon teaches us how to create a world-of-information using the binary based computer. Gordon Moore describes how semiconductor scaling makes the computer grow exponentially more capable over time. Those great ideas have shaped the microelectronics and changed our society as a whole. Surprisingly, not a single penny has been charged by those great men when individuals and organizations use them for creating hundreds of thousands of products and making profits of billions of dollars. In this world, the most precious things can be free.

The shift from the space-dominant paradigm to the time-oriented one could be an idea of such kind. It is a big vision that we shall delve deeper into the domain of time to get more processing power, to circumvent the space obstacle and to advance further to next stage. It is a firm belief that, in order to use time more efficiently, we need to first settle the two long-standing problems of arbitrary-frequency-generation and instantaneous-frequency-switching. This vision and belief are invaluable in their power of guiding future course. They are free of charge, available to anyone who is unsatisfactory of status quo and has the desire to innovate.

On the other hand, to materialize the solution to the two long-standing problems, it requires ingenious skill of engineering besides pioneering cogitation. This engineering effort however might come with some monetary price. In the pursuing of good solutions to practical problems, various engineering approaches can be attempted, often accompanied by certain financial expense. Eventually, some of the endeavor would be proven solid. To use any of those solutions, a particular kind of compensation might be needed to reimburse the developer (so that they can have motivation for more innovations). Solution to practical problem is classified as method and apparatus. They are intellectual property that is developed with financial sponsorship. It is hence reasonable to ask for return on this type of investment. As said by some sage, science is an act of using money to gain knowledge while technology is to use knowledge for making wealth.

In the case of this paradigm shift from space to time, most of the knowledge and skill accumulated so far in the old space-dominant paradigm, such as the EDA tools, the design methodologies, the test instruments, the manufacture equipment and etc., remain effective since the task of signal processing is still voltage based (the collective effect of a large number of electrons). The vital change will happen primarily inside people's brains. What matters most is the mindset that governs how time is viewed and treated in our cogitation. This change-of-mindset does not necessarily involve money. Comparing to the outcome resulted from this paradigm shift, this monetary price, if any, is trivial. Therefore, this transition from old paradigm to new one is essentially free of financial burden, at least in principle.

References

Henri Bergson, Leon Jacobson, "Duration and Simultaneity: Bergson and the Einsteinian Universe", *Clinamen Press Ltd.*, 1999.

R. Best, Phase Locked Loops 6/e: "Design, Simulation, and Applications," 6th edition, *McGraw-Hill Professional*, 2007.

Jimena Canales, "The Physicist and the Philosopher: Einstein, Bergson, and the Debate that Changed Our Understanding of Time", Princeton, NJ: *Princeton University Press*, 2015.

W. F. Egan, "Frequency Synthesis by Phase Lock," 2nd edition, *Wiley-Interscience*, 1999.

F. M. Gardner, "Phaselock Techniques," 3rd edition, *Wiley – Interscience*, 2005.

Y. Ma, X. Wei, and L. Xiu, "A Novel Spread Spectrum Clock Generation Technique: Always-On Boundary Spread SSCG," *IEEE Transactions on Electromagnetic Compatibility*, vol. 62, no. 2, pp. 364–376, April 2020.

Mair, H. and Xiu, L., "Architecture of high-performance frequency and phase synthesis," *IEEE J. Solid-State Circuits*, vol. 35, no. 6, pp. 835–846, 2000.

B. Razavi, "Phase-Locking in High - Performance Systems: From Devices to Architectures," *Wiley-IEEE Press*, 2008.

P. Sotiriadis, "Theory of Flying-Adder frequency synthesizers, Part I: modeling, signals' periods and output average frequency," *IEEE Trans. on Circuits and Systems I*, vol. 57, pp. 1935–1948, Aug. 2010a.

P. Sotiriadis, "Theory of Flying-Adder frequency synthesizers, Part II: time and frequency domain properties of the output signal," *IEEE Trans. on Circuits and Systems I*, vol. 57, pp. 1949–1963, Aug. 2010b.

P. Sotiriadis, "Exact spectrum and time-domain output of Flying-Adder frequency synthesizers," *IEEE Trans. on Ultrasonics, Ferroelectrics, and Freq. Control*, vol. 57, pp. 1926–1935, Sep. 2010c.

P. Sotiriadis, "Spurs-Free Single-Bit-Output All-Digital Frequency Synthesizers With Forward and Feedback Spurs and Noise Cancellation," in *IEEE Transactions on Circuits and Systems I*: Regular Papers, vol. 63, no. 5, pp. 567–576, May 2016.

S. A. Talwalkar, "Quantization error spectra structure of a DTC synthesizer via the DFT axis scaling property," *IEEE Trans. on Circuit And System I*, vol. 59, pp. 1242–1250, June 2012.

S. A. Talwalkar, "Digital-to-Time synthesizers: separating delay line error spurs and quantization error spurs," *IEEE Trans. on Circuit And System I*, vol. 60, pp. 2597–2605, Oct. 2013.

"TMS320DM816x DaVinci Digital Media Processors Technical Reference Manual," Texas Instruments Inc., 2015, avialble: http://www.ti.com/lit/ug/sprugx8c/sprugx8c.pdf.

L. Xiu, "VLSI Circuit Design Methodology Demystified: A Conceptual Taxonomy", *John Wiley IEEE-press*, Piscataway, 2007a.

L. Xiu, "A Flying-Adder Based on-chip Frequency Generator for Complex SoC," *IEEE Trans. on Circuit And System II*, vol. 54, pp. 1067–1071, Dec. 2007b.

L. Xiu, "A Flying-Adder PLL Technique Enabling Novel Approaches for Video/Graphic Applications," *IEEE Trans. on Consumer Electronic*, vol. 54, pp. 591–599, May, 2008a.

L. Xiu, "The Concept of Time-Average-Frequency and Mathematical Analysis of Flying-Adder Frequency Synthesis Architecture," *IEEE Circuit And System Magazine*, 3rd quarter, pp. 27–51, Sep. 2008b.

L. Xiu, "A Novel DCXO Module for Clock Synchronization in MPEG2 Transport System," *IEEE Trans. on Circuit And System I*, vol. 55, pp. 2226–2237, Sep. 2008c.

L. Xiu, "A Fast and Power-Area Efficient Accumulator for Flying-Adder Frequency Synthesizer," *IEEE Trans. on Circuit And System I*, vol. 56, pp. 2439–2448, Nov., 2009.

L. Xiu, "Nanometer Frequency Synthesis Beyond the Phase-Locked Loop," *John Wiley IEEE-press*, Piscataway, 2012.

L. Xiu, "From Frequency to Time-Average-Frequency: A Paradigm Shift in the Design of Electronic system," *John Wiley IEEE-press*, Piscataway, 2015a.

L. Xiu, "Direct Period Synthesis for Achieving Sub-PPM Frequency Resolution through Time Average Frequency: The Principle, The Experimental Demonstration, and Its Application in Digital Communication," *IEEE Trans. on VLSI*, vol.23, no.7, pp.1335–1344, 2015b.

Xiu, L., "Circuit And Method For Adaptive Clock Generation Using Dynamic-Time-Average-Frequency", US 9,118,275, August 25, 2015c.

Xiu, L., "Microelectronic System Using Time-Average-Frequency Clock Signal As Its Timekeeper", US 9,143,139, Sep. 22, 2015d.

L. Xiu, "Spectrally Pure Clock vs. Flexible Clock: Which One Is More Efficient in Driving Future Electronic System?" chapter 13, "Mixed-Signal Circuit", CRC press, 2016.

L. Xiu, "Clock Technology: The Next Frontier", *IEEE Circuit And System Magazine*, vol. 17, no. 2, pp. 27–46, 2017.

L. Xiu, L., and P. L. Chen, "A Reconfigurable TAF-DPS Frequency Synthesizer on FPGA achieving 2 ppb Frequency Granularity and Two-Cycle Switching Speed," *IEEE Trans. on Industrial Electronics*, vol. 64, pp. 1233–1240, Feb. 2017.

L. Xiu, and Z. You, "A 'flying-adder' architecture of frequency and phase synthesis with scalability," *IEEE Trans. Very Large Scale Integr. Syst.*, vol. 10, no. 5, pp. 637–649, 2002.

L. Xiu, Z. You, "A New Frequency Synthesis Method Based on Flying-Adder Architecture," *IEEE Trans. on Circuit And System II*, vol. 50, pp. 130–134, March 2003.

References

L. Xiu, Z. You, "A Flying-Adder frequency synthesis architecture of reducing VCO stages," *IEEE Trans. on VLSI, vol.13*, pp. 201–210, Feb., 2005.

L. Xiu, W. Li, J. Meiners and R. Padakanti, "A Novel All-Digital PLL with Software Adaptive Filter," *IEEE J. Solid-State Circuit*, vol. 39, no. 3, pp. 476–483, March 2004.

L. Xiu, C. W. Huang, P. Gui, "The Analysis of Harmonic Energy Distribution Portfolio for Digital-to-Frequency Converters,", *IEEE Trans. Instrum. Meas,* vol. 59, pp. 2770–2778, Oct., 2010.

L. Xiu, M. Ling, H. Jiang, "A Storage Based Carry Randomization Techniques for Spurs Reduction in Flying-Adder Digital-to-Frequency Converter," *IEEE Trans. on Circuit And System II*, vol. 58, no.6, pp 326–330, June. 2011.

L. Xiu, Kun-Ho Lin and M. Ling, "The Impact of Input-Mismatch on Flying-Adder Direct Period Synthesizer," *IEEE Trans. on Circuit And system I*, vol. 59, pp. 1942–1951, Sep. 2012.

Xiu, L., Lin, W. T. and Lee, T., "A Flying-Adder Fractional-Divider based integer-N PLL: the 2nd generation Flying-Adder PLL as clock generator for SoC", *IEEE J. Solid-State Circuits*, vol. 48, pp.441–455, Feb. 2013.

L. Xiu, Xiangye Wei, Yuhai Ma, "A Full Digital Fractional-N TAF-FLL for Digital Applications: Demonstration of the Principle of a Frequency-Locked Loop Built on Time-Average-Frequency", *IEEE Trans. on VLSI*, vol. 27, no.3, pp. 524–534, March 2019.

F. Yang and T. Haider, "MPEG-2 transport stream packet synchronizer," *US patent 8249171*, 2012.

Chapter 5
Time Moore Strategy: Bridging the Old and New Paradigms

5.1 The Essence of Moore's Law and the Principle of Least Action

The entire edifice of microelectronics is built on the foundation of matter and energy, which are settled on the background framed by space and time. This chain of structures can be appreciated from Fig. 5.1. In essence, the sole purpose of any electronic device is to process information. This is analogous to the reality that the very existing of an individual human being is to consume energy and survive while the ultimate goal of survive is to process and enjoy information. The human brain is the hardware and the mind is the software of a vast biological computer. Everything it does is information processing. As a formidable assistant, electronic devices can help us become more efficient in performing this task of information processing and make us healthier (hopeful happier as well) and live longer.

Taking a look around us, modern society is full of such electronic devices and hence is a world of electrized information, which is generated and consumed by those devices. Nowadays the amount of information, when quantified in bit, is exploding rapidly on a daily basis. In the coming years and decades, it is foreseeable that every person on the earth will have more than a few electronic devices attached to him or her 24 hours a day and 7 days a week. Those devices are constantly generating and consuming data, which is presumably useful information for helping the devices' owner make decisions and hopefully manage his or her life better. This pleasant, and at the same time somehow fearful, picture requires the consumption of a huge amount of matter and energy for it being realistic.

"Nature is thrifty in all its actions" is the notion that is at the core of modern physics and mathematics. This statement captures the soul of *the principle of least action* (Feynman 1965), which is formulated by Pierre Louis Maupertuis in 1774 as commonly believed. Essentially, this principle tells us that all the occurrences in nature tend to happen in the way that requires the least effort. For instance, the beam of light travels in a straight line because that is the shortest path between two points. If you drop a ball, it will go straight toward the center of the earth. This principle

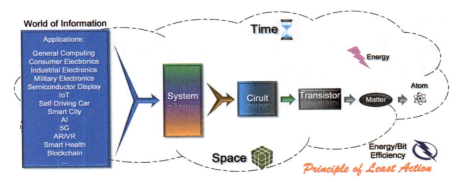

Fig. 5.1 Applications built on matter and energy against the background of space and time

has been broadly applied in various fields such as thermodynamics, fluid mechanics, the theory of relativity, quantum mechanics, particle physics, and string theory. It is therefore believed that this principle must also hold true for the development of microelectronics. As a matter of fact, although Moore's Law has been formulated in a variety of formats in the past several decades, its essence lies in the application of this "principle of least action" in the practice of electronic engineering.

In the original presentation of Moore's Law, it is stated that the number of components per integrated circuit doubles every year (Moore 1965). In the following developments around Moore's Law, this observation or predication is materialized by the continuous miniaturization of transistor. Although electronic engineering seems intricate due to the large number of circuits developed and a full suite of tools used, its focus is simply on mere one thing, namely the electrized information. Or, its work focuses solely on processing electronic data. The real impact from the practice of Moore's Law is therefore the continuous reduction in the amount of matter and energy consumed when each bit of information is processed. This is very much in compliance with the spirit of "the principle of least action". In microelectronics, it is mandatory for us to make the very best effort in using as less energy and matter as possible when processing each bit of data. To achieve this goal of "least action" in electronic world, as the potential from space dries up, we definitely need to pay more attention to time, or use time better, when practicing this art. At current stage of a space-related crisis, we need to develop a new strategy for playing this game. This strategy will be developed around the thing we called time and its embodiment, namely, the clock signal.

5.2 The Importance of Clock Technology: From Application Point of View

As stressed many times in previous chapters, information processing is the key concern of electronic devices. Inside devices, information takes the shape of electrical signals which in turn is embodied as bitstream in the majority of applications (i.e., the digital signal processing). Only in special occasions, signal processing is performed in pure analog fashion. Although limited in its scope of utilization, analog signal processing is an indispensable part in the art of microelectronics. The task of information processing, digital or analog, is supported by several underpinning circuit technologies. In a bird's-eye view, the entire field of semiconductor integrated circuit design can be roughly classified into four major technologies: analog/RF technology, processor technology, memory/storage technology and clock technology, as illustrated in Fig. 5.2.

In the past several decades, tremendous effort has been spent on processor, memory/storage and analog/RF technologies. As a result, significant advances have been made in those areas. Among the four, clock technology however has fallen behind. In Sect. 3.4, the concept of clock signal is introduced. In Sect. 4.4, the most important attribute of clock signal, namely its frequency, is discussed. In Sect. 4.5, the traditional way of generating clock signal is reviewed. An observation is made that, due to the ineptitude of the mainstream clock generation technology PLL, there are two long-standing problems, all related to clock frequency, that have not been solved to our complete satisfaction. This is the justification that, compared to the other three areas, clock technology has not advanced as much.

As can be clearly apprehended from the illustration in Fig. 5.2, clock technology is fundamentally important. It serves as the driver of all the other three since all of them require a sense-of-time to be established within their respective domains. For clock technology itself, it has three subjects of study: clock usage, clock distribution and clock generation. Clock generation further includes its own four key issues, as shown in the right-hand side of the figure. The task of clock generation is often referred as frequency synthesis since the most distinguished characteristic of a clock signal is its frequency. In the history of frequency synthesis, there are three major approaches: Direct Analog Synthesis (the mix-filter-divide approach) (Popiel-Gorski

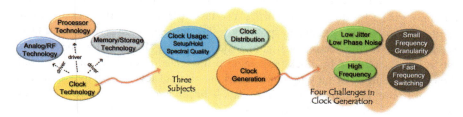

Fig. 5.2 The four major technologies in circuit design and the importance of clock technology

1975), Direct Digital Synthesis (the look-up-table approach) (Kroupa 1998), and Phase Locked Loop (PLL, the compare-then-correct approach, the indirect approach) (Best 2007; Egan 1999, 2007; Gardner 2005; Goldman 2007; Razavi 2008). Among them, the first and second ones are routinely found as stand-alone solutions. For on-chip clock generation, PLL is the most widely used method due to its easy integration with other on-chip circuits. As discussed in Chap. 4, TAF-DPS is now emerging as a new alternative with great up-rising potential due to its attempt in addressing the two long-standing problems.

From a user perspective, when a clock signal is evaluated, the concerns mainly focus on output frequency range (tuning range), frequency resolution (frequency granularity), frequency stability (phase noise, jitter) and frequency switching speed (settling time). From a hardware design perspective, unfortunately, these requirements are often mutually contradictory. These requirements can be itemized as four design targets: high frequency, low phase noise/jitter, small frequency granularity and fast frequency switching as depicted in Fig. 5.2. In the past, most clock-generation-related work has been focused on the first two targets (the green ones), and it is fair to say that those two issues have been well understood and decent solutions are available. The third and fourth targets (the brown ones), however, remain to be the problems that have not been solved to our complete satisfaction. In other words, small frequency granularity and fast frequency switching are the two long lasting problems in the field of on-chip clock generation as discussed in Sect. 4.5. In the upcoming time-oriented paradigm, clock technology needs to catch up the pace of the others. This will surely help improve the bit/energy efficiency, rising it further to next level. This plan can be symbolized as "play with frequency whenever and wherever possible". TAF-DPS is a tool for turning this agenda into reality. But let's first look at the deleterious consequence of the two long-standing problems.

5.3 The Deleterious Consequence of the Two Long-Standing Problems

As argued, information processing is the centerpiece of all modern applications. As a matter of fact, the sole purpose of most chips is to handle information, including the tasks of generating and receiving data, processing data, moving data and, eventually, consuming data for directing actions. Those tasks are being carried out at various levels such as circuit, system and application. The subject-of-study in all those tasks is data, which is embodied in bitstream made of "0" and "1". The movement of this bitstream is controlled by clock frequency, as illustrated in the left side of Fig. 5.3. Among those tasks of moving data at various abstract levels, one foundationally basic operation is the transportation of data between two different physical places, symbolically labelled as TX (transmitter) and RX (receiver), respectively, in the right-hand side of Fig. 5.3.

5.3 The Deleterious Consequence of the Two Long-Standing Problems

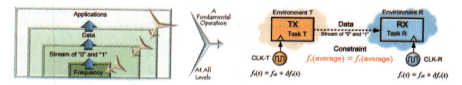

Fig. 5.3 Transferring bits is an omnipresent operation in microelectronics

In this operation that underpins the edifice of electronic system, clock signal controls the data flow, clock frequency is its pace. In most cases, each module of TX and RX has its own clock source with its intrinsic frequency. Moreover, each module works in its own environment including the conditions of voltage, temperature, loading and etc. For those reasons, the instant frequencies of the two parties, $f_t(t)$ and $f_r(t)$, most likely will not be equal. For the successful transfer of a stream of bits, however, the paces of the data flows on the two sides (the transmitting stream from the TX's output port and the receiving stream at RX's input port) must be equal in average so that no data-lose or cycle-slip occurs. This demands a mechanism for adjusting the $f_t(t)$ and $f_r(t)$, to match them averagely over a targeted time frame. For different level of performance, the time frame used for "calculating the average" is different. It could be in the range of second, millisecond, microsecond, or even nanosecond. The shorter this period of "averagely-achieving-equal-data-flow" is, the higher the overall information processing efficiency will be.

This basic operation is so ubiquitous that it can be found everywhere in microelectronics design: between small circuit blocks, between functional modules, between chips and even among nodes in a network. Frequency is absolutely the key factor in this operation. At the very bottom level (i.e., inside electrical circuits), data is materialized through bitstream. A designated bitstream traveling from one place to another is the electrical signal (digital signal in this case) that has been discussed in Sect. 3.2. In principle, the characteristic of digital signal and the issue of signal-transfer can be judged from the viewpoint of clock frequency. This reasoning can be explained with the help of the illustration in Fig. 5.4.

When an individual signal is concerned, it can be characterized by the time interval used for covering one bit of data. From this perspective, there are two types of digital signal, isochronous and anisochronous, as shown in the left of Fig. 5.4. In the case of isochronous, all the bits use same length for their time intervals (i.e., the time span

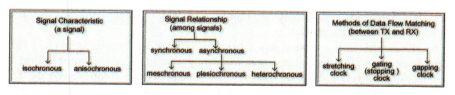

Fig. 5.4 The characteristic of electrical signal and the issue of matching-signal-flows are discerned by frequency

occupied by any bit is equal to that of any other bit). Otherwise, it is anisochronous wherein the time interval separating any two significant instants of an anisochronous signal does not necessarily relate to the time interval separating two other significant instants of the same signal. In general, data driven by TAF clock belongs to the category of anisochronous while data driven by conventional clock is isochronous signal, as can be appreciated from the TAF pulse waveform depicted in Fig. 4.8.

When more than one signal is considered, the relationship among signals will become important. As depicted in the middle of Fig. 5.4, the relations of signals can be classified as two types: synchronous and asynchronous. Two or more signals, if they are synchronous, must be driven by same clock, or by clocks sharing a common reference (i.e., they have exactly the same frequency but might have phase difference of fixed value). On the other hand, asynchronous signals are driven by clocks that are independent of each other, which naturally bear different frequencies. Inside asynchronous category, there are three subclasses of meschronous, plesiochronous and heterochronous. For meschronous, the two driving clocks have same frequency but with a random and unknown phase difference. In plesiochronous case, the driving clocks have different instantaneous frequencies but their frequencies over a given time frame are equal in average. Finally, for heterochronous, even the average frequencies are different.

When data is transferred from TX to RX, most likely, the mode will be asynchronous since TX and RX usually locate in different places and use their own local clocks. To match the data flows of both sides for avoiding bit loss, the frequencies of the clock signals on both sides therefore have to be adjusted from time to time. In the right-hand side of Fig. 5.4, three methods of matching the signal flows are listed. As expected, all the three are operated on clock: stretching clock, gating clock and gapping clock. Stretching prolongs or shortens an individual clock pulse at selected times. Gating (sometimes also referred as stopping clock) stops the clock pulse train when necessary. In the case of gapping, one or more clock pulses are intentionally removed at appropriate times.

In all the cases, the goal is to match the TX data flow to that of RX. Except in the straightforward case of synchronous data transfer where both TX and RX share the same clock, the goal of data-flow-matching can only be achieved in an average sense. In achieving this, there must be something in between the TX and RX that can function as a temporary place to hold the data so that nothing is lost (that is the reason of dot line connection instead of solid line between the TX and RX in the right-hand side illustration of Fig. 5.3). This scheme is illustrated in the right-hand side of Fig. 5.5. The purpose of bringing in this storage is to guarantee that, within a specified time frame, all TX data are faithfully moved to RX with nothing dropped or repeated (extra data added).

From the discussion around this scheme, it is therefore immediately clear that "ample supply of frequency" and "fast frequency switching" are the most desirable features in performing this data transfer task since adjustable frequency on the TX and RX clocks can simplify the design of the storage located between TX and RX. In other words, the more flexible the $f_t(t)$ and $f_r(t)$ values are, the smaller the size of the storage can be. Reflecting this desire in hardware design, the more flexible the

5.4 Whenever and Wherever: It's Time to Use Time

Fig. 5.5 A storage must be inserted between TX and RX for ensuring no data loss

clock generators are, the simpler the structure of the temporary storage will be, and the higher the efficiency of the data transportation process will be. Flexible clock generator will make the designing of the rest of circuits in the chip easier.

Practically speaking, data transport is one of the most critical tasks in signal processing, as important as the tasks of arithmetic and logic operations (i.e., the datapath design). The deleterious consequence of the two long-standing problems, as discussed intensively in Sect. 4.5 and (Xiu 2016, 2019), is that this problem of data transport cannot be solved efficiently since the clock frequency cannot be easily changed. Putting it in another way, the doctrine of fixed-frequency clocking is not effective. In the past when people work in the space-dominant paradigm, the lack of flexible clock generator (i.e., the incompetent in lower hardware level) leads to this unfortunate situation. This is why the awkward and inefficient solutions of stretching clock and gapping clock are invented in the first place. They are used to compensate the ineptness of clock generator in handling frequency. This problem of inefficient-data-transfer caused by the relatively impotent clock generator lies at the very bottom level in the field of electronic system design. Its damaging influence is immense and cannot be overstated since this issue is omnipresent throughout the entire microelectronics.

5.4 Whenever and Wherever: It's Time to Use Time

In past practice (i.e., space-dominant paradigm), a fixed-frequency clock signal would very likely be sufficient for most applications to perform their normal functions. For this reason, clock generators of the old times have focused their attention on just generating, for a given application, a few selected frequencies with high frequency stability. Moreover, fast frequency switching speed is not considered as a high design priority. Instead, low phase noise or small jitter is the primary goal. Therefore, it is reasonable to denominate this kind of clock signal as *rigid* clock. After more than a half century's evolution, the sophistication of electronic design has reached a formidable height. Moving forward, the challenges at the forefront of future electronic system design can be recapitulated as below. They are all related to clock signal in one way or another. Therefore, logically speaking, something has to be done on the clocking side so that further progress in the entire field of microelectronics can be possible.

- Connect every electronic device to a network (Internet of Things)
- Move ever-increasing amount of data among heterogeneously clocked systems
- Reduce power consumption, improve energy efficiency.
- Alleviate EMI (electromagnetic interference) problem
- Improve network time synchronization accuracy (e.g., to nanosecond range)
- Enhance electronic device's sensing capability, improve measurement accuracy.
- Promote reconfigurable hardware, improve hardware programmability
- Embed hardware-based security into electronic devices
- Deal with miniaturization challenge
- Manage interoperability challenge
- Cope with heterogeneous system architecture, multi-core architecture and reconfigurable computing
- Operate under rough environment (larger manufacture process variation, larger swing on supply voltage, bigger change on temperature and higher degree of EM interference)
- More …

As can be appreciated from the discussion around aforementioned challenges, the design of future electronic system is more complex and its working environment is harsher. Therefore, the help from the clocking side will be surely welcomed. For this reason, in this transition from space-dominant paradigm to time-oriented one, now, it is time to treat time more seriously. This spirit is appropriately reflected in this slogan: *It's time to use time*. In this new paradigm, it is mandatory for system designers to possess the capability of changing clock frequency whenever and wherever they desire, and the capability of setting the clock signal frequency to whatever value they want. In particular, ample-supply-of-frequency and quick-frequency-switching are the two features that must be supported by clock circuit designers from lower level and be available to system architects at higher level. Those two features are no longer "nice-to-have" but "must-have". A clock signal of such characteristics can be called *flexible* clock, in contrast to the rigid clock in space-dominant paradigm. For system architects, when they envisage the design of a whole system from high level, the deleterious effect originated from the two long-standing problems must be cleared from the apprehension. In this respect, a clock generator capable of generating such flexible clock signal can be appropriately named as Field Programmable Frequency Generator (FPFG) since people working in the field can now do whatever they want when clock frequency is concerned (Xiu 2015a, b).

When FPFG becomes available, the ideal scenario of interacting between clock signal and its environment can be realized in a fashion as illustrated in Fig. 5.6. In principle, the environment is the driver and it forces the clock generator (the FPFG) to tune its operating frequency so that the chip's working pace can be optimally adapted to the environment. The work need-to-be-accomplished and the energy-must-be-consumed thus can potentially reach an optimal balance. A more detailed description to achieve such goal is depicted in Fig. 5.7. A sensor or monitor is constantly watching the environment. Its output, an analog to digital converter (A/D) is needed if the sensor's output is in analog format, is fed into a situation assessment

5.4 Whenever and Wherever: It's Time to Use Time

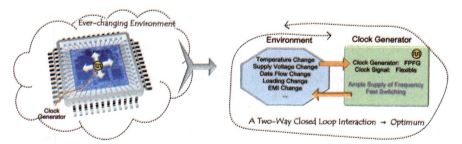

Fig. 5.6 The flexible clocking strategy

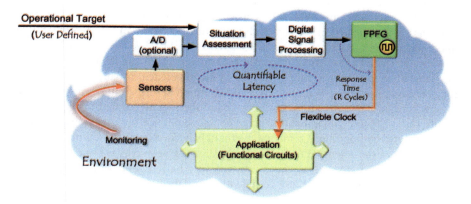

Fig. 5.7 The feedback scheme in the doctrine of flexible clocking

unit which analyzes both the environmental data and the operational target defined by user. The output of the assessment unit is passed to a digital signal processing unit for controlling the flexible clock generator (the FPFG), to generate the new clock frequency that is more suitable for current environmental condition. The new clock frequency will drive the functional circuits to achieve the optimal result when the ratio of bit/energy is used as the measuring metric, as discussed previously around Fig. 5.1.

As seen, this is a typical closed-loop-action of negative feedback that targets for balance and stability. It is a self-regulating process that aims for smoothing future activity so that fluctuations can always be brought back to a tolerable level. This is most likely what we want to achieve when using an electronic device for processing information. Notably, this feedback loop must operate in a time-predictable fashion. The loop latency needs to be quantifiable in terms of clock cycle. In achieving this, one of the enabling factors is the quantifiable behavior of the clock generator (the FPFG). When a command of frequency-update is received, the clock generator must respond quickly and time-quantifiably such that the digital signal processing unit it drives can be designed appropriately. Unfortunately, PLL is not a capable contender for being such clock generator since it responds to command-of-change neither quickly nor time-quantifiably. The TAF-DPS clock generator, described in Sect. 4.6, is ideal for playing this role because its response time is quantifiable, typically two cycles in most designs (i.e., R = 2 in Fig. 5.7).

The scenario of Fig. 5.6 is inspired by the Gaia hypothesis formulated by James Lovelock in the 1960s (Lovelock 1972, 2005). This hypothesis likened the earth to a self-regulating living organism. While this does not imply that our world is actually animate, it does entail complex and connected interactions between life and the physical environment (e.g., the atmosphere, the oceans, the polar ice sheets and the rock beneath our feet). The picture painted in Fig. 5.6 can be considered as a Gaia electronic world where information is the life. An electronic device is a self-regulating living organism wherein information (the bits), as the basic form of life, is generated, propagated and eventually consumed (which is equivalent to death). New information emerges from the ash of the old ones. Someday in future, human-like intelligence might arise from this electronic Gaia world. This scenario however can nowhere be close to a reality if we continue to operate following the fixed-frequency clocking ideology. This is simply due to its incompetence in handling of the two long-standing problems. With the TAF-DPS solution described in Sect. 4.6, the era of new time-oriented paradigm has arrived and the picture painted in Fig. 5.6 can finally start to become imaginable. This of course requires effort of serious technical work. But more importantly, a mental overhaul is needed as discussed in Chap. 4.

Let's return our attention to the two long-standing problems for one more time. In the first step, it is worth to check how the very basic operation of data transfer can be handled differently (or more efficiently) after the two long-standing problems are cleared. As can be appreciated from Fig. 5.3, frequency is absolutely the key in achieving the lossless bitstream transfer. The more flexible the clock sources CLK-T and CLK-R are, the more quickly they can adjust their frequencies to match each other. This adaptiveness can be supported by the features of arbitrary-frequency-generation and instantaneous-frequency-switching from circuit level. It then can be exploited by system architect at system level, as illustrated in Fig. 5.8. As said, a temporary storage must be inserted between TX and RX. This is simply due to the reason that CLKT and CLKR are usually independent of each other (they don't share a common reference). Their frequencies are denoted as $f_t(t)$ and $f_r(t)$, respectively. Those values vary with time due to factors such as environmental disturbances, component aging and etc. Therefore $f_t(t) = f_{t0} + \delta f_t(t)$ and $f_r(t) = f_{r0} + \delta f_r(t)$ are

Fig. 5.8 Flexible clocking strategy for the basic operation of data transfer

not equal most of the times, where f_{t0} and f_{r0} are the stationary nominal values and $\delta f_t(t)$ and $\delta f_r(t)$ are their variations, respectively.

To ensure successful data transmission, the average rate-of-flows (the frequencies) of the two sides have to be equal, namely, $f_{t\_avg} = f_{r\_avg}$. This demand leads to the scenario that RX must track TX. With this point in mind, the key issues concerning the TX clock's frequency are therefore accuracy, stability and precision. Accuracy is the extent to which a given measurement agrees with the definition of the quantity being measured. Precision is the extent to which a given measurement of one sample agrees with the mean of a measurement set that includes a multiple of measurements of the same sample. Stability describes the amount of change that occurs to certain thing (e.g., accuracy and precision) as a function of some parameters such as time, temperature, and shock. In short, it is preferable for the TX clock frequency to be precise, accurate and stable so that the RX has an easy target to track. For receiver RX, adaptability and error-toleration are the desirable qualities. They in turn demand that the features of ample-supply-of-frequency and fast-frequency-switching be available from the clock generator inside the RX (as a matter of fact, this is preferred in TX as well). TAF-DPS described in Sect. 4.6 is ideal for serving this purpose. When a good solution to this key basic operation is in sight, we can expect that many problems associated with the scenario described in Fig. 5.6 can be dealt with in a smoother fashion (Xiu 2015a, b). The key here is to use the dimension of time more often and more wisely. Or, "it is time to use time", whenever and wherever possible.

5.5 Time Moore Strategy as the Bridge

The space induced problem investigated in Chap. 3 leads to the recognition of a crisis and consequently motivates the paradigm shift from space-dominant to time-oriented. The movement from the old to new paradigm is initially motivated by the clearance of the deleterious influence of the two long-standing problems. It then grows into a grand campaign of using time in a more sophisticated fashion whenever and wherever possible, namely, the campaign symbolized by the slogan of "it's time to use time". Throughout this campaign, all the operations will be guided by a central strategy named "Time Moore strategy". It will serve as the bridge between the two worlds of old and new, as illustrated in Fig. 5.9. In the old one of space-dominant paradigm, the task of signal processing for the most part is carried out by using one arm since our attention is predominantly focused only on space. In the new time-oriented paradigm, the weight will be shared by two arms. From the old to the new, Time Moore strategy will guide the way. What then exactly is the Time Moore strategy? Let's take a quick review on its chain-of-development.

Figure 5.9 is the illustration of the role that Time Moore strategy will play in this campaign. It starts with the ideology of "using as less energy and matter as possible for processing each bit", as argued in Sect. 5.1 when the "principle of least action" is applied on microelectronics. This is then translated into the doctrine of using time more meticulously. This doctrine in turn demands the clock signal

Fig. 5.9 Time-Moore strategy as the bridge connecting the paradigms

to have the features of ample-supply-of-frequency and fast-frequency-switching. Perplexed by those two challenges (the so called two long-standing problems), it forces us to investigate the reason why they remain unsolved for such a long time. This investigation eventually points to the very concept of clock frequency. By breaking the incantation of "all the cycles in a clock pulse train must have same length-in-time", a more innate interpretation of clock frequency, Time-Average-Frequency, is surfaced. This new clock frequency concept enables us to create a novel clock signal generation technology, TAF-DPS, as discussed in Sect. 4.6. TAF-DPS provides a solution to the two long-standing problems. The rationale behind this Time Moore strategy now shall become clear. It is a new circuit design philosophy including a series of components, from concept, theory, and circuit to application. Fundamentally, deep in the heart, it requires a change-of-mindset from the view of uniform-time-flow to the nonuniform-time-flow. This is a tough battle against a formidable enemy of a firm belief brewed in a long tradition.

Figure 5.10 vividly paints a picture of logical chain behind the Time Moore strategy. The breaking of the incantation, the change-of-mindset from uniform-time-flow to nonuniform-time-flow, is the very crucial first step. It naturally leads to the TAF concept. From this new concept, a new set of theories is developed (please refer to Sect. 4.6 and references Sotiriadis 2010a, b, c, 2016; Talwalkar 2012, 2013). This group of theories can be used to direct the exploration of various applications. At the same time, it can help the design and creation of novel circuits at various levels.

There are four development paths that the TAF-DPS technology can advance the field of circuit design. The first one is the circuit structures at the fundamental level.

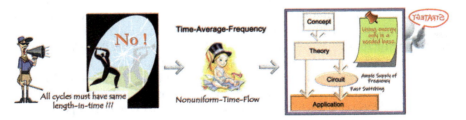

Fig. 5.10 The logical chain of Time Moore strategy

5.5 Time Moore Strategy as the Bridge

Electrical signal is in essence the movement of voltage level over time (please refer to Sects. 3.1 and 3.2). In most of the cases, this movement of voltage level (a large collection of electrons) is embodied in the form of electrical pulses. Being a technique of direct waveform synthesis, TAF-DPS is naturally good at producing all sorts of pulses. Therefore, it is a powerful weapon in circuit designer's arsenal for producing all sorts of signals (e.g., a high caliber pulse width modulator). The second one is a dedicated group of circuits that are specialized in generating clock signal, for the purpose of driving processing circuits, especially digital circuit (Xiu 2015a, b). The third path is along the line of system architectures that address certain specific problems on a scope larger than individual circuits. On this front, the features of ample-supply-of-frequency and fast-frequency-switching are utilized in the investigation of design problems from architectural level, as illustrated in Figs. 4.12 and Fig. 5.11. The last type of development deals with large-chip implementation problems. As chip size grows bigger and process feature size becomes smaller, physical implementation of large complex chip becomes an extremely difficult challenge. For instance, clock distribution is one of the toughest issues in large chip implementation. In this hard battle, TAF-DPS provides a means for attacking such type of problems from a new perspective. Figure 5.11 shows the roadmap of those four development paths in relation to the concept, theory and etc.

Table 5.1 summaries the Time Moore Strategy. Its influence will reach the entire spectrum of electronic design. This is simply due to the fact that everything in electronics has something to do with clock (i.e., the flow of time). Chapter 6 will be dedicated to the demonstration of the power of Time Moore Strategy, through its handling of some specific technical problems.

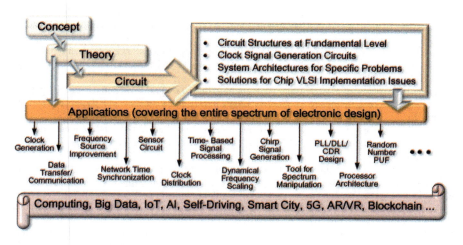

Fig. 5.11 Time Moore strategy exerting influence on the entire spectrum of microelectronics

Table 5.1 Summary of Time Moore strategy

What	Implication
Motivation	Using as less energy and matter as possible for processing each bit
Governing law	Principle of least action
Slogan	It's time to use time
Doctrine	Flexible clocking → arbitrary-frequency-generation and instantaneous-frequency-switching
Change-of-mindset	Uniform-flow-of-time → Nonuniform-flow-of-time (Time-Average-Frequency concept)
Tool	TAF-DPS technology
Development paths	Fundamental circuit on manipulating electrical pulse, clock generation circuit, system architecture, VLSI implementation issue
Outcome	The turn of Moore's Law from space to time → Paradigm shift from space-dominant to time-oriented

References

R. Best, Phase Locked Loops 6/e: Design, Simulation, and Applications, 6th edition, *McGraw-Hill Professional*, 2007.

W. F. Egan, Frequency Synthesis by Phase Lock, 2nd edition, *Wiley-Interscience*, 1999.

W. F. Egan, Phase - Lock Basic , 2nd edition *Wiley - IEEE Press*, 2007.

Richard Feynman, "The Character of Physical Law," *Modern Library*, 1965, ISBN 978–0–679–60127–2.

F. M. Gardner, "Phaselock Techniques," 3rd edition, *Wiley – Interscience*, 2005.

S. J. Goldman, "Phase Locked Loops Engineering Handbook for Integrated Circuits," *Artech House Publishers*, 2007.

V. F. Kroupa, "Direct Digital Frequency Synthesis," *IEEE Press*, 1998.

J. E. Lovelock, "Gaia as seen through the atmosphere", *Atmospheric Environment*. 6 (8): 579–580, 1972.

J. E. Lovelock, "Gaia: And the Theory of the Living Planet," *Gardners Books*, 2nd Edition, 2005.

G. E. Moore, "Cramming More Components Onto Integrated Circuits," *Electronics*, vol. 38, no. 8, pp. 114-117, 1965.

J. Popiel-Gorski, "Frequency Synthesis: Techniques and Applications," *IEEE Press*, 1975.

B. Razavi, "Phase-Locking in High - Performance Systems: From Devices to Architectures," *Wiley-IEEE Press*, 2008.

P. Sotiriadis, "Theory of Flying-Adder frequency synthesizers, Part I: modeling, signals' periods and output average frequency," *IEEE Trans. on Circuits and Systems I*, vol. 57, pp.1935–1948, Aug. 2010a.

P. Sotiriadis, "Theory of Flying-Adder frequency synthesizers, Part II: time and frequency domain properties of the output signal," *IEEE Trans. on Circuits and Systems I*, vol. 57, pp.1949–1963, Aug. 2010b.

P. Sotiriadis, "Exact spectrum and time-domain output of Flying-Adder frequency synthesizers," *IEEE Trans. on Ultrasonics, Ferroelectrics, and Freq. Control*, vol.57, pp. 1926–1935, Sep. 2010c.

P. Sotiriadis, "Spurs-Free Single-Bit-Output All-Digital Frequency Synthesizers With Forward and Feedback Spurs and Noise Cancellation," in *IEEE Transactions on Circuits and Systems I*: Regular Papers, vol. 63, no. 5, pp. 567-576, May 2016.

S. A. Talwalkar, "Quantization error spectra structure of a DTC synthesizer via the DFT axis scaling property," *IEEE Trans. on Circuit And System I*, vol. 59, pp.1242–1250, June 2012.

References

S. A. Talwalkar, "Digital-to-Time synthesizers: separating delay line error spurs and quantization error spurs," *IEEE Trans. on Circuit And System I*, vol. 60, pp.2597–2605, Oct. 2013.

Xiu, L., "Circuit And Method For Adaptive Clock Generation Using Dynamic-Time-Average-Frequency", US 9,118,275, August 25, 2015a.

Xiu, L., "Microelectronic System Using Time-Average-Frequency Clock Signal As Its Timekeeper", US 9,143,139, Sep. 22, 2015b.

Xiu, L., chapter 13 *"Spectrally Pure Clock vs. Flexible Clock: Which One Is More Efficient In Driving Future Electronic System?"* in book of *"Mixed-Signal Circuit"*, 2016, CRC press.

Xiu, L., chapter 6 "Exploiting Time: The Intersection Point of Multidisciplines and the Next Challenge and Opportunity in the Making of Electronics" in book of "Low-Power Circuits for Emerging Applications", 2019, CRC press.

Chapter 6
Old World and New Insight: Solving Problem with a Gestalt Switch

6.1 Recapitulation: Data Processing Investigated from Two Perspectives

"Gestalt Switch" is a process whereby a person's perspective on certain thing changes from one to another. As we all have experienced at some time in life, something strange can happen when we are looking at a picture. At first you see one image, and then suddenly your eyes and brain recognize the second one and store it in your memory. The next time you look at this picture you will see both images immediately. This capability of being able to read both sides of the story will last forever after that magic moment. It indicates a fact that our mind has been renewed. This not only applies to reading picture but to our experience in life in general.

The "Rubin Vase", which is presented in the left of Fig. 6.1, is a classic case showing the effect of Gestalt Switch. The proposition behind it is that it demonstrates the way in which our brain makes sense of the world, or, rather, tries to make sense of the world. The Rubin vase belongs to a type known as bi-stable image, meaning that it creates two different stable perceptions: one of a vase, and one of a pair of faces. The picture at the right-hand side of Fig. 6.1 presents another example of tree and animals. From vase to faces, from tree to animals, it might not be easy to recognize them. But once our brain has registered both perceptions, we will not look at those images in the same way as before. Instead of one, we will always see both perceptions.

This idea that the brain can switch from seeing the world from one way to a completely different way is quite what we suggest to do when promoting the paradigm shift from space-dominated to time-oriented. The task of data processing can be carried out from a pure space perspective, meaning that we pack more transistors into a given space as best as we could, just as we relentlessly did in the past six or seven decades. Or, the very task can be investigated from a new perspective of using both space and time (please refer to Fig. 4.14). The benefit of this gestalt switch may not be recognizable at first. But, when the brain does connect the dots, the image unrecognizable before will emerge and we will never want to go back to seeing the

86 6 Old World and New Insight: Solving Problem with a Gestalt Switch

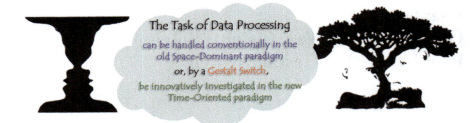

Fig. 6.1 The task of data processing can be investigated from two different perspectives

world in just the old way. In this chapter, we will provide some real examples of engineering problem as such type of dots. It is our hope that readers of this book can connect those dots and make a full new image out of it.

In the persuasion work of chapter four and five for a paradigm shift (i.e., the turn of Moore's Law from space to time), some illustrations (e.g., Figs. 4.12, 4.15, 5.2, 5.5, 5.6 and 5.9) have been used to elucidate the point and paint the whole picture from a high-level. To materialize this transition of paradigms from old to new, however, battles have to be fought one by one, individually carried out on each and every technical problem. This will be a two-front war with old issues and new challenges, as illustrated in Fig. 6.2. The power of applying the Time-Moore strategy in this war of paradigm shift will be demonstrated through two approaches: solving old problems with new technology and dealing with new challenges by fresh insight. This campaign will have a wide influence on the entire spectrum of electronic design, as will become clear after the examples presented in this chapter. This is simply due to the reason that sense-of-time is an omnipresent issue embedded deeply inside all the problems that human and machine must deal with. In this chapter, some real cases will be presented as inspiring examples for others to follow. The topics in this chapter are not organized in any specific order and reader is encouraged to pick up any subject that is interesting to him/her at any given time.

Before getting into any specific topics, it is helpful to go over some key features of TAF-DPS since they will be used in all the subsequent discussions. In Sect. 4.6, the principle of TAF-DPS technology has been explained. For detailed circuit level description, readers are referred to Xiu (2012a, 2015f, g). To assist the discussions

Fig. 6.2 A two-front war: solving old problems with new technology and dealing with new challenges by fresh insight

6.1 Recapitulation: Data Processing Investigated from Two Perspectives

in following sections, the TAF-DPS frequency transfer function needs to be briefly reviewed here. As argued in previous chapters, TAF-DPS is a novel clock generation technique that is fundamentally different from PLL. It is a direct period synthesizer that synthesizes the waveform of each of its output pulse directly. From a base-time-unit Δ, the synthesizer first creates two types of cycles T_A and T_B. Their length-in-times are given by (6.1.1) where I is an integer. When synthesizing a particular frequency (period) f_s, it uses T_A and T_B in an interwoven fashion. The output frequency (period) is expressed in (6.1.2) where $F = I + r$ is the frequency (more precisely speaking, period) control word. The fraction r controls the occurrence probability of T_A and T_B. The frequency resolution can be derived from (6.1.2) and is expressed in (6.1.3).

$$T_A = I \cdot \Delta, \quad T_B = (I+1) \cdot \Delta \tag{6.1.1}$$

$$T_{TAF} = \frac{1}{f_s} = (1-r) \cdot T_A + r \cdot T_B = (I+r) \cdot \Delta = F \cdot \Delta \tag{6.1.2}$$

$$df_s = -\frac{dF}{F^2 \Delta} = -\frac{dF}{F} \frac{1}{F\Delta} = -\frac{dF}{F} f_s, \quad \frac{df_s}{f_s} = -\frac{dF}{F} \tag{6.1.3}$$

$$f_1 = \frac{1}{x} = \frac{1}{x_0 + \delta x}, \quad \frac{f_2 - \frac{1}{x_0}}{\delta x} = \frac{d}{dx}\left(\frac{1}{x}\right)\bigg|_{x=x_0} = -\frac{1}{x_0^2}, \quad \frac{\delta y}{y} = \frac{f_1 - f_2}{f_1} = \left(\frac{\delta x}{x_0}\right)^2 \tag{6.1.4}$$

The left side of Fig. 6.3 is the f_s veersus F curve. Its output frequency is inversely proportional to control word F. From the equation and the curve, it is clear that monotonicity is mathematically guaranteed. In any given small area, the relationship is almost linear. Linearity can be studied and defined by the help of tangent line. In the right-hand side, a zoom-in view is provided for a selected point x_0. Selecting another point x located $\delta x = x - x_0$ away from x_0, the $1/x$ curve and the tangent line

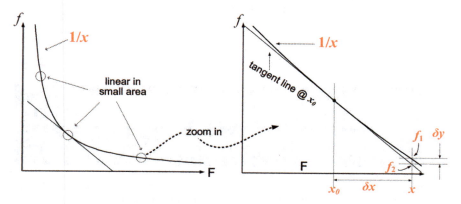

Fig. 6.3 TAF-DPS transfer function f versus F curve

have their respective values f_1 and f_2. If we denote $\delta y = f_1 - f_2$, the linearity at point x can be defined as $\delta y/y$ where y is the frequency value f_1 at point x and f_2 is the expected value from the tangent line. Using some simple algebra, linearity can be derived as $\delta y/y = (\delta x/x_0)^2$ as shown in (6.1.4). Equation (6.1.1) to (6.1.4) will be referenced frequently in the following sections.

One particularly important piece of information is how the TAF-DPS responds to the change of the control input (i.e., frequency changing command). Figure 6.4 is the demonstration of its response speed using real data obtained from experiment. It is a collection of measured frequency points, which are generated from a real chip by TAF-DPS using Eq. (6.1.2). In the top of the figure, the changing-pattern of frequency control word F is displayed. This pattern is fed into a TAF-DPS circuit and its output period (frequency) is measured using a frequency counter. The resultant measured period data is displayed in the middle plot. As seen, the output period faithfully follows the F as predicted by (6.1.2). In the very bottom of the figure, the corresponding waveform is displayed, captured from an oscilloscope. As seen, the waveform indeed varies as F changes. More importantly, the output frequency changes instantly after the command is received as evidenced from those plots. As a matter of fact, the output waveform (and thus its frequency) changes after precisely two cycles of receiving the command.

Equations (6.1.1–6.1.4) is the underpinning support for all the innovations that will be discussed in the following sections. Readers are encouraged to come back here and visit them as often as desired.

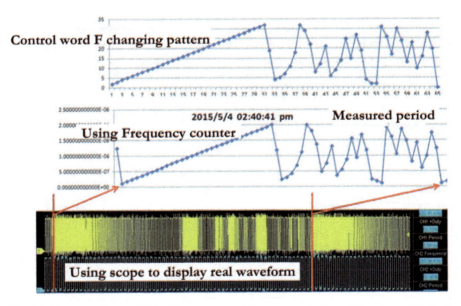

Fig. 6.4 Demonstration of TAF-DPS output period T versus control word F: pattern of F (top), trend of output period (middle) and corresponding output waveform (bottom)

6.2 A Novel Frequency-Tunable Clock Source: TAF-DPS DCXO

6.2.1 Review on Tunable Frequency Source

Frequency source is a foundational component in electronics. It is the base for creating clock signal, which is the most important electrical signal. Clock is used to coordinate events inside electronic world. For example, clock is a key factor in time synchronization process which is the foundation for myriad applications from military, metrology, industrial, consumer, communication networks, automotive, to power grid, banking, scientific project and etc. In applications, clock signal originates from reference frequency standards that require high quality factor (Q factor) resonating element, such as micro-machined or piezoelectric crystal that acts as an electromechanical resonating module. As an instance, crystal in oscillator circuit behaves like a high Q-tuned network, permitting vibrations at its resonant frequency.

Among the many issues around frequency source, long term frequency stability is one of the key concerns (Gonzalez 2007; Vig 2000). Stability of a source is defined as the degree to which the oscillator generates a constant frequency throughout a specified period of time. In many applications, there is however demand for such type of a frequency source whose frequency can be slightly adjusted through an external control variable. Typical examples include VCXO (voltage-controlled crystal oscillator) where an analog voltage is the control and DCXO (digital-controlled crystal oscillator) with digital value as the control. The performance of such sources is usually judged by frequency step, pulling range, monotonicity and linearity. For a particular source, unfortunately, frequency adjustability and frequency stability are two requirements that contradict each other.

Figure 6.5 depicts several types of frequency adjustable sources. All frequency adjustable sources can be considered as made of virtually two parts, namely, the primary frequency source and the frequency tuning module. Figure 6.5a is the case of VCO (voltage-controlled oscillator) and DCO (digital-controlled oscillator) where variable capacitor diode or MOS varactor array is incorporated inside oscillator circuit. For VCO, frequency tuning is accomplished through an analog voltage. A voltage change leads to a corresponding change in capacitance value which in turn induces a change in its oscillation frequency (Kent 1965; Helle 1975; Ishii et al. 1996). For DCO, the tunable capacitance is realized by switched capacitor array. The array is controlled by a digital control word which turns on or off each individual capacitor. Therefore, the oscillation frequency is controlled through a digital word (Tsai et al. 2008; Griffith et al. 2010; Tran et al. 2017). Besides crystal, the resonator could also be LC tank based (LCO), RO (ring oscillator) and etc.

In recent years, MEMS (Microelectromechanical) oscillator becomes a popular solution for replacing crystal oscillator in some markets thanks to its features of lower power, smaller form factor, and lower price (Partridge et al. 2013). There are two application drivers wherein quartz addresses with mechanical process but MEMS addresses electronically: (a) providing a wide range of application frequencies, and

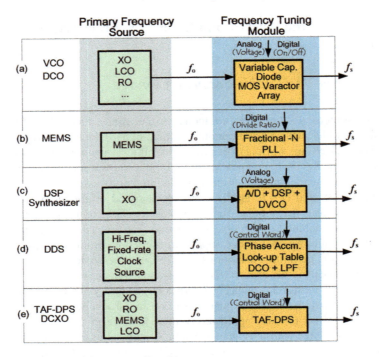

Fig. 6.5 Commonly used frequency-adjustable sources

(b) trimming the resonator frequency over production tolerances. The frequency tuning of MEMS oscillator is achieved not through adjusting its resonant frequency but by a post-processing electrical circuit, which is a fractional-N PLL as shown in Fig. 6.5b. The desired frequency adjustment is represented as divide ratio and is fed into the PLL (Chance et al. 2014).

Another approach for tuning frequency is based on digital signal processing. In Kobayashi et al. (2014), two crystals are formed as a temperature sensor. The difference in their f versus T (temperature) characteristics is converted into digital information representing temperature, which is then processed by a DSP + DDS (Direct Digital Synthesis) block. The output of this block is fed into a PLL as its input. Thus, the frequency tuning occurs from the input side of the PLL. In Petrowski and Clark (2006), the frequency tuning is accomplished through a digital controlled on-chip oscillator. The digital control is generated from a DSP processor as shown in Fig. 6.5c, to form a DCO. The DCO is then used in a frequency feedback control loop with extensive digital processing units. The resultant circuit is the so-called DSPLL (also called any-rate any-output clock generator). DDS is another important and mature technology (Analog Devices 1999; Goldberg 1999; Kroupa 1998). It is a very popular frequency synthesis technique. Typically, it is embodied as a standalone solution and is mostly found in high-end instruments. It is especially applicable for arbitrary waveform generation. Due to its strong capability in generating frequency, it can also be considered as tunable frequency source as shown in Fig. 6.5d. Figure 6.5e shows the structure of our TAF-DPS based solution, the TAF-DPS DCXO (Xiu and Wei 2019), which will be described in the following text.

6.2.2 Architecture of TAF-DPS Based Frequency Tuning

Figure 6.6 depicts the architecture of using TAF-DPS for frequency tuning (Xiu 2020). Being a frequency synthesis engine, TAF-DPS requires some kind of fuse (i.e., the primary frequency sources shown in Fig. 6.5) to fire the electrical oscillation. The frequency source could be any of the XO, MEMS, LCO or RO (ring oscillator). Its output is denoted as f_c. As shown, after the frequency adjustment performed by all the circuit elements in between, f_o is the final output. The key element in this architecture is the TAF-DPS wherein the process of frequency tuning actually takes place. The Integer-N PLL is used to boost the output frequency when needed.

As the fuse for initialing electrical oscillation, the output of the frequency source is used to drive the block of base time generator. If the frequency f_c is high enough, a divider chain can be used to generate the plurality of signals. For example, a Johnson counter of K/2 stages can produce K signals with frequency $f_\Delta = f_c/K$ from its K/2 flip-flops (an example is available in Xiu and Chen 2017). In the cases when f_c is low, an integer-N PLL of divide ratio M can be used to boost f_c to $f_m = M \cdot f_c$. If this PLL has the capability of producing multiple outputs (an example is available in Xiu et al. (2013)), it can function as the base time generator and its output can be fed into the TAF-DPS directly.

In general, the frequency at the output of base time generator can be expressed as $f_\Delta = C \cdot f_m$ where C is a constant, C = 1 if a multi-phase PLL is used as the time base generator and C = 1/K if a K/2-stage Johnson counter is used. This plurality of signals is then fed into the TAF-DPS where $\Delta = T_\Delta/K = 1/(K \cdot f_\Delta)$. From (6.1.2), it is derived that $f_s = (K/F) \cdot f_\Delta$. Therefore, the final output f_o is related to the original f_c as shown in (6.2.1).

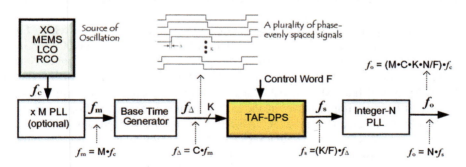

Fig. 6.6 Architecture of TAF-DPS based frequency tuning

$$f_o = (M \cdot C \cdot K \cdot N/F) \cdot f_c \qquad (6.2.1)$$

$$F = I + r = M \cdot C \cdot K \cdot N \cdot (1 + x) \qquad (6.2.2)$$

$$(1 + r/I) \cdot [I/(M \cdot C \cdot K \cdot N)] = (1 + x) \qquad (6.2.3)$$

In (6.2.1), the values of K, M, N, and C are design parameters controlled by user and f_c is a known value. Therefore, the frequency value of f_o can be tuned by adjusting the digital value of F. From (6.2.1), it is clear that this DCXO has a frequency tuning characteristic f_o versus F which is precisely describable by using this formula. In certain case, when situation arises, one might want to tune the frequency from its current value f_c to a new target value f_{target}. This target frequency can be expressed as $f_{\text{target}} = f_c/(1 + x)$ where x represents the amount of frequency adjustment. The goal is to make $f_o = f_{\text{target}}$. Feeding these two conditions into (6.2.1), (6.2.2) and (6.2.3) can be derived. Since x is the desired adjustment amount which is known by user, I and r can be chosen to make $f_o = f_{\text{target}}$. Usually, I is first chosen. After that, r can be used to fine tune the frequency.

6.2.3 TAF-DPS DCXO Demonstration on FPGA

The architecture presented in Sect. 6.2.2 is an operational DCXO. It is therefore termed as TAF-DPS DCXO. It can be used in many applications for various purposes. In this section, through a particular example, we demonstrate its capability of correcting the frequency error of a frequency source on-the-fly. When a frequency source is used on the field, its frequency value (the so-called "accuracy on shipment") is usually not aligned precisely with the number specified in its specification. At this stage, it is however often too late to modify the structure of the frequency source to correct the problem. The cost is prohibitively too high. TAF-DPS DCXO provides an elegant solution to this problem. This will be demonstrated next through an implementation on FPGA.

Xilinx KC705 Evaluation Board for Kintex-7 FPGA is selected as a test vehicle for this purpose. The key component of this system is the XC7K325T-2FFG900C FPGA chip. Regarding clock resource, it has a 200 MHz MEMS oscillator from SiTime and an I^2C programmable oscillator from Silicon Labs. In addition, the FPGA chip has 10 MMCM (Mixed Mode Clock Manager) modules. Each MMCM module has a built-in PLL. In our implementation, the LUT (look up table) and flip-flops on board of the FPGA are used to realize the TAF-DPS circuit. The implementation style in this work is HDL coding → simulation → synthesis (mapping to FPGA). A sample VHDL code is available in Appendix 4A of Xiu (2012a).

Referring back to Sect. 4.6 and Fig. 4.9, the Flying-Adder circuit can virtually be split into three parts: the signal path consisting of the two K → 1 multiplexes and one 2 → 1 multiplex, the adder at the top and the accumulator at the bottom. Those

6.2 A Novel Frequency-Tunable Clock Source: TAF-DPS DCXO

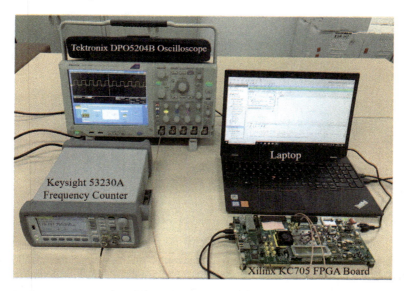

Fig. 6.7 Platform for evaluation of TAF-DPS DCXO

parts are mapped to FPGA hardware separately for better control. The multiplexes are all realized by logic cells. The adder and accumulator are realized by the LUTs and flip-flops. Flying-Adder circuit of K = 16 is configured in this case. We reserve 32 bits for representing the fraction r and 5 bits for the integer I. Figure 6.7 is the photo of the test platform. In experiment, we have evaluated the performance of TAF-DPS DCXO in two frequency points: 26 MHz and 622.08 MHz. Reference source of 26 MHz is often used in telecommunication while 622.08 (155.52) MHz is used in SONET OC-192, SDH STM-64.

The evaluation is carried out with the focus on frequency granularity (resolution), pulling range and linearity. For high precision study, a good frequency source is required as the base. For this reason, the 10 MHz OCXO from Keysight 53230A frequency counter is used as the source as shown in Fig. 6.7. Since the source frequency f_c is low at 10 MHz, it is first boosted by an integer-N PLL with setting M = 32. An 8-stage Johnson counter, driven by this PLL's output, is used to generate a plurality of 16 phase-evenly-spaced signals. We hence have C = 1/16 (please refer to Fig. 6.6). The setting for the PLL after TAF-DPS is N = 1 since final output is low at 26 MHz. Plugging in all the numbers into Eq. (6.2.1), it becomes $f_o = (32/F) \cdot f_c$. Since f_c = 10 MHz and we want f_o = 26 MHz, this leads to a nominal value for F = 12.30769230769. By adjusting F around this value, the output frequency of TAF-DPS DCXO can be tuned, either in large or small step. Figure 6.8 shows the measured result when F varies in a range of [12.18583396801, 12.4320124320]. It corresponds to a frequency range of ±1% around 26 MHz. The crosses are the measured data points and the red line is from calculation. The linearity calculated from measured data is 0.005% while the predication calculated from (6.1.4) is 0.01%.

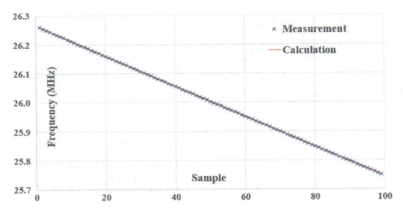

Fig. 6.8 TAF-DPS DCXO f versus F curve around 26 MHz, range of $\mp 1\%$ measured

Figure 6.9 shows the result of fine frequency adjustment, when F makes 100 steps around 26 MHz, each step is 2^{-32} (the LSB of the register holding the value of r). In the left of the figure, fine measurement result around the central part of Fig. 6.8 is shown. From (6.1.3), $df/f = dF/F = 2^{-32}/12.30769230769 = 1.89\text{e}^{-11}$, which is about 0.019 ppb in each step. The measured data confirms this predication. To our best knowledge, 0.019 ppb/step is the finest result reported so far in the literatures. On the right-hand side of this Fig. 6.9, the screen snapshots of several measurements are displayed. The fine resolution is clearly evident.

With a slight modification on the configuration, we test the DCXO on another important frequency of 622.08 MHz. Instead of N = 1 in previous case, now N is set as N = 8. Equation (6.2.1) now becomes $f_o = (32 \cdot 8/F) \cdot f_c$. Since $f_c = 10$ MHz and we now want $f_o = 622.08$ MHz, this leads to F = 4.11522633745. The plot in the left of Fig. 6.10 is the measured f versus F curve when F makes 2^{-32} increment at each step (the measured signal is a divide-by-4 version since the frequency counter's up-limit is 300 MHz). The measurement in the right-hand side of that figure is its corresponding df/f, measured at each step. As seen, the measured

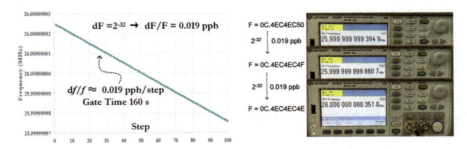

Fig. 6.9 Fine measurement around the central part of Fig. 6.8: movement step on F is 2^{-32} (left), screen snapshots of several measurements (right)

6.2 A Novel Frequency-Tunable Clock Source: TAF-DPS DCXO

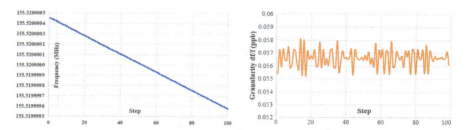

Fig. 6.10 Measured f versus F curve around 622 MHz with step of 2^{-32}: measured signal is its divided-by-4 version (left), corresponding df/f measured at each step (right)

data shows ~0.057 ppb/step. The calculation is dF/F = 2^{-32}/4.12 = 0.058 ppb per step, which agrees very well with the measured.

6.2.4 In Comparison to Commercial Products

The operation of frequency tuning in TAF-DPS DCXO is accomplished through the adjustment of the weighting factor between T_A and T_B cycles, as illustrated in the drawing on the right-hand side of Fig. 6.11. Compared to the physical processes of changing the motional capacitance of quartz crystal (left drawing in Fig. 6.11, Texas Instruments Inc. (2007)) or turning on/off a small capacitor in an on-chip switched-cap array (middle drawing in Fig. 6.11, Silicon Labs Inc. (2018)), the TAF-DPS approach of tuning frequency is virtually "free" since it is mere a mathematical operation. More significantly, the tuning control is rigorously precise. This leads to a much predictable performance, as evident from Figs. 6.8, 6.9 and 6.10.

Compared to the pulling range of just several hundred ppm in commercial products (Texas Instruments Inc., 2007; Silicon Labs Inc., 2018), the tuning range in TAF-DPS DCXO is almost unlimited. For example, in previous case of 26 MHz with F = 12.30769230769, the whole set of measurement datum can be regenerated around 27 MHz (an important reference frequency for video applications) by simply changing F to 11.85185185185. On another front, when studying the output

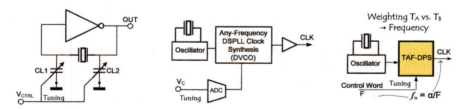

Fig. 6.11 Comparison of frequency tuning mechanisms: changing motional capacitance (right), turning on or off a small capacitor (middle) and adjusting weight factor (left)

frequency versus control, linearity refers to the maximum deviation away from the best straight line drawn through the tuning curve, usually expressed as a percentage. For TAF-DPS DCXO, the linearity is well controlled. Taking Fig. 6.10 for example, the linearity in this group of measured data is 3×10^{-9}. This level of performance is largely own to the fact that there is no physical or electrical process involved when tuning the frequency in TAF-DPS, only mathematical operation. In term of tuning resolution (frequency granularity), some results on frequency resolution reported in literatures are 0.003 ppm/step in Tsai et al. (2008), 0.07 ppm/step in Liu et al. (2017a), 2 ppb/step in Griffith et al. (2010) and 5 ppb/step in Chang et al. (2012). The datum presented in 6.2.5 and Fig. 6.10 is the hard evidence to support the claim of TAF-DPS DCXO superiority (Xiu 2008a). The reason behind this is due to its use of the TAF concept.

6.2.5 Application Example #1: Correcting Frequency Error On-The-Fly

The Xilinx KC705 Evaluation Board uses a chip of SiT9102AI-243N25E200 from SiTime as its system clock. This is a MEMS resonator oscillator. From official data sheet, its frequency tolerance is in the range of ±10 to ±50 ppm. The actual output frequency hence can deviate from its specified 200 MHz in a large amount. At this stage, there is nothing that can be done conveniently to fix this problem of inaccuracy on frequency. In this experiment, we will use the TAF-DPS to correct the frequency error on-the-fly and bring the output back to the target.

Our first step is to measure the MEMS oscillator's output. Based on the measured data of error, we then use the architecture of Fig. 6.6 to compensate it. For the particular SiT9102AI-243N25E200 on this FPGA system board we used, the measured output is $f_c = 200.000584249699$ MHz, which is the mean value out of 600 measurements. According to the definition $f_c = (1 + x) \cdot f_{target}$ of Sect. 6.2.2, it is derived that error $x = -2.921248495 \times 10^{-6}$ ($f_{target} = 200$ MHz). Referring back to Fig. 6.6, due to the reason that f_c of ~200 MHz is high enough, the output from this source is therefore used to directly drive the base time unit made of a 16-output Johnson counter. Thus, M = 1 and C = 1/16 (and K = 16). Feeding those parameters into Eq. (6.2.3), it gives us $(1 + r/I) \cdot I/N = (1 + x)$. We can further pick design parameters for I and N as I = N = 16. As a result, we have r = $x \cdot I = 4.673997592 \times 10^{-5}$. This leads to F = I + r = 16.0000467.

The top curve (blue) in Fig. 6.12 shows the 600 measurement points of the SiT9102AI-243N25E200 MEMS output, before the correction is applied. It is seen that its frequency varies in a range of ~5 ppm with a mean value of 200.000584249699 MHz. This is off-target by 584 Hz (2.92 ppm). The bottom brown curve shows 600 measurement points after the correction is applied. Its mean value now is 200.000033609561 MHz. The center is moved down by 550.6 Hz toward the

6.2 A Novel Frequency-Tunable Clock Source: TAF-DPS DCXO

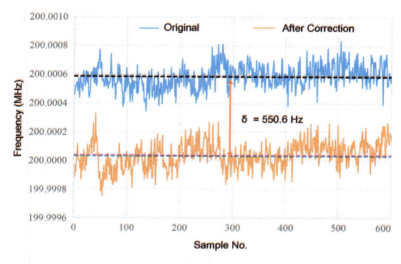

Fig. 6.12 Correction of frequency error on SiTime 200 MHz source

target. After the correction, the off-target amount is about 34 Hz (error is reduced from 2.92 ppm to 0.17 ppm).

In another example, using the architecture presented in Fig. 6.6, the accuracy of a frequency source of 100 MHz crystal is improved from 99.999723266537 MHz (2.77 ppm) to 100.00000057985 (5.80 ppb). The snapshot of the lab measurement is presented in Fig. 6.13.

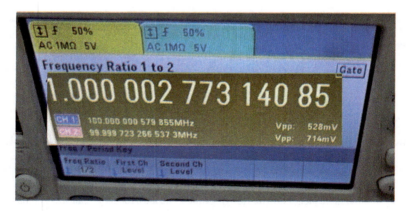

Fig. 6.13 A 100 MHz crystal's frequency accuracy is improved from 99.999723266537 MHz (2.77 ppm) to 100.00000057985 (5.80 ppb)

6.2.6 Application Example #2: Counteracting Aging-Induced Frequency Drift

Aging is a major concern in electronic product's life cycle. One of the problems caused by aging is the frequency drift in frequency source's performance. In this section, an imaginary frequency source having an aging rate of 0.5 ppb/day is assumed and the TAF-DPS is used to demonstrate that this aging-related frequency drift can be compensated. Referring back to Fig. 6.6 and Eq. (6.2.1), we can make a configuration with a M = 16 PLL and a 16-output Johnson counter (C = 1/16). Under this setting, the frequency f_s at TAF-DPS output is $f_s = (16/F) \cdot f_c$. When F is set as F = 16, we have $f_s = f_c = 10$ MHz.

To mimic aging effect, a small value dF can be added/subtracted to F periodically. To make this plan feasible in laboratory setting, we have to however accelerate the aging process. For this reason, the aging rate is accelerated from 0.5 ppb/day to 0.5 ppb/minute. In the experiment, starting from F = 16, a small value of dF = 0.000000008 is subtracted from the value of F once every minute. The blue curve in Fig. 6.14 shows the emulated frequency drift due to aging. To counteract this aging problem, we can occasionally (for example, once every few months in real situation) add a calculated amount to the F. This can bring the output back to the specified value. The red curve in Fig. 6.14 shows the measured result after applying this scheme. A correction of dF = 0.0000008 is added back to F once every 100 min. As seen, the output frequency is brought back to the target every time the correction takes place. Without compensation, the drift will make the frequency off-target by 0.5 ppm after 1000 min. With compensation, the drift is controlled within 0.050 ppm all the time.

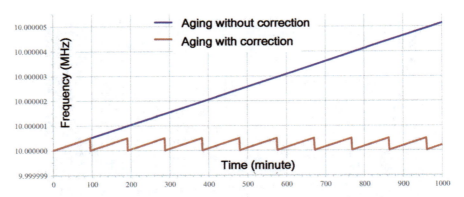

Fig. 6.14 Measured frequency drift due to aging (blue), and the frequency trend after compensation (red)

6.2.7 Application Example #3: On-Chip Emulation of Si570 Function

The Xilinx KC705 evaluation board has a programmable clock source Si570 designed by Silicon Lab. Its output can be programmed in the range of 10 to 810 MHz with resolution of 50 ppm (Silicon Labs Inc. 2014). It is believed that the function of this Si570 can be emulated by the TAF-DPS DCXO in terms of frequency resolution, range and etc. The power of TAF-DPS on achieving fine frequency resolution has already been demonstrated in Sect. 6.2.3. The frequency range is not an issue neither since, as can be appreciated from previous discussion, TAF-DPS has a large operating range (for the high end, the on-chip PLL can offer some help since it can reach 1.4 GHz). For those reasons, we want to spend some effort on investigating only the issue of frequency switching.

Si570 has a constraint that its output clock must be stopped if clock frequency is switched more than ∓ 3500 ppm (Silicon Labs Inc., 2014). TAF-DPS has an advantage of seamless frequency switching thank to its instantaneous frequency switching capability. This is an important feature that can be crucial in system application since it allows for dynamically switching frequency without interrupting system operation. To illustrate this point, a test circuit of multiplication-then-division has been created as shown in Fig. 6.15. Functionally, it outputs a logic "0" in every cycle when working correctly, otherwise "1". For this experiment, this circuit is setup-constrained to 120 MHz.

A continuous frequency switching pattern is generated from TAF-DPS. This is accomplished by a configuration of a M = 64 PLL and a 16-output Johnson counter (C = 1/16) in Fig. 6.6. Thus, the frequency at TAF-DPS output is $f_s = (64/F) \cdot f_c = 640/F$ MHz. The control word F is switched in the pattern of $32 \to 4 \to 32$ in step of one. For each F value, a time frame of 50 cycles is allocated. This pattern is shown as the brown curve in the top potion of Fig. 6.16. The TAF-DPS output frequency f_s is measured and displayed as the blue curve. As expected, the frequency sweeps from 20 to 160 MHz. This TAF-DPS output is used to drive the aforementioned test circuit.

The bottom potion of Fig. 6.16 is the screen capture of the test circuit output waveforms. At the top is the waveform of TAF-DPS output. As seen, the waveform

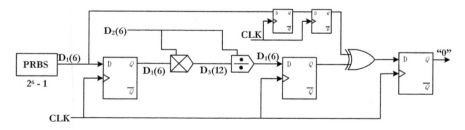

Fig. 6.15 A test circuit for evaluating dynamic frequency switching

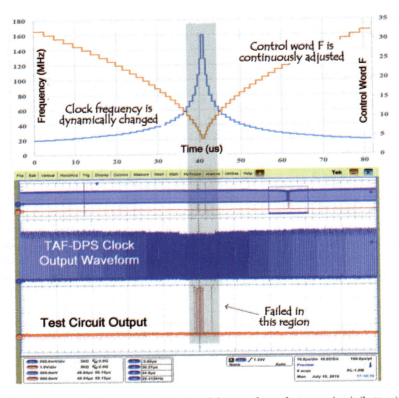

Fig. 6.16 The measured frequency trend (top) and the waveforms from test circuit (bottom)

is continuously varying since its frequency is constantly changing. The red curve at the bottom is the test circuit output. It fails (output becomes "1") in the region when frequency is high. This is because the test circuit of Fig. 6.15 is constrained at 120 MHz and the set-up constraint is not met when the frequency of its driving clock, TAF-DPS output, is higher than that value. In all the other areas of lower clock frequency, the circuit works correctly.

Two points can be made out of this experiment. First, it is important to point out that the test circuit functions normally under the environment that its driving clock changes its frequency continuously in a non-stop fashion. In other words, the system operation is not interrupted. Secondly, also worth mentioning is that this test circuit is representative of all digital circuits regardless of their functionalities. To our best knowledge, this feature of seamlessly frequency switching has not been reported from other type of clock generators. This feature will be an important enabling factor for innovation in future electronic system design (Xiu 2017a).

6.2.8 Application Example #4: A DCXO Module for Clock Synchronization in MPEG2 Transport System

One of the major advantages of digital transmission is the reliable delivery of information from one location to another with high fidelity. In this transmitting and receiving process, timing information can be embedded in the data stream. This technique of embedding timing signal within data has been used in many applications, such as telecommunication, digital TV broadcasting, digital Audio, ADSL and Set-Top-Box. The extraction of this embedded timing signal from the data is often referred as clock recovery. The key issue in clock recovery is the clock synchronization between the transmitter and the receiver. It requires constant adjustment on the local clock (receiving side) based on the timing information extracted. Traditionally, this task of clock adjustment is performed by either external VCXO chip or on-chip VCXO or DCXO module.

The block diagram in the left of Fig. 6.17 depicts a scheme for MPEG2 data stream transmission and reception used in digital TV satellite broadcast. The pictures-for-send are first converted into digital data stream by a video encoder. Then the data stream is transmitted through a communication channel to the receiving side, which could be a Step-Top-Box or a TV set. In this data transmission process, the data stream at the encoding side is generated based on its 27 MHz reference clock (possibly with some rate multiplication). This data stream then is latched by the decoder with its own local clock. Without clock matching between the two sides, the pictures generated from the video decoder will not be synchronized with that of the source. In MPEG2 transport mechanism, a timestamp called program clock reference (PCR) is inserted into the data stream at the rate of at least once in every 100 ms (>10 Hz). This PCR timestamp can be used by the receiver to retrieve the source clock information. As shown, the information generated from this timestamp processing block in receiver is used to drive a VCXO, to slightly adjust its own local 27 MHz reference. In MPEG2 specification, the maximum allowable variation on the two 27 MHz references is 27 MHz ± 810 Hz.

The local clock frequency adjustment is typically achieved by an external VCXO chip, which is a dedicated chip separated from the main processing chip as shown in the left-hand side of Fig. 6.18. There are several commercial vendors offering such VCXO chips as discrete component. Another approach of performing this kind of local clock synchronization is to design a VCXO or DCXO block on chip. This

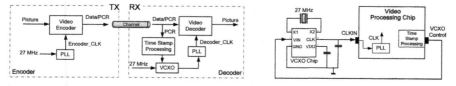

Fig. 6.17 MPEG2 data stream transmission and reception scheme (left), a VCXO chip used for clock recovery (right)

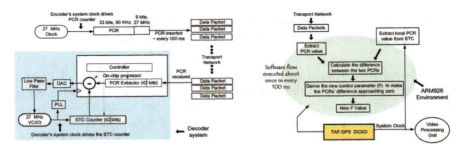

Fig. 6.18 The VCXO chip used in the MPEG clock synchronization process (left) and the scheme of using on-chip TAF-DPS DXCO (right)

approach can eliminate the off-chip VCXO chip and potentially reduce system cost. From the discussion carried out so far in this section, it is clear that TAF-DPS DCXO is ideal for this application. An implementation of such idea has been used in a commercial chip. The scheme of using on-chip TAF-DPS DXCO to carry out the MPEG clock synchronization is illustrated in the right-hand side of Fig. 6.18. As a result, the system cost is greatly reduced. And the overall performance is also improved since the TAF-DPS DCXO is digitally controlled and thus friendly for programming. One important point worth mentioning is that the speed of frequency switching on TAF-DPS DCXO is predictable. It is precisely two clock cycles. This is extremely helpful to system level planning, which is crucial in this case of MPEG2 clock synchronization. For more details, please refer to Xiu (2008b).

6.2.9 Application Example #5: A Method of GPS Disciplined Clock for Improving Frequency Accuracy and Steering Frequency

The idea of using GPS signal for assisting time transfer and frequency syntonization has been explored extensively in the past. In all those methods, the frequency tuning tools involved are complex and expensive. In the left of Fig. 6.19, the key functional blocks used in all the past works are summarized and depicted. The 1 PPS (pulse per second) signal locked to GPS is used as a gate signal to count the number of pulses of a to-be-disciplined clock. The error message, after being processed by some control algorithm, is used to drive the frequency tuning circuit and generate future frequency. The resulting clock is therefore disciplined since its long-term frequency stability is now aligned with that of atomic clock in GPS satellites.

The principle of TAF-DPS DCXO based solution is illustrated in the right-hand side of Fig. 6.19. Our aim is to simplify the tuning module and significantly lower the cost and engineering effort so that a clock disciplining tool can be available to a much larger audience. In Fig. 6.20, more technical details are shown. On the top

6.2 A Novel Frequency-Tunable Clock Source: TAF-DPS DCXO

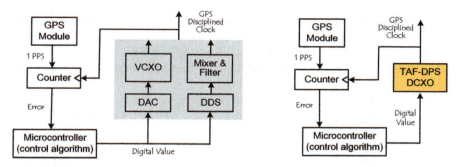

Fig. 6.19 Using GPS to discipline clock: traditional approaches (left) and TAF-DPS DCXO based solution (right)

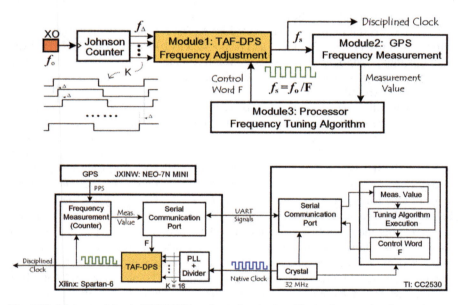

Fig. 6.20 Major modules in TAF-DPS based solution (top) and its implementation (bottom)

is the block diagram that includes major modules. The key block is the TAF-DPS frequency synthesizer. It performs the heavy lifting of tuning frequency. The GPS module is used as a tool for measuring frequency. The TAF-DPS output frequency is measured against the 1 PPS signal. Based on the measured result, a tuning algorithm is executed on the processor module. Its output is a frequency control word that will be fed into the TAF-DPS for directing future output.

We want to develop a method that can be easily applied by ordinary user in the field, not necessarily frequency-control or circuit-design expert in research and development. For this reason, the TAF-DPS used here will be implemented purely in digital fashion. For any electronic system equipped with a commercial-grade primary frequency source (such as a regular XO), the TAF-DPS can be configured as its companion. The interface between XO and TAF-DPS is a Johnson counter whose purpose is to generate a group of K phase-evenly-spaced signals. Those signals are then fed into the TAF-DPS. This guideline has been introduced in the discussion around Fig. 6.6.

On the bottom of Fig. 6.20, the functional modules depicted in the top are realized by three discrete components, all available commercially off-the-shelf for DIY (do it yourself): Spartan-6 FPGA from Xilinx, CC2530 SoC of "Solution for 2.4-GHz IEEE 802.15.4 and ZigBee Applications" from Texas Instrument, and NEO-7 N MINI from JXINW. The NEO-7 N is a GPS receiver that provides 1 PPS signal. The Spartan-6 is the FPGA used for implementing the TAF-DPS. Its output is the disciplined clock. As shown, the crystal output with a nominal value of 32 MHz signal is fed into the Spartan-6. This clock is then trained by the 1 PPS signal through the TAF-DPS. The output from the Spartan-6 is the disciplined clock. In implementation, the 32 MHz crystal output is first divided by 4. The resulting 8 MHz signal is fed into an in-house PLL on Spartan-6 and be boosted to 100 MHz. The 100 MHz PLL output is then used to drive a 16 stage Johnson Counter that produces a plurality of 16 same-frequency-phase-evenly-spaced signals (K = 16). This group of signals is fed into the TAF-DPS.

Following the generic architecture presented in Fig. 6.6 and using Eqs. (6.2.1)–(6.2.3), the TAF-DPS output frequency can be tuned. This process can be constantly executed under the guidance of the 1 PPS signal locked to GPS. As a result, the clock is disciplined. The disciplining target can be the specified value in product datasheet (e.g., the 32 MHz in this case) or a user-defined new value for frequency steering (for example, the 6.78 or 13.56 MHz used in ISM bands). Figures 6.21, 6.22 and 6.23 are some of the results obtained from real tests in the field. Using a system described in Fig. 6.20, the data plotted in Fig. 6.21 shows the improvement on frequency accuracy after applying the disciplining scheme. The data on Fig. 6.22 is generated from 20 such systems. Those systems are configured as a cluster of nodes in a network. The improvement on precision among the 20 nodes is clearly seen after the disciplining scheme is applied. Figure 6.23 is the result for frequency being steered to 13.56 MHz (from 32 MHz), including both cases of without and with the disciplining scheme.

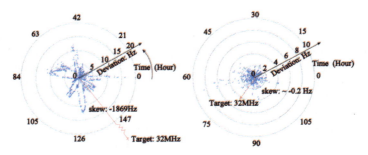

Fig. 6.21 Improvement on accuracy: without (left) and with (right) clock disciplining

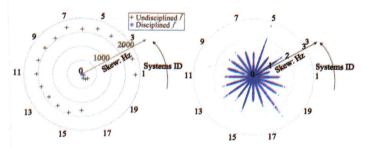

Fig. 6.22 Precision-related data obtained from a group of 20 systems, full scale (left) and detailed view (right)

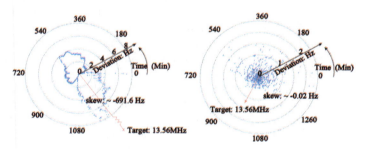

Fig. 6.23 Frequency steering to 13.56 MHz, without (left) and with (right) clock disciplining

For more details on TAF-DPS based clock disciplining scheme, please refer to Li et al. (2021).

Gestalt Switch:

The resolution of the two long-lasting problems (arbitrary frequency generation and instantaneous frequency switching) enables new treatment on old problems. In this section, the elucidation of this spirit is though the example of tunable frequency source, which is a widely used component in microelectronics.

6.3 On-Chip Syntonistor for Assisting Time Synchronization in Network

In architecture planning and engineering construction of telecom network, computer network or network of any other types of electronic devices, synchronization is an important issue (Lévesque and Tipper 2016). It is the bedrock of the information superhighway. It is the enabler for successful transfer of voice, data and video over a network. It can optimize the use of a given bandwidth, increasing throughput in the case of a fixed spectrum band. Synchronization is also recognized as the base for Cyber Physical System (CPS) (NIST 2016) and IoT framework (Al-Fuqaha et al. 2015). Further, synchronization is critical in industrial automation and data acquisition, to improve precision, productivity, and quality (Harris 2008; Mach et al. 2007; Glass and O'Donoghue 2008; Wang et al. 2011; Yan et al. 2013). Moreover, it plays key role in many other applications (Koskiahde et al. 2008; Estrela et al. 2014; Wu et al. 2009).

In the past several decades, network architecture has evolved from synchronous TDM style (time division multiplexing) (Donegan 2011, Latremouille et al. 2007) and bus-driven and fieldbus architectures (Sauter 2010) toward asynchronous packet-based style. The aim is for more efficient use of precious spectrum and network hardware (Subrahmanyan 2007 and Pinchas and Cohen 2007). In TDM based network, there is a physical link for reliable frequency transfer and thus the network can coordinate its activities using this common frequency. In packet-based network, however, the direct physical connection is nonexistence. Instead, both the data and timing-information have to be conveyed through cell packet. This presents a different and more difficult challenge for time synchronization.

In the development of old-time telecom-network (solely for voice communication) and the currently popular packet-network (mainly for data transfer), there are three concepts associated with timing signal transportation: frequency transfer, phase transfer and time transfer. They correspond to three types of systems: frequency-locked system, phase synchronized system and time synchronized system. In frequency-locked system, all the nodes' frequencies remain bounded all the times by a given value. In phase synchronized system, the distribution of reference timing signals ensures that their significant events occur at the same instant (within an accuracy qualification). Time synchronized system is similar to the phase synchronized one in that time synchronization simply consists of naming the significant instants (e.g., time-of-day information). In phase and time synchronization, a common notion of time needs to be established within a network so that activities occurred in the network can be coordinated. This task of establishing a common notion of time within a given network is also termed as clock synchronization.

Clock is a physical device that realizes a theoretical principle: the states of a system repeat at a constant rate. Time generated from a clock is defined by counting the states as they repeat themselves. The transition from connection-oriented network optimized for voice traffic (TDM) to connection-less network optimized for data (packet network) enables a suite of new TAACCS (Time-Aware Applications, Computers, and Communication Systems) applications that calls for innovation on the design of clock device (Weiss et al. 2015). From the aspect of timing signal, the optimal use of data in computing and in networking unfortunately is in the opposite direction. In one hand, computer hardware, software and networking are all operating in isolated timing processes (e.g., the layered approach in OSI (Open Systems Interconnection) model), allowing data to be processed with maximum efficiency due in part to asynchrony. On the other hand, coordination of processes, time stamping of events, latency measurement and optimal use of precious spectrum are all enabled by timing which needs to be accessible by all the involved parties. Thus, the fundamental

6.3 On-Chip Syntonistor for Assisting Time Synchronization in Network

breakthrough enabling future TAACCS will most likely come from an interwork of timing and data signals that break through the existing barriers. One of the solutions that can positively impact the task of synchronization is to directly control clock hardware's frequency, not just the value of time registers as currently done in practice. This is a work that requires effort of expertise from multidiscipline: computer architecture, network architecture, clock and oscillator circuit design, VLSI system design, frequency and timing control, algorithm development, software engineering and etc. (Weiss et al. 2015).

As discussed extensively in previous texts, TAF-DPS is an emerging frequency synthesis technology. It enables a new type of clock device that bears the features of arbitrary frequency generation and instantaneous frequency switching, achieved simultaneously for a given design. Its output frequency is precisely predictable with mathematical rigor and is controlled through a digital word. This novel clock device provides us with a means to attack the time synchronization problem from a new angle. In this section, TAF-DPS circuit is illustrated as a tool to improve network time synchronization accuracy and precision. It answers to the challenges raised in TAACCS (Weiss et al. 2015).

6.3.1 Notion of Time

In the field of time synchronization, there are two key concepts regarding time: real time and clock time. Real time refers to the Newtonian time frame that is not directly observable. Clock time is the reading value from certain man-made time-keeping devices, such as the wrist watch, wall clock, computer clock and etc. Following Lamport and Melliar-smith's convention (Lamport and Melliar-Smith 1985), we use lowercase t to represent real time and uppercase T for clock time. Further, we use upper case C to define a clock that maps from real time to clock time $C(t) = T$ and lower-case c for mapping from clock time to real time $c(T) = t$. The two mapping functions $C(t) = T$ and $c(T) = t$ have fundamentally different implications. $C(t) = T$ means that at real time t the clock's reading value is T. This function is used to describe the scenario where clocks are used to measure the real time when an event occurred in real world, for example, to record the moment-of-time when a rocket is launched from the launching pad. If t is the real time at which the rocket is launched, then $|C_1(t) - C_2(t)|$ represents the difference in the recording values of two different clocks C_1 and C_2 (For any two physically separated clocks, there always exists a nonzero $|C_1(t) - C_2(t)|$ for an event happening at a particular physical location. This statement is underpinned by Einstein's theory of relativity).

On the other hand, $c(T) = t$ is used for the scenario of describing an event happened inside a network. Suppose an event E occurs at time T (clock time) inside a network, which is made of a plurality of nodes. From each node's perspective, this very event happens at its own local time t, which could be different from that of other nodes due to the variation caused by physical configuration and electrical characteristic of the network. The mapping $c(T) = t$ serves to reflect the nature of this phenomenon.

For instance, if a command of all-power-down is issued at clock time T for all the nodes in the network, $c_q(T)$ and $c_p(T)$ will be the real times for this command being issued at node q and p, respectively, where q and p are two nodes in the network. $|c_q(T) - c_p(T)|$ will be the difference in real time when the two nodes are actually shut down. A network is said to be synchronized to a precision of ζ when $|c_q(T) - c_p(T)| \le \zeta$ is true, where q and p are any two different nodes in this network. Real time t is unobservable while clock time T is realized through an oscillator (or a clock circuit) and thus is observable.

In the implementation of clock time T, there are two more concepts of physical clock and logical (or virtual) clock. Logical clock is solely intended for the purpose of clock synchronization. Those concepts of real time and clock time, physical clock and logical clock are illustrated in Fig. 6.24 (exaggerated for illustration purposes). In this section, physical clock refers to the hardware oscillator (optionally plus a clock circuit) with a running frequency f. Its reading value is generated from a counter driven by the oscillator. Logical clock is also driven by the oscillator. Its value however can be arbitrarily adjusted by some operations, usually controlled by an algorithm. Logical clock is used to synchronize all the nodes' time registers in a numerical fashion. This notation of physical and logical clocks is in compliance with Lamport and Melliar-Smith (1985).

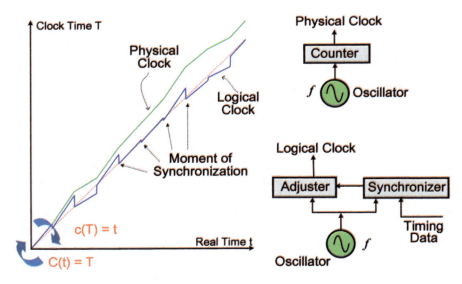

Fig. 6.24 Real time and clock time, physical clock and logical clock

6.3.2 Frequency, Phase and Time Synchronization

As mentioned previously, synchronization can be deployed in network at three levels: frequency, phase and time. Among them, frequency synchronization is the most straightforward one since an accurate frequency can exist stand-alone without the concern of time. Figure 6.25 illustrates two major architectures for all the nodes in a network to share a common frequency (i.e., the action of synchronizing in frequency, or syntonization). On the left is the plesiochronous distribution hierarchy (PDH) where multiple PRCs (primary reference clock) are distributed independently among the nodes (Ferrant et al. 2013). The PRCs are frequency sources of high frequency stability, typically at $\sim 10^{-11}$. On the right is the master–slave method where a high-quality PRC functions as the source and is distributed to slave nodes through electrical channels (i.e., the action of passing an electrical signal, which is a train of electrical pulses). This task is usually accomplished by using a special kind of frequency lock circuit (e.g., phase locked loop, or PLL). This approach is used in network of Synchronous Digital Hierarchy (SDH) (Ferrant et al. 2013).

In contrast to the direct frequency connection, phase and time synchronizations however require a method of transfer. As network infrastructure gradually migrates from TDM-based to packet-based architectures, synchronization becomes increasingly difficult. While TDM networks have timing transfer capability inherently built into them, packet networks (e.g., Ethernet, IP, MPLS) are naturally asynchronous and are not designed with timing transfer in mind. Using TDM-like timing techniques, Synchronous Ethernet (Sync-E) has emerged as a powerful and simple technology for accurate timing transfer over Ethernet networks. In packet network, there are two ways to transport timing information. In Fig. 6.26, a dedicated digital bit stream of CBR (constant bit rate) is used to link the PHYs in transmitter and receiver. A CDR (clock data recovery) circuit is then used in the receiver to extract the timing information (i.e., frequency) embedded in the data stream (Aweya 2014). As another method, in Fig. 6.27, a high-level feedback control loop (similar to PLL in principle) is used to extract the frequency from the incoming packet containing timing (Mills 1991). After the frequency link is established over the packet network, phase and time are derived from it using protocols and algorithms.

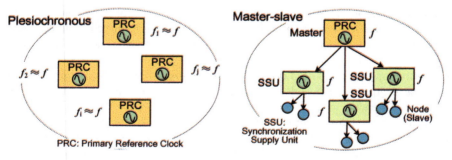

Fig. 6.25 Frequency synchronization: plesiochronous (left) and master–slave (right)

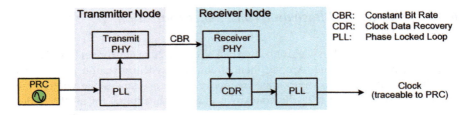

Fig. 6.26 Frequency transport in packet network through CDR

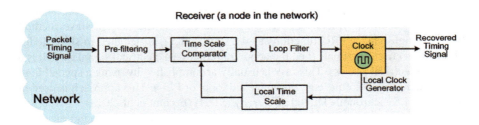

Fig. 6.27 Frequency transport in packet network through timing cell packet

Besides the aforementioned approaches, a special type of method for delivering frequency, phase and time to nodes in a network is through Global Navigation Satellite System (GNSS, GPS in North America). It can provide synchronization in high accuracy. However, it is expensive and requires line-of-sight of satellites. Moreover, the low signal strength makes it vulnerable to unintentional and intentional interferences.

6.3.3 Establishing a Common Notion of Time

Although time is a common notion for scheduling activities in all kinds of organizations, an explicit notion of time is not an absolute necessity for operation. For example, a notion of time is not necessarily a must-have for an isolated system. The clock pulse train from an oscillator is good enough to drive a digital system since the circuit's output is deterministic as long as its internal states evolve step by step. The sequential order inherently built inside the clock pulses is sufficient for supporting operation, no need of notion of time. It is also worth mentioning that, in this case, the quality of the pulses (i.e., jitter and wander) does not affect the function of event-sequencing as long as the circuit operates safely (Xiu 2012a). For the case of centralized systems, there is no need for synchronized time either since there is no time ambiguity. As in a single-processor computer with OS, a process can get the time by simply issuing a system call to the kernel. When another process tries to get time, it will get either an equal or a higher time value. There is a clear ordering of

6.3 On-Chip Syntonistor for Assisting Time Synchronization in Network

events and the times at which these events occur (Lamport 1978). Therefore, it is not always necessary for a clock to keep synchronized to a reference.

In distributed systems, however, the story is different. There is no global clock or common memory. Each processor has its own internal clock and its own notion of time. These clocks can easily drift seconds per day due to frequency instability, thus accumulating significant error over time. Worst yet, because of the fact that different clocks drift at different rates, these clocks most likely will not remain synchronized although they might be at start. This poses problems to applications that depend on a synchronized notion of time. For most applications and algorithms that run in a distributed system, we need to feel time in one or more of the following senses.

1. The relative ordering of events that happened on different machines in the network.
2. The time interval between two events that happened on different machines in the network.
3. The time of the day at which an event happened on a specific machine in the network.

Those three issues lead to the need of seeking a common notion of time for distributed system. For applications in which a notion of time can be substituted by causality-based relationship, clocks are relatively synchronized to each other because the requirement is only to provide an ordering of events, and not the exact real-world time at which each event occurred (issue 1 and 2 listed above). For such kind of clocks that provide only relative synchrony, phase synchronization is sufficient. For clocks that must not only be synchronized with each other but also have to adhere to a reference time (issue 3 in the list), time synchronization must be performed to time-stamp each significant instant using an accurate real time standard, such as the universal coordinated time (UTC). One strategy for keeping clocks synchronized is to give each node a GPS receiver and all nodes then use time signal sent by satellites. With this scheme, there are of course concerns of reliability and cost. An alternative approach is software based: to use protocol and then to design synchronization algorithm for seeking a common notion of time. The underlying foundation for such algorithms to work is the frequency and time transfer mechanisms discussed in last Sect. 6.3.2.

6.3.4 Applications of Packet-Switched Synchronization

Mobile Cellular Networks: Synchronization is important for synchronizing the base stations and monitoring the network performance. Traditionally, synchronization in mobile cellular networks is supported by TDM technologies since TDM can provide sufficient timing performance. However, it has low efficiency in utilizing the bandwidth. It is envisioned that the traditional TDM synchronization architecture will be replaced by packed-based synchronization scheme due to the cost efficiency and the convergence of packet-based services provided by an all-IP backhaul network (Mock et al. 2000 and Mirabella et al. 2010).

Industrial Applications: Synchronization is also critical in industrial applications of automation and data acquisition, to improve precision, productivity, and quality. In the case of data acquisition of synchronized inputs from sensors, for instance, events measured in different areas of an object need to be precisely time-stamped so that the whole picture of correlated events can be reconstructed (Koskiahde et al. 2008). As another example, navy shipboards require precise time measurements in order to detect, locate, and identify moving targets (Glass and O'Donoghue 2008).

Smart Grid: Smart grid is an enhancement of the legacy power grid. It consists of certain level of intelligence. One of its important features is the integration of two-way-flow of energy and information between the premises of generation and customer. It supports emerging features such as distributed renewable energy resources and demand response. It improves grid's efficiency and reliability. Timing for the emerging smart grids is a critical requirement (Wang et al. 2011; Yan et al. 2013). For example, one significant modification to the grid with a stringent synchronization requirement is the wide area monitoring system, supporting the use of synchrophasor measurements of time-stamped harmonic power components for real time grid control. With the arrival of new applications such as demand response, distributed energy resources (e.g., wind and solar) and advanced metering infrastructure, new communications and synchronization requirements will emerge. Synchronizing data in a fully integrated manner is expected to become a challenging requirement in this domain, as a variety of wired and wireless communications technologies will be used in those applications.

Other Applications: Many other types of applications exist or are emerging that require tight synchronization of devices (Harris 2008; Wu et al. 2009), such as those IoT applications that need highly accurate time-based devices to be deployed in large-scale. For instance, synchronized sensors are important in law enforcement, military and crisis applications (Koskiahde et al. 2008). Accurate time is a key element in such applications since the processing of collected information would be significantly impaired if timestamps are not accurate. Financial industry, where PTP is currently used, is another example that time synchronization is critical since inaccurate timing can lead to disastrous outcome (Estrela et al. 2014).

6.3.5 Clock Synchronization in Network Standards

Clock synchronization can be achieved either by hardware or by software, depending on required precision (from millisecond to nanosecond) and geographic spread of the distributed system (from a few meters to thousands of kilometers). Moreover, cost is also an important factor. In actual implementation, the task of clock synchronization has to be carried out according to protocols defined in industrial standards. The most popular standards are Network Time Protocol (NTP), Precision Time Protocol (PTP, IEEE 1588) and Sync-E (Synchronous Ethernet).

NTP is one of the oldest protocols that is mostly known due to the fact that it is used in the Internet (Mills 1991; Ullmann and Vögeler 2009; Neagoe et al. 2006). It is a highly robust protocol, especially for distributed unreliable networks. It is widely deployed throughout all the branches of the Internet. It is generally regarded as the state of the art in time synchronization protocols. It can reach synchronization accuracy to the order of a few milliseconds in public Internet and to sub-millisecond level in local area networks. NTP is cost-effective since it does not require any specific hardware. It is also scalable due to its hierarchical structure. Compared to the solution of using a GNSS or atomic clock at every node, it is very cost effective. Its primary drawbacks are the extra communication overhead associated with the request-response messages and the relatively low level of accuracy.

PTP is a master–slave protocol for delivery of highly accurate time over local area networks (Ferrant et al. 2013; IEC Technical Committee 65 and IEEE Standards Association (IEEE-SA) 2008; Eidson 2006). It enables precise synchronization of clocks over heterogeneous systems with accuracy in the order of microsecond to submicrosecond range. The supported protocols are UDP, DeviceNet, ControlNet, and PROFINET. It targets at large scale distributed applications requiring microsecond synchronization, such as smart grid, industrial automation and telecom systems. Unlike NTP which does not require any special device between slave and master nodes, PTP requires dedicated hardware at intermediate nodes for the introduction of transparent and peer-to-peer clocks. It is worth to mention that PTP potentially provides a syntonization method to correct slave clock's rate.

Sync-E is based on and compatible to its TDM predecessor SONET/SDH. They share the common architecture of synchronization, clock specifications, and network elements (i.e., PRC and SSU). Synchronization networks can be built as mix of Sync-E and SDH/SONET elements. Sync-E and SDH/SONET share Synchronization Status Message (SSM) principle and it allows interworking (Aweya 2014).

6.3.6 Key Components in Clock Synchronization Algorithms

In centralized systems, mutual exclusion and inter-task communication problems are generally solved using methods of semaphores and monitors. For distributed systems connected through a network, the situation is much complex. They in general bear the following characteristics: multiple concurrent computation threads, interconnections for inter-thread communication and globally shared state that all individual computers cooperatively maintain. Clock synchronization provides an internal consistency in the flow-of-time for all the involved nodes. This can help ease the implementation of the aforementioned issues. During the past decades, much research has been conducted in developing algorithms for seeking a common notion of time in fault-tolerant distributed systems. They can be software, hardware or hybrid solutions.

From performance point of view, clock synchronization algorithms can be classified as deterministic, probabilistic and statistical. Deterministic algorithms assume an upper bound on transmission delays and guarantee a maximum difference between any two simultaneous clock readings. Probabilistic algorithms guarantee a maximum deviation between synchronized clocks. At any time, a clock knows if it is synchronized or not with the other, but there is a non-zero probability that a clock will become out of synchronization when things go wrong. Statistical algorithms assume that the expectation and standard deviation of the delay distributions are known. Clocks do not know how far apart they are from each other. But a statistical argument is made that, at any time, any two clocks are within some constant maximum deviation with a certain probability.

Software algorithms use standard communication networks and exchange synchronization messages to get the clocks synchronized. They are frequently used in situations where a loose synchronization in the range of some milliseconds is acceptable (such as the Internet). Pure hardware clock synchronization achieves tight synchronization through the use of special synchronization hardware at each node and the use of a separated network solely for clock signals (such as Sync-E). Due to the cost and size of the additional hardware, it is in most cases only affordable when the system spread is within a limited range. Hybrid solutions are based on software algorithms with moderate hardware support. They can achieve reasonably tight synchronization and are still cost effective in comparison to pure hardware approaches. The supporting hardware usually carries out the tasks of maintaining the local clock, applying the required corrections and providing some facilities to ease the exchange of synchronization-related messages (such as in the case of PTP).

In the development of clock synchronization algorithm, following factors must be taken into consideration.

- Network configuration: broadcast or point-to-point or bus, fully-connected or partially-connected.
- Internal or external: internal synchronization only considers the clock states of the nodes in the network while external further requires the synchronization to an external time standard (e.g., UTC). The criterion for internal synchronization is precision while accuracy is additionally considered for external.
- Clock model: clock synchronization algorithm executes on local processor and takes the clock readings of local and remote clocks as inputs. The computed correction is then applied to correct the local clock time. Most nodes are equipped with a clock of oscillator-plus-counter. Synchronization on all those local hardware clocks (i.e., the physical clocks) is not possible since directly tuning the oscillator is difficult if not impossible. Thus, the concept of logical clock is introduced, whose value is determined by adding an adjustment term to the physical clock (please refer to the illustration on the right side of Fig. 6.24). The adjustment term can be either a discrete value changed at each re-synchronization, or a linear function of time. The discrete adjustment may cause a logical clock to instantaneously leap forward or be set back, and then continue to run at the speed of its underlying hardware clock. Such behaviors of negative time and abrupt time-jump cannot be tolerated by most applications. Therefore, a linear function of time for clock adjustment is often mandatory.

6.3 On-Chip Syntonistor for Assisting Time Synchronization in Network

- Clock rate variation: this is caused by oscillator's frequency instability. In real world, frequency drift is one of the key reasons for network's gradually-out-of-sync behavior.
- Message exchange method: this defines the way that nodes exchange clocking-related messages. The first type is asymmetric approach (master–slave structure) where one dedicated node is designed as master and provides the time to the other nodes designed as slaves. The second one is symmetric approach where all the nodes participate in an active manner and execute the whole clock synchronization algorithm. Another one is hierarchical approach where the synchronization is spread at different levels and within each level there may again be categorized either as asymmetric or symmetric. Asymmetric scheme is a low-cost solution in terms of message traffic but it bears the risk of single point failure. Symmetric method is more robust but having a higher traffic cost.
- Network uncertainty: it refers to the fact that, in packet-switched network, the time that it takes for a network to deliver synchronization message is nondeterministic.
- Fault type: this describes the faults that a synchronization algorithm must deal with. There are clock-related faults (clock byzantine failure and clock timing failure), processor-related faults (processor crash failure and processor performance failure) and link-related fault (link omission failure and link performance failure).

In operation, clock synchronization algorithm needs to perform three key tasks: synchronization event detection, remote clock estimation and clock correction. Due to oscillators' frequency instability, clocks must be re-synchronized periodically to guarantee precision and accuracy. Clock synchronization algorithms are usually round-based, each round being devoted to the re-synchronization of all the clocks. Thus, the function for resynchronization-event-detection must be activated periodically. When a new round initializes, every node needs to get some knowledge of the values of remote clocks. Due to indeterministic communication delays and clock drifts, only estimates can be made. It is essential that these estimations closely resemble the remote clock values since those clock readings will form the input for directing the subsequent clock correction. The action of clock correction can be implemented in two different fashions: clock state correction where only clock's current state is modified and clock rate correction in which the clock rate can be adjusted as well.

6.3.7 Motivation for On-Chip Syntonistor: Eliminate Frequency Offset and Drift from a Hardware Perspective

From the intensive review on the field of time synchronization in previous texts, it can be understood that one of the key reasons for network's tendency of gradually-out-of-sync is the mismatch among the frequencies of the oscillators of the nodes.

It is therefore almost mandatory to control the frequency variation occurred inside the nodes if high quality synchronization is demanded. Oscillator's frequency can be influenced by deterministic factors (such as temperature, aging, pressure, shock) and stochastic factors (white noise, flicker noise, random noise and etc.), as discussed in Sect. 6.2. Frequency instability is often categorized as short-term noise (jitter, higher than 10 Hz) and long-term noise (wander, below 10 Hz) (Poore 2001; Tektronix 2017; Sullivan et al. 1990). Short-term noise impairs the operation of electrical circuit. Its impact on network clock synchronization (which operates at a higher level than individual circuit) however can be ignored since short-term noise is already taken care of by circuits and it cannot reach system level. Frequency wander (a long-term effect) is usually caused by systematic reasons, such as temperature difference, manufacture imperfection and component aging. Practically speaking, compared to the frequency variations caused by temporary environmental factors, frequency wander is the leading cause of network's trend of gradually-out-of-sync (Gaderer et al. 2011; Mahmood et al. 2017).

Frequency wander is typically caused by systematic reasons, such as manufacture nonideality and component aging. Atomic oscillator (Rubidium and Cesium) has the highest frequency stability of 10^{-11} to 10^{-12} ($\tau = 1$ s). OCXO (oven-controlled crystal oscillator) is next in line at $\sim 10^{-10}$ to 10^{-11}. TCXO (temperature-controlled crystal oscillator) is in the range of $\sim 10^{-8}$ to 10^{-9} (Lombradi 2002 and Vig 2000). These high-quality oscillators are bulky and expense and usually only used as PRC at high stratum level. For most nodes in a network, regular XO (crystal oscillator) is used. This type of oscillator's stability is in the range of $\sim 10^{-5}$ to 10^{-6}. It also has significant aging rate in the range of tens of ppm per year (Vig 2000).

Low grade frequency sources, such as XO, are widely used in most of today's networks as the base of local clock. When XO of such low stability is used in communicating nodes as time origin, some methods of compensation is certainly desired. Before proposing a solution, it is necessary to dig deeper into the issue of time error. Time error is literately defined as $X(t) = T(t) - T_{ref}(t)$ in ITU-T Recommendation G.810 (https://www.itu.int/rec/T-REC-G.810-199608-I/en), where $T(t)$ is the time obtained from the clock-under-study and $T_{ref}(t)$ is that of a reference. In practice, depending on the cost versus performance trade-off of the network, reference time can come from GPS, WWVB radio station operated by NIST, RDS from FM radio, RT-Link of AM radio or even the 60 Hz of AC power line. For NTP users, the stratum 0 level grand master clock comes from "The U.S. Naval Observatory Alternate Master Clock at Schriever AFB, Colorado". In principle, $T_{ref}(t)$ can also come from a node in the network that is selected as the reference for other nodes.

$$X(t) = x_0 + (y_0 - y_{0,ref})t + \frac{(D - D_{ref})}{2}t^2 + R(t) \qquad (6.3.1)$$

The time error $X(t)$ can be expressed using a digital value when the times $T(t)$ and $T_{ref}(t)$ are materialized through time counters and read out from clock registers. Or, it can be represented as measured phase difference of the two clocks when the clocks are embodied as electrical signals. The latter expression is often used by clock circuit designers while the other one is usually adopted by network system professionals when developing their algorithms. Time error can be deconstructed into constant error (DC component) and dynamic error (AC component). In general, time error $X(t)$ can be described by using Eq. (6.3.1), where x_0 is constant time error, y_0 and $y_{0,ref}$ are frequency offsets (e.g., manufacture error) of the two respective clocks, D, D_{ref} are frequency drifts (e.g., component aging) of the clocks and $R(t)$ represents random variations (e.g., caused by temperature change).

In most of current practices, it is assumed that the two clocks are locked in phase and, therefore, the terms associated with frequency offset y and drift D can be eliminated. For remaining terms, x_0 corresponds to the constant time error (or time offset) and $R(t)$ is the dynamic time error that is time dependent. In ITU-T G.8273.2/Y.1368.2 (https://www.itu.int/rec/T-REC-G.8273.2-202010-I/en), the term cTE is created to formally represent Constant Time Error. It is virtually a measure of accuracy since it reveals the average magnitude of the time error. The term dTE is defined to express Dynamic Time Error. It more or less represents stability since it shows the dynamic change of the time error. The terms cTE and dTE are only meaningful when frequency offset y and drift D are not presented in the time error analysis. This is a reasonable expectation in the environment of connected TDM networks where frequency is transferred in the form of electrical pulse train through the physical layer (e.g., SyncE). For packet-based network (e.g., wireless sensor networks), this is however hardly the case since the frequency of reference clock is not conveyed in any physical way. Frequency offset y and drift D are such slow fluctuation affecting the phase precision of clock signal. When frequency wander is nonignorable in Eq. (6.3.1), the magnitude of time error can grow unbounded.

As mentioned, PTP provides a method of syntonization to correct slave clock's rate. With this method, in principle, the frequency offset y and drift D on local clocks of participating nodes can be compensated, relatively to the reference node. If properly done, the two terms associated with y and D in Eq. (6.3.1) can be gradually diminished, eventually eliminated. In current practice, this method is however implemented on logical clock (i.e., time register), not on actual clock circuitry. In other words, it is a software-based solution. Therefore, it is not a true syntonization since physical clock is not altered (the clock circuit is not touched). Its performance is thus limited. To advance the art, it is desirable to create a hardware based syntonization method. For this reason, it is desirable to adopt a clock circuit having following features.

- fine frequency granularity and fast frequency switching
- frequency-tuning being done in a highly quantifiable fashion (so that high-level processing unit can change node's frequency precisely in real time)
- frequency adjustment being digitally performed and in a fashion that is transparent to algorithm (no extra computation-load and message-traffic burdening the main processor)
- can be on-chip integrated into nodes with low cost.

Our motivation is to incorporate such a clock circuit into the clock synchronization operation. By providing this capability of directly adjusting the physical clock, it is believed that some difficult problems that cannot be efficiently dealt with previously at algorithm level, such as the rate adjustment and the linear time function, can be solved from a new perspective. This can eventually lead to a higher precision and accuracy in time synchronization. This type of clock circuit for performing the synchronization on clock frequency is termed syntonistor. In short, we need an on-chip syntonistor to be incorporated into the time synchronization process. The word syntonistor is defined as a clock circuit whose frequency is tunable and its purpose is for assisting the synchronization of frequency (not time) among nodes in a network.

6.3.8 Suitability of Frequency Sources as On-Chip Syntonistor

From the discussion in previous sections, it can be appreciated that high quality network time synchronization requires frequency-tuning capability that must be incorporated in all the communication nodes. In Sect. 6.2.1, a variety of frequency-tunable sources have been reviewed. Based on that study, Table 6.1 compares the aforementioned sources for their suitability as on-chip syntonization tool for network time synchronization, or their aptness as on-chip syntonistor. From this comparison, it is believed that TAF-DPS DCXO is a good candidate as on-chip syntonistor for this task (Xiu 2019b; Xiu and Wei 2020).

6.3.9 TAF-DPS as On-Chip Syntonistor for Assisting Time Synchronization

$$f_s = \frac{C}{F} f_o \qquad (6.3.2)$$

$$dF = -\frac{df_s}{f_s} F_c = -x \cdot F_c, \quad F_{new} = F_c + dF \qquad (6.3.3)$$

On the left of Fig. 6.28, the circuit diagram of Flying-Adder frequency synthesis is depicted. It is an implementation of TAF-DPS. A plurality of K phase-evenly-spaced signals at frequency f_Δ is fed into the circuit. Δ is a base time unit, it is the time span between any two adjacent signals of said plurality of K signals. Its output frequency f_s is controlled by the frequency control word F, which is a rational number having both integer and fractional parts. On the right-hand side of Fig. 6.28, the scheme of integrating it on main chip (e.g., a SoC implemented in ASIC style) is illustrated. For high-performance applications, the circuit shall be designed and layouted using transistor-level custom design method. For low-cost applications, the implementation can be done using digital design approach. In both cases, the size of this circuit is

6.3 On-Chip Syntonistor for Assisting Time Synchronization in Network

Table 6.1 Comparison of sources for suitability in assisting clock synchronization

Source	Frequency tunability method	Switching speed	Circuit cost (Power and size)	Programmability (Interface with software)
XO	Mechanical trim	N/A	N/A	None
VCXO	Tuned by voltage, frequency granularity is limited by voltage resolution, sub-ppm	~Hundreds of clock cycles	High cost, high power usage, hard to be fully integrated on-chip	Difficult to program, inconvenient for being used in algorithm since its response time is long and unquantifiable
DCXO	Tuned through digital word, granularity is limited by varactor, sub-ppm	~Hundreds of clock cycles	Large size, medium power usage	Easy to program, but not ideal for being used in algorithm since its response time is long and unquantifiable
MEMS	Tuned by digital word, granularity is limited by the PLL circuit, sub-ppm	~Hundreds of clock cycles	Medium cost, and power, hard to be fully integrated	Easy to program, unsuitable for being used in algorithm since its response time is long and unquantifiable
DSP Synthesizer	Tuned through voltage, frequency granularity is limited, sub-ppm	~Hundreds of clock cycles	Large size, high power usage, hard to be fully integrated	Difficult to program, unfriendly to algorithm since the hardware response time is long
TAF-DPS DCXO	tuned through digital word, Frequency granularity sub-ppb	Two clock cycles	Small size, low power usage, can be easily integrated	Easy to program, friendly to algorithm since the hardware responds very fast and quantifiable

equivalent to several hundred NAND2 gates. The actual size is largely influenced by the number of bits used in the fractional part (the size of the accumulator).

As an emerging frequency synthesis technique, TAF-DPS is a circuit level enabler for system level innovation. Network clock synchronization is one of the areas that it can play innovative role and incite some novel solutions. Figure 6.29 illustrates the idea of using TAF-DPS as oscillator's companion circuit for assisting the task of clock synchronization. The plurality of phase-evenly-spaced signals needed by the TAF-DPS comes from the oscillator, such as a XO type oscillator. The TAF-DPS's

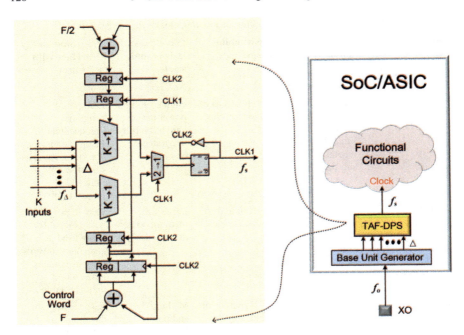

Fig. 6.28 Flying-Adder circuit for realizing TAF-DPS (left) and the scheme of integrating it on-chip (right)

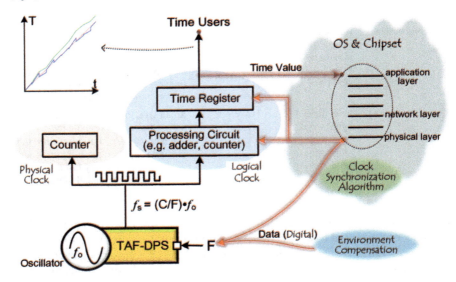

Fig. 6.29 TAF-DPS as a companion circuit for assisting oscillator in clock synchronization

6.3 On-Chip Syntonistor for Assisting Time Synchronization in Network

output is used to drive both the physical and logical clocks. Physical clock is simply a counter driven by this clock while logical clock could have some more sophisticated operation whose input comes from the data processed by algorithms executed in higher level. This companion circuit's output frequency can be expressed in (6.3.2), where f_s is the final output frequency and f_o is the oscillator's native frequency. C is a constant, a user-controlled design parameter. F is the frequency control word. Equation (6.3.2) is in the same form as (6.2.1) since they both originate from (6.1.2). It is the same TAF-DPS being used in different applications.

By changing the current value F_c, frequency f_s can be adjusted. If f_x is the desired new frequency: $df_s = f_s - f_x$ and $df_s/f_s = (f_s - f_x)/f_s = x$ where x represents a small change, then (6.3.3) can be used to calculate the new value F_{new}. Most likely, slightly adjusting the fraction r can produce the desired result. Experimental data is available for the proof that frequency granularity (resolution) can reach sub-ppb range (Xiu and Chen 2017; Xiu and Wei 2019). Further, frequency switching can be accomplished very quickly (two clock cycles after the command is received). Those two features of arbitrary frequency generation and instantaneous frequency switching are extremely valuable for the task of clock synchronization. They provide great flexibility in algorithm design. Those features make it possible that high-level communication layers can access this frequency adjustment capability when necessary. Moreover, data intended for environment compensation (e.g., temperature) can also be inputted to TAF-DPS for corresponding frequency adjustment. For all those reasons, this TAF-DPS DCXO is suggested as a competent on-chip integrated syntonistor for periodically fine-tuning clock frequency in real time, to assist the task of time synchronization.

6.3.10 *A Method of Using TAF-DPS Syntonistor for Tuning Physical Clock*

A. *Definition*

t	Real time, Newtonian time, unobservable
k	A natural number, its increase represents the forward movement of real time t
t[k]	The value of t at k, unobservable
R[k]	A golden time reference, observable (e.g., the UTC)
Δt	$= t[k+1] - t[k]$ or $= R[k+1] - R[k]$
$T_c^l[k]$	A common notion of time established within a network, logical clock, calculated by algorithm
$f_i[k]$	ith node's hardware clock frequency at k
$T_i^p[k]$	ith node's physical clock at step k, influenced only by $f_i[k]$
$T_i^l[k]$	ith node's logical clock at step k, can be adjusted by algorithm
$m_i[k]$	Rate of $T_i^l[k]$ increment at step k, $T_i^l[k+1] = T_i^l[k] + m_i[k]$

N	The number of ticks, in unit of k, between two consecutive synchronizations, also called one interval, or one round
M	The number of ticks, in unit of k, between two consecutive syntonizations
Accuracy	A parameter quantifying a relationship between two objects of a target and an object-under-study: the distance between the target and the object-under-study, measured using certain unit. Accuracy is the degree of conformity of a measured or calculated value to its definition
Precision	A parameter quantifying a relationship among multiple objects-under-study: the standard deviation of the distances of all the objects to the mean value
Stability	A parameter quantifying the degree of variation over a given time interval of a quantifiable parameter, such as accuracy or precision. Stability doesn't indicate whether the parameter, such as the accuracy, is "good" or "bad" but only whether it stays the same

This group of definitions establishes the base for the following discussion. k is a natural number which is independent of time. Its monotonic increase corresponds to the arrow-of-time: the forward movement of real time t. In hardware, the value of k can represent the number of oscillations occurred in clock circuit since the start. The movement of k is also called tick or step. $f_i[k]$ is an attribute of physical clock $T_i^p[k]$ (e.g., the value of f_s in Fig. 6.29 at step k). $T_i^l[k]$ is logical clock, the reading from time register at step k. The movement in its value is controlled by $m_i[k]$ (i.e., the virtual clock frequency derived from $T_i^l[k]$). For clock-state algorithms, only the value of $T_i^l[k]$ can be adjusted. For clock-rate algorithms, the value of $m_i[k]$ can be adjusted as well. When implementing rate adjustment on $m_i[k]$, there are two possibilities. The first one is to design algorithm for advancing the value of logical clock $T_i^l[k]$ (Loy 1996; Horauer 2004), but the physical clock's rate $f_i[k]$ is not altered. The second one is to adjust the value of $f_i[k]$. This second approach requires support from clock hardware (Buevich et al. 2013). Using the TAF-DPS syntonistor introduced in Sect. 6.3.9, we will develop a more powerful and universal tool for manipulating $f_i[k]$.

Naturally, for a particular node i, the value of $m_i[k]$ should be set as $m_i[k] \equiv 1/f_i[k]$ so that physical and logical clocks move coherently. In real case, however, the value of $f_i[k]$ can only be estimated, based on the number specified in specification. This is due to the constraint that we cannot continuously measure the value of $f_i[k]$ in real time. Hence, its true value cannot be determined with high confidence due to the frequency error introduced in manufacturing and, equally likely, the frequency variation caused by environment. On another note, in the environment of a connected network where multiple nodes try to establish a common notion of time, the value of $m_i[k]$ needs to be constantly adjusted by algorithm so that $T_i^l[k]$ can achieve better consistency with $T_c^l[k]$ or $R[k]$, a value that is shared by all nodes. In other words, the value of $m_i[k]$ is affected by the states of other nodes. Therefore, $m_i[k]$ is typically not equal to $1/f_i[k]$ and physical clock $T_i^p[k]$ and logical clock $T_i^l[k]$ usually move in different paces, as illustrated in Fig. 6.24.

6.3 On-Chip Syntonistor for Assisting Time Synchronization in Network

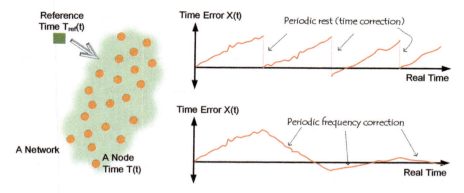

Fig. 6.30 The methods of phase reset (top) and rate adjustment (bottom) to correct time error

To judge the quality of time-keeping activity of a particular node, a criterion of $|T_i^l[k] - R[k]| \leq \zeta$ can be used, where ζ is a bound of accuracy. To achieve this, the most direct way is phase-reset, namely, letting $T_i^l[k_j \cdot N] = R[k_j \cdot N]$ at each synchronization moment (when $k = j \cdot N$). For the very next tick this relationship however is generally not maintained (i.e., $T_i^l[k_j \cdot N+1] \neq R[k_j \cdot N+1]$) due to the variation of $f_i[k]$ (i.e., the source of the problem, the clock rate, has not been touched). This approach of using phase reset to keep time error from growing out of bound is illustrated in the top of Fig. 6.30. Although simple, phase-reset can cause problems of negative-time and abrupt-time-jump as described in Lamport and Melliar-Smith (1985). This can result in severe system level issues since clock's monotonicity and continuousness are violated.

Another approach is rate adjustment. Instead of simply resetting $T_i^l[k_j \cdot N]$, the compensation for the difference of $|T_i^l[k_j \cdot N] - R[k_j \cdot N]|$ is amortized (i.e. spreading out) over a time interval by adjusting the $m_i[k]$ of that interval (i.e., increasing or decreasing the speed of logical clock) (Schmuck and Cristian 1990). The aim is to implement a linear time function and avoid the problems of negative-time and time-jump. This idea is shown in the bottom of Fig. 6.30. In current practice, this adjustment on $m_i[k]$ is carried out on logical clock. It is accomplished by a simple digital circuit (e.g., adder plus counter, driven by $f_i[k]$), as demonstrated in Loy (1996), Horauer (2004). It however has a serious drawback. Although all driven by $f_i[k]$, the logical and physical clocks' trajectories can be completely out-of-sync, as illustrated in Fig. 6.24, when the value of $m_i[k]$ is forced to accommodate the golden reference $R[k]$ (or $T_c^l[k]$) while $f_i[k]$ is left to free run. This fact impairs the performance of synchronization. A more effective way is to directly modify the value of $f_i[k]$, through the interaction between $R[k]$ and $T_i^p[k]$. In other words, the current values of $R[k]$ and $T_i^p[k]$ are used to direct the adjustment on future value for $f_i[k]$. The implementation of this idea will involve clock hardware and is made feasible by the TAF-DPS syntonistor, as shown in Eqs. (6.3.2) and (6.3.3) and Fig. 6.29. It is worth to note that this plan is only possible with TAF-DPS circuit since there is no other clock circuit, as elucidated through Table 6.1, bearing the features

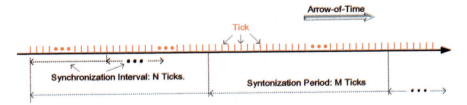

Fig. 6.31 Tick, synchronization interval N and syntonization period M

of arbitrary frequency generation and instantaneous frequency switching, achieved simultaneously on a given design.

B. *TAF-DPS Syntonistor*

With a golden reference R[k], the advance of real time t is expressed as $\Delta t = R[k+1] - R[k]$. Ideally, for the ith node, $f_i[k]$ is supposed to materialize this Δt. However, as said, frequency error can impair the integrity of $f_i[k]$, leading to the deviation of its value from specification. In this study, a first order approximation that lumps together all the error-inducing factors into one parameter is adopted. A bias x is introduced for representing this lumped error $f_i[k_q \cdot M] = f_0 \cdot (1 - x)$ where f_0 is the frequency value specified in product specification and $f_i[k_q \cdot M]$ represents the real frequency value at each syntonization moment (i.e., when $k = q \cdot M$). The relationships of tick, N and M are illustrated in Fig. 6.31. The goal of the proposed syntonization method is to counteract this error x through adjusting control word F and make Eq. (6.3.4) true.

$$\Delta t \equiv 1/f_0 = (1 - x)/f_i[k_{q \cdot M}] = 1/m_i[k_{q \cdot M}] \qquad (6.3.4)$$

The value of x is estimated from the values of R[k] and $T_i^P[k]$. For convenience, we assume that x is a constant in each syntonization period and its estimation is performed once in every syntonization period. The assumptions are summarized as follow.

1. The task of time synchronization is split into a series of intervals. Each interval consists of N ticks.
2. The task of frequency syntonization is split into a series of periods. Each period consists of multiple intervals, total M ticks.
3. From the position of a particular node, the values of R[k], $T_i^P[k]$ and $T_i^1[k]$ of other nodes can only be observed at the moments of synchronization (i.e., when $k = j \cdot N$, j is an integer).
4. All the factors contributing to clock hardware's frequency error is lumped into a parameter x.
5. The method used for estimating x is the syntonization recommended in PTP standard (IEEE, 802.1AS-2011 2011).

According to IEEE 802.1AS-2011, frequency ratio of master and slave clocks can be estimated as the ratio of elapsed times between two consecutively received

6.3 On-Chip Syntonistor for Assisting Time Synchronization in Network

timestamps of the two clocks (IEEE, 802.1AS-2011 2011). The value of x can then be calculated from (6.3.5) and (6.3.6) at the beginning of every syntonization period. Practically speaking, x ≪ 1 for real world applications. By using (6.3.3), this value of x will be applied to the equation and the next control word F of TAF-DPS circuit can be derived.

$$f_0 = \frac{R[k_{q \cdot M}] - R[k_{(q-1) \cdot M}]}{M \cdot \Delta t}, \quad f_i[k_{q \cdot M}] = \frac{T_i^P[k_{q \cdot M}] - T_i^P[k_{(q-1) \cdot M}]}{M \cdot \Delta t} \quad (6.3.5)$$

$$x = 1 - \frac{T_i^P[k_{q \cdot M}] - T_i^P[k_{(q-1) \cdot M}]}{R[k_{q \cdot M}] - R[k_{(q-1) \cdot M}]} \quad (6.3.6)$$

The action for compensating x is shown as the procedure in (6.3.7). At each of the following syntonization period, the value of x_{new} is recalculated using (6.3.5). It is expected that $|x_{new}| < |x|$ since, by using this method, frequency error is supposed to be gradually diminishing in each period. After a few steps, the clock circuit's output frequency shall approach f_0 quickly as shown in (6.3.7). This is a continuous procedure and eventually the frequency error will be practically zero. If this is not the case (i.e., $|x_{new}| \geq |x|$), it indicates that something has happened during the last period and the clock's frequency has been disturbed. Then, the procedure can be restarted from the beginning, using x_{new} as the new starting point.

$$f_i[f_{(q+1) \cdot M}] = f_i[k_{q \cdot M}] \cdot (1 + x) = f_0(1 - x^2)$$

$$f_i[f_{(q+2) \cdot M}] = f_i[k_{(q+1) \cdot M}] \cdot (1 + x^2) = f_0(1 - x^4)$$

$$\vdots \quad (6.3.7)$$

$$f_i[f_{(q+n) \cdot M}] = f_i[k_{(q+n-1) \cdot M}] \cdot (1 + x^n) = f_0\left(1 - x^{n^2}\right) \approx f_0$$

$$E \approx -\left(x + \frac{x^2}{1 - x^2}\right) \quad (6.3.8)$$

During this process, although frequency error is diminishing, the time error on physical clock $|T_i^P[k] - R[k]|/(M \cdot \Delta t)$ will still grow. However, it will be bounded since the clock hardware's frequency is being corrected continuously and the frequency error will eventually become zero. Therefore, the time error on physical clock will gradually reach a final-state value E, which is expressed in (6.3.8). We can then make an effort to improve time synchronization accuracy with the help from logical clock $T_i^l[k]$. At the beginning of each syntonization period when x becomes available, E is calculated using (6.3.8). During the next (M/N − 1) synchronization intervals, the error of E is compensated by amortization. A linear time function can be implemented to spread the error over those (M/N − 1) intervals.

C. *Simulation of TAF-DPS Syntonistor*

SimEvents® from MathWorks® is a tool for studying the effects of task timing and resource usage on the performance of distributed control systems, software and hardware architectures, and communication networks. It is used in this work for developing and verifying the proposed method. For this particular study, a network system made of ten nodes is constructed using this tool. A golden reference R[k] representing real time t is controlled by a monotonically increasing counter k. In simulation, R[k] is given an ideal rate of 1. For each of the ten nodes-under-study, its rate is assigned as $f_i[k] = 1 - x + r_e$, where x is the rate-error introduced in Sect. 6.3.10B and r_e is a number representing random frequency variation. Their values are constrained in the ranges of $x \in [-0.1, 0.1]$ and $r_e \in [-0.02, 0.02]$, respectively. The $\mp 10\%$ rate offset and $\mp 2\%$ random variation are not realistic. They are several orders larger than the values observed in real cases. We want to use those unrealistically large values to amplify the effects-under-study. In simulation, $\Delta t = 0.01$ is advanced for each increment of k (i.e., one step forward of k), or one tick. Synchronization interval is selected as N = 1000 ticks and syntonization period is $M = 10 \cdot N = 10{,}000$.

Figure 6.32 shows the behavior of one of the nodes when there is no synchronization algorithm activated (i.e., free-run scenario). The top graph is the trajectories of physical clock $T_i^p[k]$ and reference R[k]. The graph in the middle is the $|T_i^p[k] - R[k]|$, representing time error. As seen, the error grows uncontrolled since there is

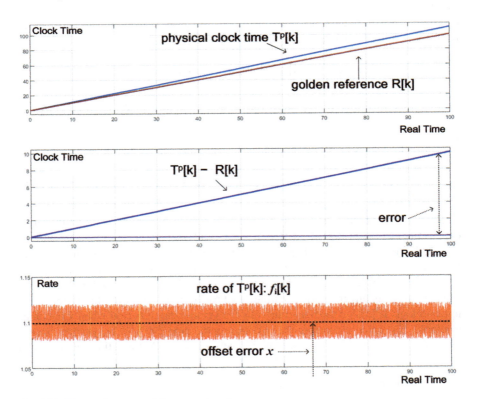

Fig. 6.32 Time trajectory and time error in a free-run case

6.3 On-Chip Syntonistor for Assisting Time Synchronization in Network

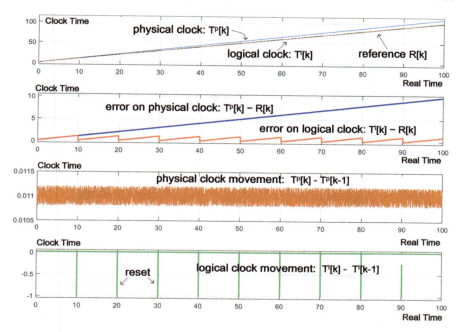

Fig. 6.33 Time trajectory, time error and clock movement in the case of using phase-reset

no action of synchronization. The bottom graph (red trace) shows the trend of $f_i[k]$ versus time. Both the offset error x and random variation r_e are displayed (ideal rate is 1, the rate of R[k]).

Figure 6.33 shows the case of phase-reset algorithm. In this figure, the trend of logical clock $T_i^l[k]$ is added. At the moment of each synchronization, logical clock's value is simply reset to the value of R[k]. Now, the synchronization error (the logical clock, the red trace) is under control, as shown in the second graph from top. However, this algorithm produces problems of negative-time and time-jump. The 4th graph of Fig. 6.33 (the green trace) is the clock movement of logical clock $T_i^l[k] - T_i^l[k-1]$. It becomes negative immediately after the moment of synchronization. This is due to the hard-reset on the clock state. In contrast, the movement of physical clock, the 3rd graph (the brown trace), is always positive since it represents the advance of real clock pulses. The effect of random variation, controlled by r_e, is also visible. Figure 6.34 compares the synchronization precisions between the free-run and phase-reset. The data is collected from all the ten nodes. The synchronization error grows unbounded in the case of free-run while it is controlled in the phase-reset case (the scale of the left graph is 10 times larger than that of right).

To eliminate the problems of negative-time and time-jump, the amortization method is used. Instead of a simple reset, the frequency error indicated by x is gradually compensated over a period of 10 intervals (one syntonization period). As a result, the negative-time issue is nonexistence as shown in the 4th graph of Fig. 6.35

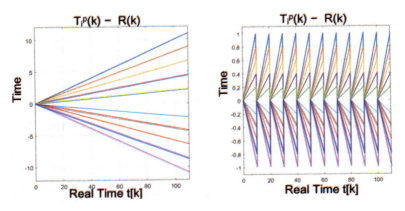

Fig. 6.34 Time errors: free-run (left) and phase-reset (right), simulation of 10 nodes

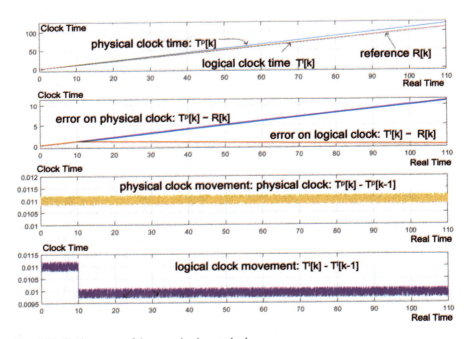

Fig. 6.35 Performance of the amortization method

(please compare it to the 4th graph of Fig. 6.33). And $|T_i^1[k] - R[k]|$ is reduced to zero at the end of the 10th intervals (the brown trace in the 2nd graph).

In previous cases of phase-reset and amortization, actions are only taken on logical clock. From the 2nd graphs of both Figs. 6.33 and 6.35, it is seen that the error on physical clock (the blue traces) grows unbounded. The reason is that the clock circuit hardware has not been touched (i.e., the $f_i[k]$ is not tuned). To improve synchronization precision, as argued, it is desirable to tune $f_i[k]$. This frequency tuning is

6.3 On-Chip Syntonistor for Assisting Time Synchronization in Network

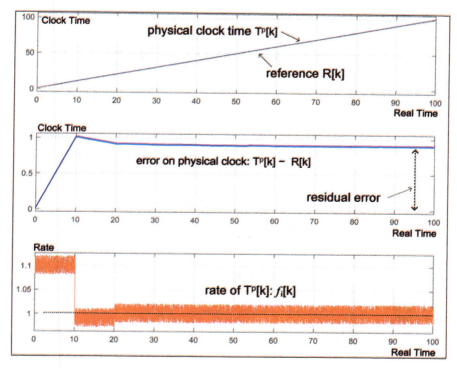

Fig. 6.36 The behavior of physical clock using TAF-DPS syntonistor for frequency correction

accomplished by the procedure introduced in Sect. 6.3.10B, using the TAF-DPS syntonistor. Figure 6.36 shows the behavior of physical clock by using TAF-DPS syntonistor. As seen from the 3rd graph (the red trace), the rate of physical clock has been corrected to the ideal rate of 1 (the mean value of $f_i[k]$ approaches f_0, the random variation is not corrected (please compare it to the 3rd graph of Fig. 6.32). However, as shown in the 2nd graph, $|T_i^p[k] - R[k]|$ bears a residual error. This is due to the reason that, initially, there is frequency error on clock. This error on time synchronization can be fixed by applying amortization to logical clock. After using it, $|T_i^1[k] - R[k]|$ is gradually reduced to zero as shown in Fig. 6.37, where the blue and red traces are the trends of $|T_i^p[k] - R[k]|$ and $|T_i^1[k] - R[k]|$, respectively (only one node is shown). As seen, the logical clock is able to approach the golden reference R[k] (amortization period is five intervals, simulation time is 500).

Figure 6.38 is more interesting. It shows the comparison between the amortization-only method of Fig. 6.33 and our approach of using TAF-DPS syntonistor. In the case of amortization-only, the time synchronization precision is up and down periodically. This is due to the fact that the physical clock is not corrected (i.e., the $f_i[k]$ is not tuned). On the contrast, our approach is free of this problem since the actual hardware clock frequency is tuned.

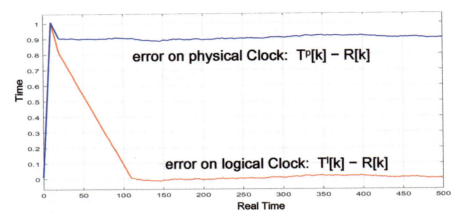

Fig. 6.37 Physical and logical clock errors using TAF-DPS syntonistor for frequency correction

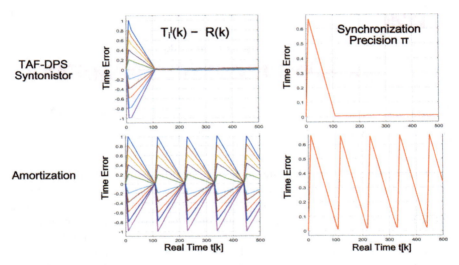

Fig. 6.38 Comparison of time errors on amortization-only (bottom) and TAF-DPS syntonistor (top), simulation of 10 nodes

The simulation works presented above are all for studying the characteristics of TAF-DPS syntonistor. The purpose is to visualize its effects with large unrealistic parameters. In next simulation, we present a case of more realistic: $f_0 = 100$ MHz (10 ns), $N = 10^8$ (synchronization interval is 1 s) and $M = 10^9$ (syntonization period is 10 s). Frequency error (manufacture imperfection) is $x = 10$ ppm and random frequency variation is $r_e = \mp 0.5$ ppm. Figure 6.39 is the simulation result. As seen, the time error $|T_i^1[k] - R[k]|$ grows to 10 us at the end of the 1st second. This is mainly due to the 10 ppm frequency offset. After that, the TAF-DPS syntonistor based synchronization process kicks in and the error starts to decrease. At the end

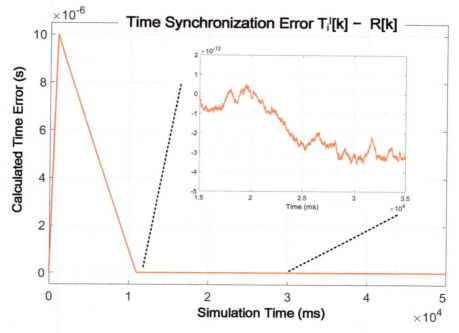

Fig. 6.39 Simulation of a realistic case: $f_o = 100$ MHz, frequency error 10 ppm, synchronization interval 1 s and syntonization period 10 s

of 1st syntonization period (after ten synchronization intervals, at 11 s), the error reaches sub-ns range.

Paradigm Shift:

In the emerging Time-Oriented paradigm, the classic problem of network time synchronization can be investigated from a new perspective. This opens the possibility for improving the time synchronization precision to next level.

6.4 Spread Spectrum Clock Generation: Always-On Boundary Spread SSCG

Spread spectrum clock generation (SSCG) is an important technique invented to mitigate EMI (Electromagnetic Interference) problems encountered by electronic systems. Being the pace-setter and time-keeper for electronic system, clock is the key for driving the innumerable logic gates in a VLSI system. For any given system, clock signal owns the highest frequency and sharpest transition edges. A clean clock signal (i.e., a signal of high purity in its frequency spectrum) is a highly effective source for emitting its electromagnetic energy, which could be harmful to electronic devices operating in its neighborhood. The technique of SSCG can lower the peak power of this energy by intentionally adding "controlled jitter" to the clock signal. This mechanism can spread the clock's energy from a concentrated line into a band of frequencies, consequently lower its peak power and weaken its capability of producing harmful effects.

Regulations have been created to ensure EMC (electromagnetic compliance) of various independently designed electronic devices and communication protocols (i.e., the state of coexistence and working normally) (ISCRI Part 1-1 2015a; ISCRI Part 2-1 2015b). This principle is realized by restricting the EMC ability of specific product from both the source view (limiting the EMI influences on environment) and the victim view (robustness under EMI disturbances). SSCG focuses on the first view of solving the problem from the source. In its principle, SSCG is implemented by constantly changing the instant frequency of the clock signal (an action of frequency modulation), thus scattering its energy in the frequency domain. Although the emitted energy remains unchanged, the power spectrum at working frequency (and its harmonics) is reshaped from a sharp peak to a lower platform. The modulation can be made by adding a time-varying command for frequency control, called modulation profile, to the clock generation circuit. In operation, the changing rate must be within the bandwidth of the clock generator, which is usually a PLL. The parameters defining a unique modulation profile include four attributes, namely, type (frequency deviation direction such as up spread, down spread, and central spread), depth (frequency deviation magnitude), rate (modulation frequency or command changing rate), and waveform (command curve shape). Since the first proposal of SSCG (Hardin et al. 1994; Lin and Chen 1994), extensive researches have been made on these issues, all focusing on the performance optimization of the EMI reduction.

In practice, a variety of modulation waveforms, both periodic and random, have been discussed and tested. The most widely studied periodic waveforms are sinusoidal (Lin and Chen 1994), triangular (Hardin et al. 1994) and sawtooth. They have been analyzed and optimized both intuitively and theoretically, leading to more complicated waveforms such as Hershey-Kiss (Hardin et al. 2003), optimal continuous (Matsumoto et al. 2005) and optimal discrete (De Caro 2013). For random-based modulations, several chaotic-map and pseudo-random sequence generators claim similar or better performance on EMI reduction (Pareschi et al. 2014a, b). Besides waveform, modulation rate is another major concern since the law of its influence on EMI reduction sometimes appears contradictory due to different measurement instruments and assessment methods in the history (Pareschi et al. 2015). State of the art results have leaded to convincing criteria by in-depth analyses and modeling on the working principle of EMI receivers standardized by the regulations. Recent tests and simulations from various works conclude to an agreed view that there exists an optimal value (when detected by peak/quasi-peak detectors) for the modulation rate, depending on the corresponding RBW (resolution bandwidth) of the EMI receiver required in different frequency measurement ranges (Bendicks et al. 2018). In addition, signal processing techniques used for modeling the EMI receivers have offered new means to construct the EMI measurement instrument itself, leading to

6.4 Spread Spectrum Clock Generation: Always-On Boundary Spread SSCG

more time-efficient EMI pre-compliance tests and TDEMI test systems (Karaca et al. 2015).

Despite the proven common belief that larger modulation depth gives better EMI reduction, applying this principle in real practice however is usually conservative. Limited by several implementation difficulties associated with conventional clock generators (especially closed-loop PLL-based), the desirable large dynamic frequency range could not be realized without negatively affecting the operation of the driven circuits (Tatsukawa 2008). For deviation type, down spread is generally preferred in most circumstances since it imposes no timing-constraint violation on existing design. For this reason, no much attention has been paid on the issue of deviation type.

From the experience of all those past works, the clock circuit for SSCG is preferred to bear following characteristics: (1) Direct: easy to handle, preferably linear frequency mapping; (2) Fast: quick response to tuning command; (3) Fine: small frequency granularity for accurate tracking; (4) Simple: small area, high power efficiency. TAF-DPS, as an emerging frequency synthesis technique, is based on the novel concept of Time-Average-Frequency (Xiu 2008c). Using this concept, a clock pulse train can be made of multiple types of pulses. In this regard, TAF clock naturally spreads its energy to several frequency points already (the so-called instant frequency). As a clock generator, TAF-DPS has two distinguished features of arbitrary frequency generation and instantaneous frequency switching. The output frequency of TAF-DPS is controlled via a digital control word. Further, the output's period is linearly proportional to the value of the control word (its output frequency is inversely linear). For this reason, TAF-DPS clock generator is ideally suitable for SSCG. TAF-DPS based SSCG holds the corresponding features of (1) Direct: simple control-word → frequency relation; (2) Fast: open-loop, few transition steps; (3) Fine: frequency resolution is only limited to the number of fractional bits used; (4) Simple: all digital, PVT irrelevant.

A novel modulation type called boundary spread, which is based on the TAF concept, is created from TAF-DPS (Ma et al. 2020). Its boundary of frequency deviation is naturally formed by fully utilizing the maximum potential modulation depth under the constraint that only two types of pulse-periods are used in the generated clock signal. This leads to the fact that, when TAF-DPS clock generator is used for SSCG, no extra jitter is introduced into the system even when large modulation depth is employed. Thus, the TAF-DPS SSCG can achieve large modulation depth but still offers high quality clock signal for system operation. As a result, any system incorporating this technique can work correctly with high confidence when SSCG clocking is turned on. This is in contrast to the traditional approaches where normal function is not guaranteed with high confidence when SSCG is on. Thus, the new technique is termed Always-on Boundary Spread SSCG, or AOBS.

6.4.1 The Architecture of Always-On Boundary Spread (AOBS)

(1) *SSCG Problem Definition*

$$f_s = f_s^r\left[1 + \frac{\delta}{2}\xi(t)\right], \quad \xi(t) \in [-1,1] \quad (6.4.1)$$

$$F(t) = F_r\left[1 + \frac{\delta}{2}\xi(t)\right], \quad \xi(t) \in [-1,1] \quad (6.4.2)$$

$$f_s(\xi) = \frac{1}{F_r\left(1 + \frac{\delta}{2}\xi(t)\right)\Delta} = \frac{f_s^r}{1 + \frac{\delta}{2}\xi(t)} \quad (6.4.3)$$

The normal expression for SSCG modulation, taking center spread for example, can be noted as (6.4.1), where $\xi(t)$ is the controllable modulation waveform varying within an interval $[-1, 1]$, δ is the non-negative magnitude of modulation depth. f_s^r represents the reference frequency without spread, the r in this work holds the same meaning for superscript and subscript. For TAF-DPS SSCG, $f_s^r = (F_r\Delta)^{-1}$ according to (6.1.2) where F_r is the frequency control word used for generating f_s^r. The relation between the control word F and the TAF-DPS synthesized frequency f_s is explicitly expressed as $F = (f_s\Delta)^{-1}$, and their derivative relation is deterministic and could be derived as (6.1.3). As mentioned in Sect. 6.1, the reciprocal relation can be treated as locally linear for simplicity with trivial approximating error. Thus, for TAF-DPS, one may tend to use a more straightforward way for SSCG as illustrated in (6.4.2), called direct control word modulation in this work. Consequently, the output frequency is changed from (6.4.1) to (6.4.3).

(2) *Architecture of Using TAF-DPS For SSCG*

Figure 6.40 depicts the plan of using TAF-DPS as the core for generating spread spectrum clock signal. As shown, a frequency modulation pattern is first created in the modulation block in digital fashion. Using Eq. (6.1.2), the pattern will be mapped to frequency domain through the TAF-DPS circuit. As a result, the output frequency from TAF-DPS will bear corresponding pattern. Consequently, the energy embedded in the clock signal is spread into a band of frequencies.

As discussed in Sect. 4.6, when a clock frequency (period T_{TAF}) is required to be synthesized, TAF-DPS circuit uses two types of pulses T_A and T_B to accomplish the goal. This mechanism is illustrated in Fig. 6.41. The locations of T_A and T_B are calculated and fixed once the value of desired T_{TAF} is determined. In operation, the virtual location of the T_{TAF} (the precise frequency) is set by the weighing factor r. When TAF clock is used to drive digital circuit, only T_A matters (please refer to Sect. 4.6). In other words, instead of using T_{TAF}, T_A should be used as the setup constraint. When TAF-DPS is used as an SSCG for spreading a clock signal's energy

6.4 Spread Spectrum Clock Generation: Always-On Boundary Spread SSCG

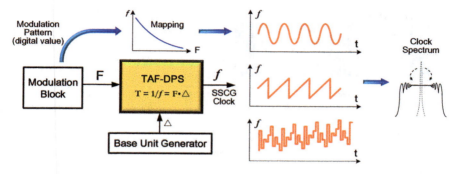

Fig. 6.40 Architecture of TAF-DPS SSCG

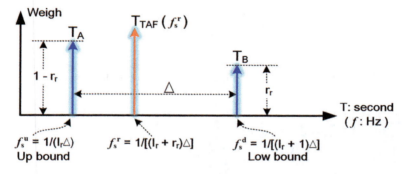

Fig. 6.41 The period (frequency) distribution of TAF clock signal

around frequency $f_s^r = (F_r \Delta)^{-1} = [(I_r + r_r)\Delta]^{-1}$, the control word $F = I + r$ can be varied using the profile defined in (6.4.2). By investigating Fig. 6.41, it is preferred that we keep the value of I fixed and only vary the r so that the setup constraint is not altered (more discussion will be given in later text). In other words, it is planned that there would be no need for changing I to get the desired large enough spread depth (i.e., $I = I_r$ is a pre-selected design parameter and r is the only actual tuning parameter during runtime). Under this guideline, we arrive at (6.4.4) which can be derived from (6.1.3).

$$\frac{df_s}{f_s^r} = -\frac{I_r + r_r}{(I+r)^2} dr \qquad (6.4.4)$$

Figures 6.42 and 6.43 are used to illustrate the difference of frequency-spread mechanisms between TAF and conventional frequency (CF). Figure 6.42 shows a case of doing spread spectrum using a sinusoid profile. In Fig. 6.42c, the period versus time trend is displayed when CF is the underlying frequency concept. The modulation range for period (frequency) is $[11\Delta, 12\Delta]$. As expected, the period varies gradually between the minimum and maximum. Figure 6.42d shows its period distribution. As

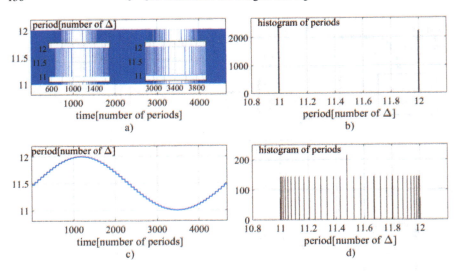

Fig. 6.42 Sinusoid SSCG implemented by TAF and CF: **a** period versus time trend of TAF, **b** period distribution of TAF, **c** period versus time trend of CF, **d** period distribution of CF

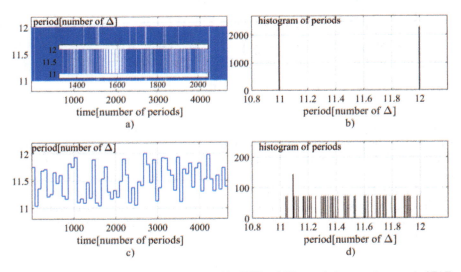

Fig. 6.43 Random fashion SSCG implemented by TAF and CF: **a** period versus time trend of TAF, **b** period distribution of TAF, **c** period versus time trend of CF, **d** period distribution of CF

seen, there are a plurality of periods (or frequencies) used in the spreading process. This is the characteristic of doing spread spectrum using CF. Figure 6.42a is the period versus time trend of doing the same sinusoid SSCG but using TAF concept. It is interesting to see that only two types of periods, 11Δ and 12Δ, are used. The sinusoid profile is accomplished through the varying local densities of the periods.

6.4 Spread Spectrum Clock Generation: Always-On Boundary Spread SSCG

Two zoom-in areas are also included to make this point clear. In Fig. 6.42b, the period distribution is displayed and, as expected, there are only two types of periods. Figure 6.43 is the case of random fashion spread profile. Using CF concept, the period is changed randomly as shown in Fig. 6.43c. Figure 6.43d shows that the periods used are indeed a series of random numbers siting within $[11\Delta, 12\Delta]$. A particular number has occurred twice. Figure 6.43a shows the scenario of using TAF for the same random fashion spread. Again, the local-period-densities are used to realize the intended profile. In Fig. 6.43b, the period distribution is displayed. As in the previous case, only two types of periods, 11Δ and 12Δ, are used. Figures 6.42 and 6.43 are helpful to the understanding of TAF SSCG mechanism.

(3) *Properties of TAF-DPS Based SSCG*

$$\frac{df_s}{f_s} = -\frac{dF}{F} = -\frac{F_r \frac{\delta}{2} d\xi(t)}{F_r[1 + \frac{\delta}{2}\xi(t)]} = -\frac{1}{\frac{2}{\delta} + \xi(t)} d\xi(t) \quad (6.4.5)$$

$$f_s^u = \left[F_r\left(1 - \frac{\delta}{2}\right)\Delta\right]^{-1}, \; f_s^d = \left[F_r\left(1 + \frac{\delta}{2}\right)\Delta\right]^{-1}, \; f_s^{u0} = f_s^r\left(1 + \frac{\delta}{2}\right), \; f_s^{d0} = f_s^r\left(1 - \frac{\delta}{2}\right) \quad (6.4.6)$$

Although range accuracy and tuning linearity are not quite emphasized for SSCG application in the past works, properties of direct control word modulation for TAF-DPS are analyzed and solutions are given here in case that strict limitations in future applications have to be met. Combining (6.4.2) with (6.1.3), when using TAF-DPS direct control word modulation, the frequency tuning gain is now dependent on current $\xi(t)$ value as expressed in (6.4.5). In (6.4.6), TAF-DPS modulation range is expressed using the boundary values (f_s^u, f_s^d). The values are compared to the ideal linear spread spectrum definition (f_s^{u0}, f_s^{d0}).

$$\frac{f_s^u - f_s^{u0}}{f_s^r} = \frac{\delta^2}{4 - 2\delta}, \; \frac{f_s^d - f_s^{d0}}{f_s^r} = \frac{\delta^2}{4 + 2\delta} \quad (6.4.7)$$

$$\Delta f_s = f_s^u - f_s^d = f_s^r \frac{\delta}{1 - \frac{\delta^2}{4}} > f_s^r \delta, \quad \varepsilon = \frac{|(f_s^u - f_s^r) - (f_s^r - f_s^d)|}{f_s^u - f_s^d} = \frac{\delta}{2} \quad (6.4.8)$$

$$\text{center spread}: \xi \in \left[\frac{-1}{1 + \frac{\delta}{2}}, \frac{1}{1 - \frac{\delta}{2}}\right], \quad \text{down spread}: \xi \in \left[0, \frac{2}{1 - \delta}\right] \quad (6.4.9)$$

Due to the reciprocal relationship, asymmetry exists for TAF-DPS modulation range as seen in (6.4.7). It can be seen that the frequency range is moved upward in both up and down boundaries, compared to the ideal linear modulation. In (6.4.8), the extents to which the range is enlarged (denoted by the absolute frequency range Δf_s) and distorted (denoted by the relative asymmetric error ε) are both proportionally correlated to the modulation depth δ. If one hopes to maintain the tuning range in accordance with the ideal linear SSCG definition, he or she can easily get the constrained tuning range for center spread and down spread using (6.4.9) by solving Eq. (6.4.6) for the desired boundary values. When the SSCG clock is used to drive a circuit, those formulas will be useful for ensuring no exceeding of working frequency out of the original design timing constraint.

$$\frac{df_s}{f_s^r} \Big/ \frac{\delta}{2} = -\frac{1}{\left[1 + \frac{\delta}{2}\xi(t)\right]^2} d\xi(t) \tag{6.4.10}$$

$$\exists E(\xi(t)), \quad \frac{dE(\xi)}{d\xi} = \left[1 + \frac{\delta}{2}E(\xi)\right]^2 \tag{6.4.11}$$

$$\frac{df_s}{f_s^r} \Big/ \frac{\delta}{2} = -d\xi \tag{6.4.12}$$

Nonlinearity of the frequency changing rate due to the TAF-DPS's reciprocal relationship of control word versus frequency can be shown as (6.4.10). As seen, the nonlinearity is quite trivial for small modulation depth. However, since the tuning relation of TAF-DPS is simple and explicit, it is expected that this nonlinearity can be compensated rigorously for general case. In other words, we hope to find a transformation E on the original modulation waveform ξ that satisfies (6.4.11). Applying this transformation to (6.4.10), a purely linear tuning relation can then be resulted in (6.4.12).

$$E(\xi) = E^* + \frac{1}{F(\xi)} \tag{6.4.13}$$

$$E(\xi) = -\frac{2}{\delta} + \frac{1}{-\frac{\delta^2}{4}\xi + C} \tag{6.4.14}$$

$$E(\xi) = \frac{\xi}{1 - \frac{\delta}{2}\xi} \tag{6.4.15}$$

$$\text{center}: E(\xi) \in \left[\frac{-1}{1+\frac{\delta}{2}}, \frac{1}{1-\frac{\delta}{2}}\right] \Rightarrow \xi \in [-1,1], \quad \text{down}: E(\xi) \in \left[0, \frac{2}{1-\delta}\right] \Rightarrow \xi \in [0,2] \tag{6.4.16}$$

Condition (6.4.11) is a polynomial differential equation in the Riccati form, whose analytical solution can be reached by following the procedure in (Zeidler and Hunt 2012). Firstly, it is obvious that the constant $E^* = -\frac{2}{\delta}$ is a particular solution to (6.4.11). Secondly, the substitution of (6.4.13) is performed to (6.4.11), resulting in (6.4.14). Finally, by assuming a constraint $E(0) = 0$, the integral constant C can be determined and we get (6.4.15) as the desired transformation function. This transformation compensates the reciprocal relation to a proportional one, $f_s = f_s^r\left(1 - \frac{\delta\xi}{2}\right)$. One could even go further by changing the sign to get a positive proportional modulation if so desired. By using the boundary values presented in (6.4.9), it can be

6.4 Spread Spectrum Clock Generation: Always-On Boundary Spread SSCG

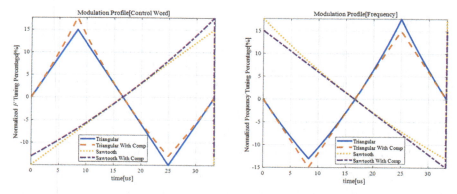

Fig. 6.44 Tuning range of control word modulation and the effect of compensation: changing pattern of normalized frequency control word (left), changing pattern of normalized output frequency (right)

easily proven in (6.4.16) that normal frequency tuning range is restored by this transformation for center spread and down spread.

The curves on the left of Fig. 6.44 depict the modulation profile and those on the right are the resulting output frequency trends. In this study, triangular and sawtooth waveforms are used in the modulation. And both the no-compensation and with-compensation cases are shown. To make the point clear, a large modulation depth of 30% is used. The values are normalized by the references to get the relative change percentages. From those curves, it can be seen that direct control word modulation causes the effects of distortion and range offset. And, they can be compensated by a simple transformation on the modulation profile.

(4) *The Issue of Clock Integrality*

The impact of spreading-clock on system operation is a serious problem that deserves supplementary attention. When a normal clock is spread, one of the most likely-to-occur problems that will degrade the system operation is the "added extra jitter" since the spreading-clock sometimes shortens the clock period. The shortened clock period, when certain limit is exceeded, can fail the system at its designated working frequency. This problem can sometimes be bypassed by lowering the operating frequency with the consequence of degraded system performance. However, in almost all the current SSCG techniques, the amount of "extra jitter" is usually unknown or uncontrollable. Thus, in many real products, the SSCG function is usually turned off in normal working condition. It is only turned on when doing EMI compliance test in lab.

TAF-DPS SSCG is a spread mechanism that is effective when dealing with the aforementioned problem. As seen from (6.4.4), we have plenty of options to choose the values of I and r for implementing the modulation profile, to achieve various spread results such as center spread, up spread and down spread. At the same time, we can also use the I and r in a controlled fashion to minimize the impact of spread on normal operation. By investigating Fig. 6.41, a natural solution is to use the region defined by T_A and T_B as the battleground for doing the frequency spread. In doing so, the setup constraint is not altered (i.e., still the T_A) so that system performance will not be degraded. Further, the frequency variation range is restricted to $[T_A, T_B]$. As a result, the clock edges only vary within a defined range. This idea leads to the essence of the proposed boundary spread method: using the region naturally created by TAF for frequency spread. This fact further leads to a positive consequence: the system's normal operation is not affected (please refer to the discussion around Fig. 6.16). In other words, AOBS SSCG can be always-on.

(5) *TAF-DPS Based Always-On Boundary Spread: The AOBS*

By investigating (6.4.2), the T_A and T_B automatically set the boundaries for the spread. If we make the clock frequency vary between T_A and T_B in certain way, the clock energy will be distributed into a frequency band set by those boundaries. At the same time, as stated earlier, its impact on system operation is none since T_A is not changed (i.e., the circuit speed constrained by setup constraint remains the same). From (6.4.2), the modulation can be restated as (6.4.17). This leads to $0 \leq r = r_r + (I_r + r_r)\frac{\delta \xi}{2} < 1$. The term $\frac{\delta \xi}{2}$ with the range $\delta > 0$, $\xi \in [-1,1]$ (center spread) reflects that the maximum possible modulation depth δ_{max} is determined by the pre-selected $F_r = I_r + r_r$. Using some simple algebra, the forms of frequency boundaries f_s^d, f_s^u are quite intuitive (corresponding to $I_r \leq F < I_r + 1$) as in (6.4.18).

$$F(t) = I + r = I_r + r_r + (I_r + r_r)\frac{\delta}{2}\xi \qquad (6.4.17)$$

$$f_s^d = f_s^r \frac{I_r + r_r}{I_r + 1} < f_s \leq f_s^r \frac{I_r + r_r}{I_r} = f_s^u \qquad (6.4.18)$$

$$f_s = f_s^c\left(1 + \frac{\delta_{max}^c}{2}\xi\right) \qquad (6.4.19)$$

$$f_s^c = \frac{f_s^d + f_s^u}{2} = f_s^r \frac{I_r + r_r}{2I_r}\frac{2I_r + 1}{I_r + 1} = f_s^r \frac{F_r}{F_r^c}, \quad \delta_{max}^c = \frac{f_s^u - f_s^d}{f_s^c} = \frac{1}{I_r + \frac{1}{2}} \qquad (6.4.20)$$

$$F(t) = F_r^c\left(1 + \frac{\delta_{max}^c}{2}\xi\right) \qquad (6.4.21)$$

$$F_{min} = I_r \leq F \leq I_r + 1 = F_{max}, \quad F_r^c = \frac{F_{max} + F_{min}}{2} = I_r + \frac{1}{2}, \quad \delta_{max}^c$$

$$= \frac{F_{max} - F_{min}}{F_r^c} = \frac{1}{I_r + \frac{1}{2}} \qquad (6.4.22)$$

$$f_s^c = \frac{F_r f_s^r}{F_r^c} = f_s^r \frac{I_r + r_r}{I_r + \frac{1}{2}} \neq f_s^r \frac{I_r + r_r}{2I_r}\frac{2I_r + 1}{I_r + 1} \qquad (6.4.23)$$

To express boundary spread in an equivalent normal center spread formation (6.4.19), the reference frequency should be adjusted in (6.4.20) to a shifted center

6.4 Spread Spectrum Clock Generation: Always-On Boundary Spread SSCG

value f_s^c, and the maximum modulation depth possible is extended to δ_{max}^c, corresponding control word can be further derived as in Table 6.3. To express boundary spread in an equivalent direct control word modulation formation (6.4.21), the shifted control word and modulation depth can be determined by (6.4.22). But the shifted center frequency here is different from the normal form (6.4.23), as well as the tuning range. This is the same problem which has been discussed in Sect. 6.4.1.(3), and could be compensated by the same transformation $E(\xi)$.

Referring back to Fig. 6.41, when normal center spread or down spread TAF-DPS SSCG is performed in the region defined by T_A and T_B, the maximum possible modulation depth is determined by the specific value of the reference frequency $f_s^r = [(I_r + r_r)\Delta]^{-1}$. Therefore, their values depend on both I_r and r_r. For boundary spread, the center is moved to a central place and consequently the full potential for modulation is utilized. In other words, boundary spread is a spread type fully deploying the nature of TAF-DPS to maximize the modulation depth (r_r does not play any role now). When original f_s^r is near center, it is equivalent to a center spread with a maximum modulation depth determined by I_r. When f_s^r is near T_A, it is actually down spread. It becomes up spread when f_s^r is near T_B. All with rigorously set boundaries.

Table 6.2 lists the maximum possible modulation depths for different spread types. Table 6.3 is the shifted central control word for AOBS. As seen, the modulation depth is automatically set at its maximum value δ_{max}^c since the TAF-DPS circuit now fully utilizes all the available TAF frequencies between the two boundary frequencies f_s^d and f_s^u in (6.4.18). It is interesting to point out that smaller I_r offers larger modulation depth but it also bears larger structure dissimilarity between T_A and T_B. Considering the fact that Δ is a design parameter that designer can control, selecting the proper I_r is a design tradeoff between modulation depth and TAF waveform irregularity.

Figure 6.45 shows a set of simulations of using TAF-DPS SSCG for down spread and boundary spread. The Δ is selected as 0.625 ns. The normal working frequency

Table 6.2 Maximum modulation depth possible for different spread types

Maximum modulation depth	With compensation	Without compensation
Center spread	$\begin{cases} \delta_{max} < 2\frac{1-r_r}{I_r+1} \\ \delta_{max} \leq 2\frac{r_r}{I_r} \end{cases}$	$\begin{cases} \delta_{max} < 2\frac{1-r_r}{I_r+r_r} \\ \delta_{max} \leq 2\frac{r_r}{I_r+r_r} \end{cases}$
Down spread	$\delta_{max} < \frac{1-r_r}{I_r+1}$	$\delta_{max} < \frac{1-r_r}{I_r+r_r}$
Boundary spread	$\delta_{max}^c = \frac{1}{I_r+\frac{1}{2}}$	$\delta_{max}^c = \frac{1}{I_r+\frac{1}{2}}$

Table 6.3 Shifted central control word for boundary spread

	With compensation	Without compensation
Central control word	$I_r^c = I_r, r_r^c = \frac{I_r}{2I_r+1}$	$I_r^c = I_r, r_r^c = \frac{1}{2}$

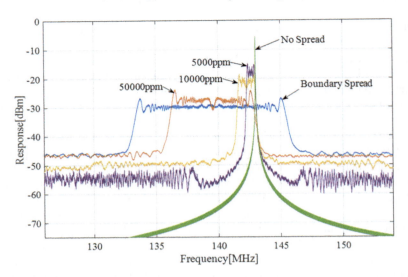

Fig. 6.45 Simulated spectrum for boundary spread and down spread

is $f_s^r = 143\,\text{MHz}$. This leads to $F_r = I_r + r_r = (f_s^r \Delta)^{-1} \approx 11.1888$. The modulation rate is chosen as 30 kHz. Using (6.4.18), the boundaries are calculated as $f_s^d = 133.333\,\text{MHz}$, $f_s^u = 145.455\,\text{MHz}$, which are agreed with the simulation result as shown. The modulation depth can be predicted by formula from Table 6.2, $\delta_{max}^c = 1/11.5 \approx 86957$ ppm. In Fig. 6.45, enveloped-FFT responses with no-spread and boundary spread using triangular profile are shown. Results from down spread using triangular profile with 5000, 10000 and 50000 ppm are also shown for comparison. Clearly, boundary spread achieves the largest distribution on energy (lowest peak power).

(6) *In Comparison to other SSCG techniques*

Table 6.4 makes the comparison between TAF-DPS based AOBS and other representative SSCG techniques.

6.4.2 Experimental Validation of AOBS SSCG

(1) *AOBS SSCG on FPGA*

The TAF-DPS based AOBS SSCG scheme has been implemented on a Kintex-7 FPGA system from Xilinx. The top-level view is depicted in Fig. 6.46. For the details of all the internal blocks, please refer to Figs. 6.28 and 6.40. The Johnson Counter is used to provide the plurality of K phase evenly spaced signals. The time span between any two adjacent such signals is the base time Δ. The output from the AOBS is fed into a Keysight N9000B signal analyzer for spectrum analysis. It is also

6.4 Spread Spectrum Clock Generation: Always-On Boundary Spread SSCG

Table 6.4 Comparison with other representative SSCG techniques

	PLL based	**Digital controlled delay line (DCDL)**	**TAF-DPS based AOBS**
Frequency concept	Conventional frequency	Conventional frequency	TAF
Frequency tuning method	Direct VCO modulation (Hsieh et al. 2008, Chang et al. 2003, Lee and Kim 2011) Divide ratio dithering (Ebuchi et al. 2009; Yang et al. 2009; Song et al. 2013)	Delay line adjustment (De Caro et al. 2015)	Adjusting weigh factor between T_A and T_B
Circuit complexity	Analog Complex Power hungry	Hand crafted digital Complex High power	Synthesizable digital Simple Low power
Max modulation depth	Less than $\mp 5\%$	Less than $\mp 5\%$	Much larger than $\mp 5\%$
Added jitter	Modulation-depth dependent	Modulation-depth dependent	Amount is fixed[*]
Modulation programmability	Low	High	High
Modulation accuracy	Low	Medium	High
Impact on normal operation	Very hard to control Unpredictable in field usage	Difficult to control Questionable in field usage	Easy to control Always-On

[*] This amount is not considered as jitter. It is the inherent $\triangle$ associated with TAF

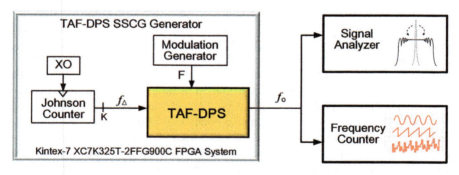

Fig. 6.46 The implementation of AOBS SSCG on an FPGA system

Fig. 6.47 The experimental platform for evaluating AOBS SSCG

fed into a Keysight 53230A frequency counter for observing its f versus time trend. A snapshot of the experimental platform is shown in Fig. 6.47.

(2) *EMI Pre-compliance Measurement Results*

To validate the effectiveness of AOBS, the circuit is implemented on the FPGA system and the outputted clock, which is spread, is studied using pre-compliance test. The results are compared with the observations reported in literature. In this study, the base time Δ is created from 16 channels of 40 MHz evenly spaced signals from a 8-stage Johnson Counter, resulting in $\Delta = 1.5625$ ns. The normal working frequency is set at $f_s^r = 120$ MHz. This leads to $F_r = I_r + r_r = \Delta^{-1}/120$ MHz $= 5.3333$. By the constraint of $F \in [5, 6]$, modulation depth is calculated as $\delta_{max}^c = 1/5.5 \approx 18\%$ per Table 6.2. From (6.4.18), the designed boundary values are $f_s^d = 106.7$ MHz and $f_s^u = 128.0$ MHz. Using the Keysight N9000B signal analyzer working in EMI receiver mode, the 120 MHz output from our TAF-DPS SSCG is studied for different RBW settings (Azpúrua et al. 2015; Matsumoto et al. 2006).

The plots on the left of Fig. 6.48 show the result. The modulation rate is 30 kHz. The modulation profile is triangular. Modulation method is boundary spread. Three tests with respective RBW settings of 9, 30 and 120 kHz are carried out. The envelop detector used is peak detector. As seen, the boundaries of measured EMI responses agree with the designed values. The AOBS is also used to compare the effectiveness of different modulation profiles. Those results are shown in on the right of Fig. 6.48. The outputs from triangular, sawtooth/ramp and random profiles are plotted in the same graph along with the no-spread result. As reported in De Caro (2013), Bendicks et al. (2018) and also observable in this figure, sawtooth profile produces the smoothest spread.

From the studies of De Caro (2013), Pareschi et al. (2014a, b), Bendicks et al. (2018), it is observed that, detected by peak/quasi-peak detector, there exists an optimal value for modulation rate when given certain RBW. This observation is also obtained by our AOBS as shown in Fig. 6.49. For periodic modulation profile,

6.4 Spread Spectrum Clock Generation: Always-On Boundary Spread SSCG

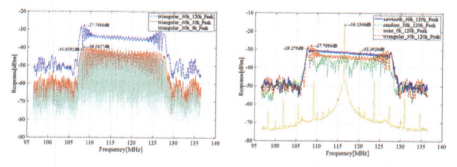

Fig. 6.48 EMI measurement responses under different RBW settings (left), comparison of EMI measurement responses under different modulation profiles (right)

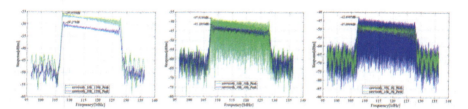

Fig. 6.49 Peak Detected EMI responses for modulation frequency 10 and 30 kHz, RBW = 120 kHz (left), 30 kHz (middle) and (9 kHz)

a modulation rate f_m that is closer to RBW can usually produce a stronger EMI reduction. In the left of Fig. 6.49, the RBW setting is 120 kHz and, clearly, f_m = 30 kHz gives better EMI reduction than f_m = 10 kHz. In the middle, the RBW setting is reduced to 30 kHz and the modulation rate of 30 kHz still produces better result. In the right-hand side of the figure, the RBW setting is further reduced to 9 kHz. In this case, the modulation rate of 10 kHz generates lower peak power than 30 kHz since f_m = 10 kHz is now closer to RBW of 9 kHz. It is worth to mention that the filter RBW is defined by −6 dB bandwidth in EMI regulation while it is usually specified at −3 dB in spectrum analyzer resolution bandwidth. Thus, results from De Caro (2013), Pareschi et al. (2014a, b) should be adjusted for comparison.

In summary, all the common laws relating the influence of RBW setting, modulation depth and frequency modulation rate on SSCG peak power reduction have been observed in AOBS. This validates the effectiveness of this technique. Furthermore, a quite large modulation depth of 18% is achieved using only two deterministic clock periods. The modulation depth is so large that descending slopes in the curves are observed. This slope is probably due to the frequency-dependent gain characteristic of the IO buffer of the FPGA. Usually, this kind of slope is hardly observable in small-range frequency tuning (i.e., 1, 2 and 5%).

6.4.3 Applying AOBS on Real System: Always-On

Figure 6.50 illustrates the various scenarios of studying the EMI related issues. The study can be categorized into two territories of aggressor domain and victim domain. On the aggressor side, the most straightforward method is to study the spectrum of the clock signal. Further, the impact of the SSCG clock signal on system operation has to be evaluated since SSCG techniques will add "extra jitter" into the clock signal. This is called aggressor function test. On the victim side, there are various standard test and evaluation methods. If a particular system is identified as victim, function test on this system also can be performed. For example, APD (Amplitude Probabilistic Detector) (Wiklundh 2006) is a method which becomes popular recently. It evaluates the BER (bit error rate) when a wireless digital communication system is operating under the influence of an EMI aggressor. In this section, two systems will be presented to verify our SSCG technique. It falls into the category of aggressor function test.

As discussed previously, in conventional SSCG techniques, spread clock can degrade system performance by demanding a lower working frequency than specified since the setup constraint is tightened due to the "added jitter". In other words, in many real commercial products, the system might fail to function normally when SSCG function is turned on, due to unforeseeable timing violation. For this reason, the common practice in industry currently is to turn off SSCG function most of the time. It is only turned on when the product is actively being tested for EMI influence in the lab. In AOBS, this is not the case since the setup constraint is not touched. Thanks to this fact, the feature of energy-spread can always be turned on when the system is in normal work state. In this section, we will demonstrate this "always-on" feature through one real examples.

A digital signal processing system is created as shown in Fig. 6.51. This system contains three RAMs for storing data and two logic blocks for performing logic and arithmetic operations. The data is generated from a PRBS (Pseudo-Random Binary Sequence) generator. The generated data passes two logic operation stages. Block #2 carries out the inverse operation performed inside block #1. Thus, after passing those

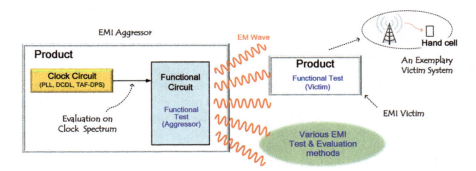

Fig. 6.50 Various scenarios of studying EMI related issues

6.4 Spread Spectrum Clock Generation: Always-On Boundary Spread SSCG

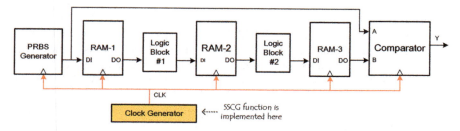

Fig. 6.51 A representative digital signal processing system

blocks, the data returns to its original state. The processed data is then compared with the original data (delayed by a certain number of clock cycles). When this circuit works correctly, the final output from the comparator shall be a logic "0". When this circuit fails, for any reason, the output is logic "1". From logic operation point of view, this circuit can represent all digital circuits regardless their functionalities. This representative system has been implemented on the Kintex-7 FPGA. When mapped to the configurable elements, the circuit is setup constrained at 260 MHz.

Figure 6.52 shows some of the captured waveforms when the system is operating and is driven by clock working in various modes. The top one is the clock waveform, which is the CLK in Fig. 6.51. The second one from the top is the comparator output, the error signal Y. The third and fourth ones are two flags: flag_speed and flag_sscg. When flag_speed is logic "1", the CLK signal's frequency is set at 300 MHz. It is at 222.22 MHz when flag_speed is logic "0". The flag_sscg is used for signaling the SSCG function. The SSCG function is turned on when its value is "1". The waveform

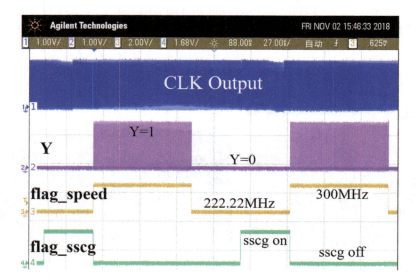

Fig. 6.52 Captured waveforms when driving clock in different modes

in this figure clearly shows that the system fails to operate correctly (as indicated by Y = "1") when the driving clock is 300 MHz. This is understandable since our circuit is set-constrained at 260 MHz. More interesting to us is the fact that, as confirmed by the waveforms (the purple and green ones), the system can function correctly when the SSCG function is turned on. This is due to the reason that AOBS does not add "extra jitter".

The test system in Fig. 6.51 is accompanied by a custom-made mini-BERT system and the BER test result is already graphically revealed in Fig. 6.52. To validate our AOBS further, the functional circuit and the TAF-DPS clock generator presented in Fig. 6.51 are brought into a Keysight BERT system M8041A. The photo in the left of Fig. 6.53 shows the test setup. The bit stream used in test is 31-bit PRBS with length of 2,147,483,647. In normal working condition (SSCG is off), the clock runs at 311.111 MHz. The PRBS stream (a portion of it is shown in the middle of Fig. 6.53) is fed into the circuit for 60 s (the stream will repeat itself when running to end) and the error rate is zero. When AOBS SSCG is turned on, the same stream has been fed into the circuit for 60 s and the resultant error rate is still zero, as shown in the left of Fig. 6.53.

Figure 6.54 shows the measurement result of the power reduction, or energy spread. Using the RIGOL RSA5065 spectrum analyzer with EMI filter bandwidth RBW set to 120 kHz (VBW set to 30 kHz) and detector set as positive peak, EMI power responses are measured for both no spread signal and boundary spread signal for comparison. The Δ is set as 1.00 ns, normal working frequency is $f_s^r = 222.22 \text{MHz}$, which leads to $F_r = I_r + r_r = \left(f_s^r \Delta\right)^{-1} = 4.5$. The modulation profile for boundary spread is a sawtooth waveform with 31.25 kHz modulation frequency. Using (6.4.18), the boundary frequencies can be predicted as $f_s^d = 200 \text{MHz}$, $f_s^u = 250 \text{MHz}$, which are validated by the result. The modulation depth for this test is $\delta_{max}^c = 1/4.5 \approx 22.2 \times 10^4 \text{ ppm} = 22.2\%$.

 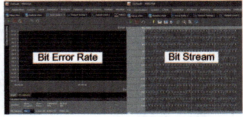

Fig. 6.53 BER test setup (left), snapshots of bit stream (middle) and error rate (left)

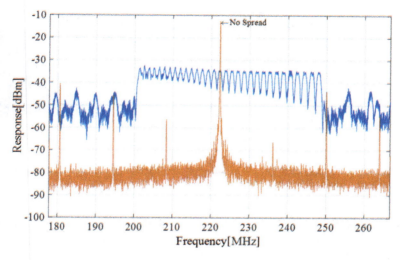

Fig. 6.54 Measured EMI peak power when AOBS is on

Paradigm Shift:

In the emerging Time-Oriented paradigm, the classic subject of EMC (electromagnetic compliance) can be investigated from a new perspective. This intricate problem becomes easier to be comprehend and dealt with.

6.5 On-Chip Chirp Signal Generator

A chirp signal is an electrical pulse train whose frequency increases or decreases with time. This type of signal is commonly used in sonar and radar. It is also found in other applications such as spread spectrum communication, resonant converters, electronic ballasts and etc. The two most common types of chirp signal are linear and exponential chirp signals. The frequency trend verse time for linear chirp signal follows $f(t) = c \cdot t + f_0$ where c is a constant (often called chirpyness) and f_0 is the starting frequency (at time t = 0). For exponential chirp signal, it can be expressed as $f(t) = f_0 \cdot k^t$ where k is the rate of exponential change in frequency. Figure 6.55 illustrates the characteristics of linear and exponential chirp signals by displaying their waveforms side by side.

Chirp signal can be generated with analog circuitry via a voltage-controlled oscillator (VCO) which has been briefly discussed in Sect. 6.2.1. If the VCO input is controlled by a linearly or exponentially ramping voltage, a linear or exponential chirp signal can be resulted. Chirp signal can also be generated digitally by using direct digital synthesizer (DDS) which consists of numerically controlled oscillator (NCO), digital to analog converter (DAC) and reconstruction lowpass filter. Moreover, it can be generated by using YIG (Yttrium Iron Garnet) oscillator. One of the

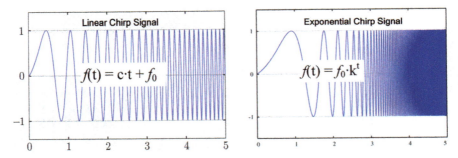

Fig. 6.55 The waveforms of linear and exponential chirp signals (By Georg-Johann, https://commons.wikimedia.org/w/index.php?curid=11330709)

major issues with all the aforementioned chirp signal generators is the implementation cost. Traditionally, the hardware cost associated with chirp signal generator is high. In many cases, discrete components have to be used to build the functional modules. The application of chirp signal is thus limited, especially so for on-chip operation.

Thanks to the linear relationship between control word and output period, TAF-DPS can be an effective tool for generating pulse-shaped chirp signal, as evident from Eqs. (6.1.2) and (6.1.3) and Figs. 6.3 and 6.4. Pulse-shaped chirp signal can be useful to many applications. If necessary, it can be converted into signal of sinusoid waveform by using a low pass filter. Figure 6.56 is the general architecture of TAF-DPS based on-chip chirp signal generator. A digital modulation block is responsible for generating the desired pattern in the form of a series of digital values. Those values are then fed into the TAF-DPS as the frequency control word F. Following the TAF-DPS working equations of (6.1.1–6.1.4), the output signal can bear a trend of *frequency versus time* defined by the modulation pattern. Thanks to the TAF-DPS mathematically traceable transfer function between the control word and the output frequency (almost linear in small region), many forms of chirp signal can be produced.

In Fig. 6.57, some measured real data from a chip are presented to show the capability of TAF-DPS on-chip chirp generator. On the left of the figure, sawtooth,

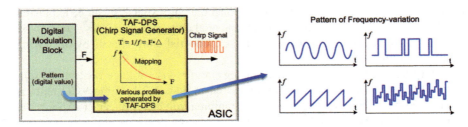

Fig. 6.56 TAF-DPS as on-chip Chirp Signal Generator

6.5 On-Chip Chirp Signal Generator

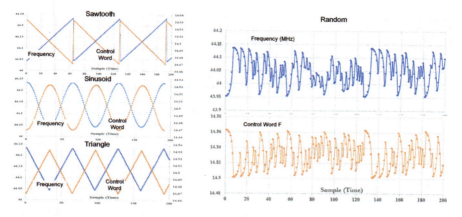

Fig. 6.57 Measured data from a TAF-DPS on-chip Chirp Signal Generator: sawtooth (top left), sinusoid (middle left), triangle (bottom left) and random (right)

sinusoid and triangle patterns are first generated by the digital modulation block, the resulting chirp signal is then frequency-measured by using a frequency counter. As seen, the trends of the output frequency follow the digital modulation patterns precisely. Those patterns of sawtooth, sinusoid and triangle have been generated in the control word F, the nominal value for F is 14.5. The range-of-adjustment is [14.46875, 14.53125], about ~0.4% of the nominal value. On the right-hand side of the Fig. 6.57, a stream of random numbers is created in the modulation block. A 7-bit PRBS generator is used to generate a stream of random number. In this case, F is varied in the range of [14.50048828125, 14.56201171875]. The corresponding frequency trend is then measured and displayed. The trends of frequency and control word clearly form a mirror-image pair which reflects the fact that TAF-DPS output frequency is inversely proportional to the magnitude of its control word. As a matter of fact, in all the plots of the four cases, the TAF-DPS output frequency responds to the control word change in an almost linear but mirrored fashion. Figure 6.58 shows

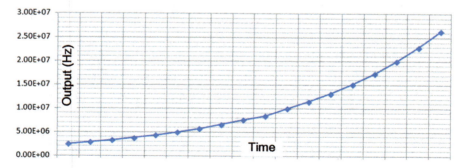

Fig. 6.58 Exponential chirp signal by TAF-DPS on-chip Chirp Signal Generator

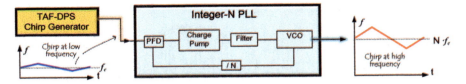

Fig. 6.59 Architecture of "TAF-DPS plus PLL" for boosting chirp signal's frequency

another case where an exponential chirp signal is produced from TAF-DPS on-chip chirp signal generator, also plotted using data from a real chip.

The simple structure of TAF-DPS circuit makes it friendly for being implemented as an IP for large ASIC chips. This on-chip TAF-DPS chirp signal generator opens the possibility for another interesting architecture. For modern ASIC-based designs, phase locked loop (PLL) is a standard IP that is available on almost all designs. Using those standard PLL, an architecture for boosting the chirp signal's frequency can be created as depicted in Fig. 6.59. A TAF-DPS chirp generator can be used to generate chirp signal at low frequency. Its output can then be fed into an integer-N PLL that will boost the input chirp signal by N times in frequency. In using this method, one issue needs to be watched closely. The chirpyness (rate of frequency change) shall be limited to a threshold that is within the PLL loop bandwidth so that the PLL can follow its input faithfully. Figure 6.60 presents a result measured from a real chip. As seen, the PLL truthfully boosts the original chirp signal (base frequency 5 MHz) to a higher frequency chirp signal (base frequency 60 MHz). During this process, the sinuous frequency trend is faithfully maintained and the frequency-spread ratio (range vs. central value) is also preserved.

This architecture of "TAF-DPS plus PLL" can be used in many places for real application. Figure 6.61 shows a case where a mini-LVDS display panel system is used to demonstrate this architecture for SSCG (spread spectrum clock generation). The system block diagram is depicted at the left of figure. The key functional blocks are DDR controller, mini-LVDS data formatter and serializer. They are driven by three clocks of 50, 81.25 and 650 MHz, respectively. The three clocks are generated from the block called clock generator. Our plan is to install the SSCG circuitry in this block and spread the clock energy associated with all those three clocks. The

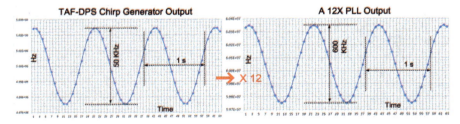

Fig. 6.60 Boosting chirp signal's frequency: measured output from TAF-DPS chirp generator (left), measured output after 12X PLL (right)

6.5 On-Chip Chirp Signal Generator

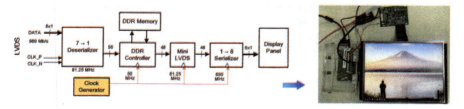

Fig. 6.61 The use of "TAF-DPS plus PLL" architecture in a mini-LVDS display panel system

display panel system is manufactured in Xilinx Spartan-6 FPGA system. Our circuit is implemented on this FPGA as well. In the right-hand side, a snapshot of the system is included.

In Fig. 6.62, the clocking plan is presented. As depicted on the left of that figure, the frequency modulation pattern can be initially created in a relatively low frequency region wherein the TAF-DPS circuit can operate comfortably. The resulting frequency modulation pattern is then fed into an integer-N PLL from its input port. In principle, the input frequency modulation pattern will be duplicated at the PLL output with its center frequency multiplied N times and its modulation depth preserved. This principle will hold as long as the input modulation rate falls within the bandwidth of the PLL.

In typical SSCG practice, the modulation rate is usually smaller than a few hundred kHz. The most popular choice is around 30 kHz. Due to the low operation frequency of the Xilinx Spartan-6 FPGA, it is difficult to directly implement the TAF-DPS circuit on the final working frequency range of 81.25 and 650 MHz. Therefore, the architecture of "TAF-DPS plus PLL" is utilized to boost the frequency in a cascaded fashion. As shown in Fig. 6.62, a triangular modulation profile of ~3.3% depth is initially created from TAF-DPS circuit, centered on ~6.6 MHz (this is a very comfortable range for the TAF-DPS being implemented in this low-cost Xilinx Spartan-6 FPGA). The first stage PLL boosts its center to ~50 MHz while preserving the modulation depth at 3.3%. The second stage PLL boosts the frequency to the final working range, centered on 81.25 and 650 MHz, respectively. During this chain of frequency boosting, the modulation depths at points A, B, C and D are all preserved at 3.3%. This scheme is implemented on this Spartan-6 system. The panel has been

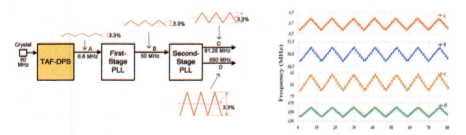

Fig. 6.62 Clock module (left) and measured frequency trends at various locations (right)

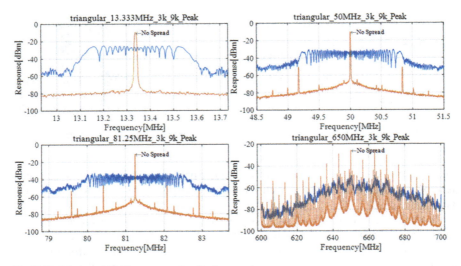

Fig. 6.63 Measured EMI responses for display panel system

putted on normal work state for hours when the TAF-DPS SSCG circuit is on. No visible video or image defect is observed.

On the right-hand side of Fig. 6.62, the measured frequency trends at various points are shown. As seen, the clock signal is frequency-boosted by the PLLs. More importantly, the modulation pattern and the modulation depth are well preserved among all the points in the chain. Figure 6.63 shows the peak power reduction measured using RIGOL RSA5065 spectrum analyzer with EMI filter bandwidth RBW set to 9 kHz (VBW set to 1 kHz) and detector set as positive peak, EMI power responses are measured for both no spread signal and boundary spread signal for comparison. It demonstrates the fact that clock signals' energies are spread as expected, peak power are reduced.

As another example, the "TAF-DPS plus PLL" architecture may be used in the FMCW (frequency modulated continuous wave) radar system. Linear chirp signal generator is a key component in FMCW radar for distance and altitude measurement, for vehicle collision warning systems and etc. In FMCW, a carrier frequency is varied linearly at a known rate over a predetermined range by using a chirp signal generator and a FM modulator. Instead of measuring the Time-of-Flight between transmission and reception directly, the change in carrier frequency is measured and divided by the rate of change. This gives the two-way travel time, which subsequently leads to the derivation of the distance/height. As shown in Fig. 6.64, the carrier frequency f_c is modulated in a triangular fashion with modulation rate f_m and maximum deviation Δf. At any moment the beat frequency, $f_b = f_t - f_r$ where f_t and f_r are the frequencies of transmitting and receiving waves respectively, is detected using a mixer. The distance/height H can be derived as $H = f_b \cdot C c/(4 \cdot \Delta f \cdot f_m)$ where C is the speed of light. In this system, one of the key components is the chirp signal generator (frequency modulator). The architecture presented in Fig. 6.59 provides a simple

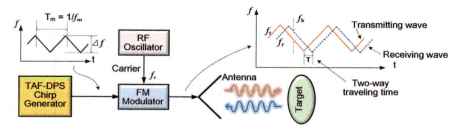

Fig. 6.64 TAF-DPS chirp generator used for FMCW radar system

and elegant solution for this application. This is especially beneficial to applications where low cost on-chip chirp generator IP block is preferred.

Gestalt Switch:

The resolution of the two long-lasting problems (arbitrary frequency generation and instantaneous frequency switching) enables new treatment on old problems. In this section, the problem of chirp signal generation is simplified by using a newer and more powerful tool.

6.6 Frequency Lock on Time-Average-Frequency: All Digital TAF-FLL

A frequency-locked loop (FLL) is an electronic control system that generates a signal whose frequency is locked to the frequency of a "reference" signal. This circuit compares the frequency of a controlled oscillator to the reference, automatically raising or lowering the frequency of the oscillator until its frequency is matched to that of the reference (or multiple of the reference). It is an example of control system using negative feedback. FLLs are used in radio, telecommunications, computers and etc., to generate stable frequencies, or to recover a signal from a noisy communication channel. It is also used in many other electronic applications such as frequency measurement, synchronization to power grid, oscillator frequency stabilization, FM spectroscopy, transceiver RF synchronization and etc. A phase-lock, or phase-locked loop (PLL), is one step further than frequency-lock. PLL is a control system that generates an output signal, in addition to frequency match, whose phase is also related to the phase of the input signal.

Both PLL and FLL are well studied circuit architectures. Traditionally, they are built mostly from analog components. The most popular monolithic PLL architecture in IC design is the third-order type II charge pump PLL. Nowadays, with the advance of digital process, digital processing of digitized analog signals becomes a popular approach of designing such frequency control loops. This new technique of processing digitized analog signal results in all-digital PLL (ADPLL) wherein the analog signal processing in loop design is replaced by digital signal processing. References (Staszewski et al. 2005; Wu et al. 2017) present the first and a recent example of all-digital PLL (ADPLL). ADPLL is achieved by innovations in the frequency/phase detection (PFD) and oscillator control (VCO). The traditional PFD is replaced by Time-to-Digital Converter (TDC) so that the error signal is expressed in a digital fashion. In such so-called ADPLL, the oscillator is typically made of on-chip inductor and capacitor. Its frequency tuning is accomplished by tuning on and off individual small capacitors in an array. This action takes place in digital domain. Thus, it is called digitally controlled oscillator (DCO). The use of TDC and DCO leads to the possibility of using digital filter in the loop design.

In some recent cases, the term ADPLL also has been used to refer to those architectures where CMOS gate-delay structures are employed for generating oscillation frequency (inductor-less). The oscillator is also digitally controlled (DCO) (Chung et al. 2016; Ho and Yao 2016). For all those ADPLLs, the all-digital nature of the loop filter enables the ADPLL to enjoy many implementation advantages, such as improved noise immunity from circuit nonidealities, compatibility with digital deep submicron CMOS process, simplified testing and calibration, and ease of integration with digital baseband circuitries. Further, the digital nature of the loop structure makes it possible for the loop feedback design to be carried out in time domain using observer-controller approach instead of pole-zero method in frequency domain (Namgoong 2010).

For FLL design, analog signal processing approach has been traditionally adopted. An unconventional analog FLL is discussed in Djemonai et al. (2001). It is based on a unique frequency-to-voltage converter. The all-digital approach is also seen in FLL design. Reference (Dean and Rane 2013) is a simple digital FLL for low frequency application of capacitance measurement where an up-down counter is used as frequency oscillator (numerically controlled oscillator, or NCO). Reference (Hazarika and Sumathi 2015) is a Moving-Window-Filter (MWF) based FLL designed for similar low frequency capacitance measurement application where a MWF is used as a frequency extractor and another MWF is used as frequency generator. In Ding et al. (2018), a RC network functions as a bang-bang type frequency detector and is used to work with an analog VCO through a DAC, to form a FLL loop. Reference (Khalil et al. 2011) is an example of ADFLL where a frequency-to-digital converter and a DAC are used to transform the analog loop into a digital one. As a result, the loop signal processing is carried out in digital domain and the DAC is used to apply the digital control to a delay-cell-based oscillator (analog in nature).

All the current PLL and FLL architectures, being called all digital or not, almost all have some kinds of analog circuitries embedded in the structure. In most cases, the frequency oscillator is analog (however, its frequency tuning can be digitally controlled). In certain scenario, it is preferred that a 100% pure digital frequency control loop be available. This can be valuable for a variety of applications when analog design approach is not feasible. Reference (Kumm et al. 2010) presents a FPGA example of such cases. It is used for high energy physics experiments. Due to volume limitation, it is uneconomical to build a FLL in ASIC style for this application. Instead, a quick and low-cost digital approach is preferred. For this particular case, all input and output signals are sampled waveforms coming from ADCs or going to DACs (the ADC and DAC are on-chip available from this FPGA). This makes it possible to analyze and simulate the whole PLL using the z transform.

As discussed previously, TAF-DPS possesses the features of small frequency granularity and fast frequency switching, achieved simultaneously for a given design (loosely termed as arbitrary frequency generation and instantaneous frequency switching). Further, its output frequency can be monotonically tuned following a changing-pattern created in digital control word. Moreover, its frequency switching speed is quantifiable. The response time from the moment of receiving control-update to the moment of frequency switching is calculable in terms of cycles. These facts make TAF-DPS a suitable circuit block for functioning as a digitally controlled oscillator (DCO). In this section, an example of TAF-DPS synthesizer is constructed entirely from digital cells and is used as a pure digital DCO. Combining a bang-bang frequency detector and this DCO with a digital control/filter block, a true digital FLL is subsequently created. Instead of conventional frequency, this loop works on Time-Average-Frequency (Xiu et al. 2004). It bears following distinguished characteristics (Xiu et al. 2019).

- Output jitter and loop design are independent from each other and the two issues can be treated separately.
- The difference between the treatments of integer-N and fractional-N is insignificant due to the use of Time-Average-Frequency.
- The response speeds of all the loop components are quantifiable in terms of DCO cycle. Hence, loop dynamic can be quantitively analyzed and precisely predicted.
- Bang-bang type frequency detector works in harmony with Time-Average-Frequency naturally, which makes the loop design simple and robust.
- The loop is 100% digitally implementable without requiring any analog component.

6.6.1 TAF-FLL Architectures

(1) *Generic Frequency-Lock Architecture on Time-Average-Frequency*

Figure 6.65 illustrates the generic frequency-lock architect which works on Time-Average-Frequency. It is a feedback loop consisting of a frequency detector (FD), a

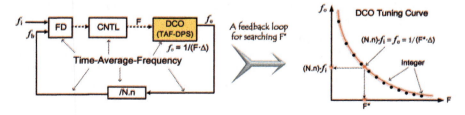

Fig. 6.65 Frequency-lock achieved on Time-Average-Frequency by using TAF-DPS as DCO

control block for receiving the FD output and generating a control word F, a DCO made of TAF-DPS and a divider of ratio N · n where N is the integer part and n is the fractional portion. Together, those components form a closed loop for searching the appropriate value for control word F^* that can make $(N \cdot n) \cdot f_i = f_o = 1/(F^* \cdot \Delta) = 1/[(I^* + r^*) \cdot \Delta]$ true. Virtually, this loop is used to implement a search algorithm as shown in the right-hand side of the figure. This search process can start from anywhere in the curve where control word takes value from $F \in [2, 2K]$. The coarse-search uses only integers. When integer boundaries are identified, fine-search is carried out into F's fractional region. The search result is judged by the FD, which is designed as a bang-bang type detector. The DCO's response time is two f_o cycles (please refer to Sect. 4.5 of Xiu (2012a)). The response times of FD, CNTL and divider are all quantifiable in unit of f_b (or f_o) cycle. The latency of this loop therefore is quantifiable. It is important to note that this feedback control loop, unlike traditional frequency loops that operate on conventional concept, works on Time-Average-Frequency. In other words, the components in this loop (all of them are digital circuits) are driven by signals of TAF (i.e., made of multiple types of cycles such as T_A and T_B). In this loop, the output from FD and CNTL are digital values and thus are shown as dotted line.

The aim of this loop is to make $T_i = T_b = (N \cdot n) \cdot T_o = (N + p/q) \cdot (F^* \cdot \Delta)$, where n is expressed as $n = p/q$ and p and q are integers and $0 \le p < q$, $\gcd\{p, q\} = 1$. This can further be expressed as $T_b = F^{**} \cdot \Delta$ where $F^{**} = F^* \cdot N + F^* \cdot p/q = I^{**} + r^{**}$. Both f_o and f_b therefore are realized in Time-Average-Frequency. In other words, generally, both use two types of cycles. This leads to the observation that TAF-FLL actually blurs the difference between the integer-N and fractional-N structures since $T_b = F^{**} \cdot \Delta = (F^* \cdot N + F^* \cdot p/q) \cdot \Delta = (I^{**} + r^{**}) \cdot \Delta$ contains both integer and fraction parts whether $p = 0$ or not. From the perspective of frequency concept, the loop treats integer-N and fractional-N the same. In other words, in the value of divider ratio N · n, whether $n = 0$ or $n \ne 0$ makes no difference when loop operation is concerned.

$$G(z) = K_D K_L K_O \frac{1}{(z-1)z^{L-1}}, \quad H(z) = \frac{1}{N}$$

$$\frac{F_o(z)}{F_i(z)} = \frac{G(z)}{1 + G(z)H(z)} = \frac{NK_D K_L K_O}{N(z-1)z^{L-1} + K_D K_L K_O} \quad (6.6.1)$$

From Fig. 6.65, it is seen that the loop is entirely digital and can be analyzed in discrete time domain. Figure 6.66 is the z-domain model of this TAF based frequency-lock loop. The DCO is modeled as a gain element instead of an integrator since our structure is a FLL not a PLL. The control block CNTL is modeled as an integrator containing one pole. This is due to the reason that it is an accumulating circuit. The nonlinear bang-bang frequency detector is simply modeled as a gain element. The model also includes a delay block with latency L for representing the time required by digital processing. The transfer function is derived in (6.6.1). Figure 6.67 includes

6.6 Frequency Lock on Time-Average-Frequency: All Digital TAF-FLL

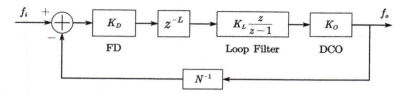

Fig. 6.66 Linear model of the discrete-time TAF based frequency-lock loop

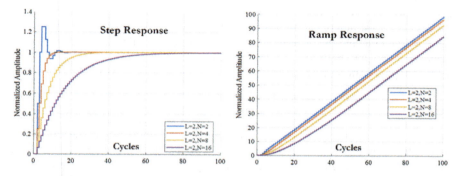

Fig. 6.67 Simulation result using the linear model: step (left) and ramp (right) responses

step and ramp responses of several settings. Latency L is chosen as two since the FD uses one $f_b/2$ cycle (will be explained shortly) and the CNTL and DCO combined processing time is less than one $f_b/2$ cycle when $N \geq 2$. The larger the N is, the less significant the CNTL & DCO processing time is and the more stable the loop is, as evident from the plots. When N is large enough, the CNTL & DCO processing time can virtually be ignored and the loop becomes a very stable first-order system.

(2) *Any rate frequency multiplier (Software FLL)*

Figure 6.68 depicts a structure of any rate frequency multiplier, called configuration A1. As shown, FD measures the frequency of the input signal f_i. Its output, a digital value representing the value of f_i, is fed into a digital signal processing block. This block carries out the tasks of filtering, frequency multiplication/division and control word generation, all in digital domain. Its output is the frequency control word F, which is used to control the TAF-DPS DCO. The DCO output bears a frequency value of $f_o = (N \cdot n) \cdot f_i$. This configuration is a one-pass forward-only operation. Thus, it is called frequency multiplier rather than FLL.

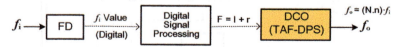

6.68 Block diagram of any rate frequency multiplier (software FLL)

Some comments can be made on A1. First, the FD in this architecture operates in a "slow but accurate" fashion since it does not need to operate all the time. A digital counter operating in a precise one-second-window is suggested since Time-Average-Frequency concept is defined in one-second-window (for higher accuracy, multiple seconds can be used, the result can be then averaged to get the value of frequency). In this structure, the search process is simplified to a straightforward calculation. Secondly, any ratio of multiplication or division can be achieved (as long as f_o is within the TAF-DPS's working range). Thirdly, the digital signal processing block can be implemented in either hardware or software, whichever is convenience. For this reason, this structure can also be viewed as a software FLL when the processing block is implemented in software. Overall, the merit of configuration A1 is that its users do not need to have any PLL/FLL related knowledge or skill. It is for general public to "play with frequency".

(3) *Frequency-Locked Loop: TAF-FLL*

Figure 6.69 depicts the core architecture of Time-Average-Frequency Frequency-Locked Loop (TAF-FLL), called A2 for the purpose of identification. The aim of this architecture is to make the loop respond quickly to the input change. In other words, we want the latency of the action-loop "frequency detection $\rightarrow$ control generation $\rightarrow$ frequency change $\rightarrow$ ..." be as small as possible. To achieve this goal, the key is to use a fast FD. For this reason, the well-known Alexander detector is selected (Alexander 1975). It uses three-point-sampling technique to make early-late decision between two compared signals. The input signal f_i is first frequency-divided by two so that it can be treated as "data" (similar to the case of CDR). Since Alexander detector requires the sampling clock being 50% duty cycle (both the rising and falling edges are used), an additional divide-by-2 process is applied to both the f_i and f_b as shown. Three transition edges that make up one cycle of $f_b/2$ are used to sample the "data". Early-late decision is made in every $f_b/2$ cycle which indicates the result of comparing f_i and f_b for fast-slow. The time required for TAF-DPS DCO update is two f_o cycles. The CNTL block is driven by f_o and its circuit can be designed very fast, taking only one f_o cycle. Thus, the latency of the whole loop is quantifiable. The loop latency L can be expressed as L = (2 + 3/N)/2 in unit of $f_b/2$ (the loop sample rate is $f_b/2$).

The search process is straightforward. The early-late decision from FD will direct the CNTL block to increase or decrease the F value by one from its current value (other more sophisticated pattern can also be used if so desired). This is an accumulation

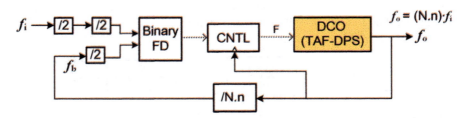

Fig. 6.69 Architecture of TAF-FLL

6.6 Frequency Lock on Time-Average-Frequency: All Digital TAF-FLL

effect and thus is modeled as an integrator. When the integer boundaries F_{up} and F_{dn} are identified ($F_{up} \leq F^* \leq F_{dn}$), the fast-slow decision will now only influence the fractional part. In real circuit, the effect of fraction is achieved by the motion of dynamic jumping between F_{up} and F_{dn}. In other words, in this A2 configuration, the DCO does not contain actual circuit for fractional part. The virtual fraction is achieved by straddling between F_{up} and F_{dn}. This fact simplifies the design and improves the circuit speed.

For the case of fractional-N, the divide ratio $N \cdot n$ can be realized by using a divider whose ratio dynamically jumps between two integer values. This type of divider has been used in many traditional fractional-N PLL and FLL architectures, often accompanied by a sigma-delta modulator to smooth the output signal's spectrum. In conventional FLL and PLL, introducing fraction into loop divider is a significant issue. The design challenge of fractional-N FLL and PLL is considerably higher than that of integer-N FLL and PLL. This is simply due to the fact that its output signal is in conventional frequency. In other words, its design goal is to make its output pulse train in compliance with the creed of "all pulse shall have same length-in-time". This however is not an issue for TAF-FLL since it uses the TAF concept (i.e., inherently uses multiple types of cycles). This change in frequency concept simplifies the design of TAF-FLL greatly.

(4) *Frequency-Locked Loop: TAF-DPS based NCO*

Traditionally, a digital controlled oscillator (DCO) can be designed from using a frequency divider. By altering the divide ratio, its output frequency is correspondingly changed. It is so popular that a name of NCO (numerical controlled oscillator) is created for it since its output frequency is controlled by a number. The largest drawback with NCO is that its frequency resolution is coarse since only integer value can be used. TAF-DPS is virtually a fractional divider (Xiu 2012a). By cascading TAF-DPS and NCO together, a fractional NCO can be created. This approach can improve its frequency resolution. TAF-FLL using this type of fractional NCO as its oscillator is illustrated in Fig. 6.70 and it is called configuration A3.

In this architecture, the frequency range is covered by the programmable divider R while the fine frequency adjustment is accomplished by the TAF-DPS. The coarse-search is now focused on setting the R value. The fine-search is carried out with the help of the fractional part of control word F. The TAF-DPS output is $1/f_{TAF} = T_{TAF}$

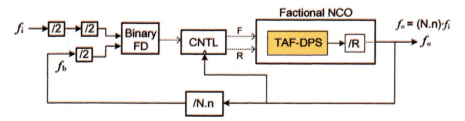

Fig. 6.70 Architecture of TAF-FLL made of fractional NCO

$= F \cdot \Delta$. The output after divider is $f_o = f_{TAF}/R$, or $T_o = R \cdot T_{TAF} = R \cdot F \cdot \Delta$. If the divider selects its divide ratio from the range of [1, R], the frequency range of the TAF-DPS output is expended in the low end from $T_{TAF\_max} = 2 K \cdot \Delta$ to $T_{o\_max} = 2R \cdot K \cdot \Delta$. Now, $T_o = R \cdot T_{TAF} = R \cdot F \cdot \Delta = F' \cdot \Delta$ where $F' = R \cdot F = R \cdot (I + r) = R \cdot I + R \cdot r = I' + r'$. This is virtually an expanded TAF-DPS circuit from K input phases to $R \cdot K$ (please refer to Sect. 4.6 of Xiu (2012a)). Compared to A2, the operating frequency range is much expanded.

(5) *Jitter or noise performance and loop design*

Referring to Figs. 6.68, 6.69 and 6.70, there are dot-line connections in all those structures. Those connections are communication channels passing digital values. There is no electrical-charge-based information being passed between FD and DCO. The DCO output is only influenced by the numerical value of F. Any "noise" on the value of F only influences *where* and *when* T_A and T_B occur. It does not have any impact on the electrical quality of the T_A and T_B pluses. Therefore, the noise performance of TAF-FLL output is not influenced by the loop design at all. It is determined only by the DCO's construction quality (mainly the quality of the layout) and the noise quality of the input K signals. This fact leads to a significant design advantage that the TAF-FLL output's jitter or noise performance can be treated separately from loop design.

(6) *TAF-FLL intended application*

TAF-FLL uses two types of cycles T_A and T_B on its output. Its intended application is to drive digital circuitries. When used as a clock signal to drive digital circuit, it behaves the same as conventional-frequency clock. The only concern is that the digital circuit driven has to be *setup*-constrained by the shorter one of the two: T_A. There is no issue regarding *hold* constrain since hold check only uses one edge. For more information regarding TAF clock and setup-hold check, reader is referred to Sects. 1.3.2, 1.3.3, 4.22 of Xiu (2012a) and Sects. 2.5, 2.6, 3.6, 3.8 of Xiu (2015a). As a matter of fact, the TAF-FLL itself is a digital circuitry and is driven by TAF clock. Its successful operation is a solid proof of TAF clock's effectiveness. When TAF clock is used for other applications, such as driving data converters or being used as carrier signal, some concerns may arise and certain techniques might be applicable to deal with them but analysis has to be carried out case by case. It is also interesting to mention that a good byproduct of using TAF is that its spectrum is inherently spread since two types of cycles occur in an interwoven way. In other words, naturally, TAF clock has less electromagnetic radiation.

6.6.2 TAF-FLL Experimental Verification

(1) *TAF-FLL implementation on FPGA*

Xilinx KC705 Evaluation Board for the Kintex-7 FPGA is selected for implementing the TAF-FLL architectures. The key component of the Xilinx evaluation system is

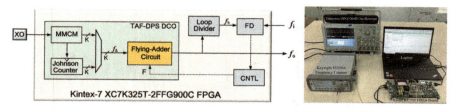

Fig. 6.71 TAF-FLL implementation on FPGA (left) and the experiment platform (right)

the XC7K325T-2FFG900C FPGA chip. It has 326,080 logic cells, 407,600 CLB flip-flops, 800 DSP slices, 500 IOs, 16,020 block RAMs, 16 high speed series transceivers and etc. For clock resource, it has a 200 MHz MEMS resonator from Si Time and an I2C programmable oscillator from Silicon Labs. In addition, the XC7K325T-2FFG900C FPGA chip has 10 Mixed Mode Clock Management Tiles (MMCM). Each MMCM can function as a PLL. In our process of implementing the TAF-FLL circuits, the plurality of K phase-evenly-spaced signals must originate from a frequency source of some type. It could be the 200 MHz MEMS resonator or the programmable clock source. This signal, after frequency boosted by MMCM, drives a K/2-stage Johnson counter made of K/2 flip flops. Their K non-inverting and inverting outputs form the plurality of K phase-evenly-spaced signals. MMCM, which has multi-phase outputs, can also function as the source of this plurality of signals. The rest of TAF-FLL circuitries are all realized by the LUT (look up table) and flip flops from the FPGA resource pool. This implementation scheme is illustrated in the left of Fig. 6.71.

The design style adopted is HDL coding → simulation → synthesis (mapping to FPGA). A sample VHDL code for the DCO (Flying-Adder) is available in appendix 4.A of Xiu (2012a). The implementation of TAF-DPS is through the Flying-Adder circuit. Referring back to Sect. 4.6 and Fig. 4.9, the Flying-Adder circuit can virtually be split into three parts: the signal path consisting of the two K → 1 multiplexes, one 2 → 1 multiplex, and the toggle flip-flop; the adder at the top; and the accumulator at the bottom. Those three parts are mapped to FPGA hardware separately for better control. The multiplexes are all realized by logic cells. The adder and accumulator are realized by the LUTs and flip-flops in the CLB blocks. In this work, Flying-Adder circuit has been constructed as K = 8. Four bits are used for the integer part and, for Configuration A1, 28 bits are reserved for the fractional part.

As stated previously, TAF-FLL is mainly used for driving digital oriented application. In this application domain, frequency accuracy is a major concern since it controls the flow rate of the digital data stream for myriad types of communications. For this reason, a Keysight 53230A Universal Frequency Counter is used as our main vehicle for evaluating the TAF-FLL performance. The experiment platform is shown in the right-hand side of Fig. 6.71. Oscilloscope is used for capturing signal waveform when loop dynamic is studied.

(2) *Frequency accuracy measurement of A1, Any rate frequency multiplier (software FLL)*

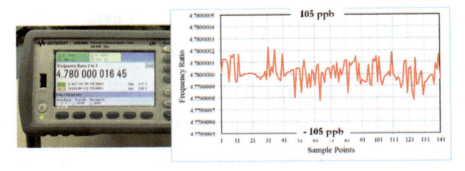

Fig. 6.72 Frequency accuracy of A1, N.n = 4.78

Figure 6.72 shows the measured output of any rate frequency multiplier (configuration A1, refer to Fig. 6.68). The multiplier ratio is arbitrarily set as 4.78. The input fed to the multiplier is 19.44 MHz, which is also fed into the frequency counter's channel 1 (CH1). The multiplier's output is fed into channel 2 (CH2). As shown in the snapshot (displayed in the left-hand side), the measured CH2/CH1 ratio is 4.78000001645. On the right-hand side, a curve containing 140 measured points are shown (taken once every five seconds). The average value is 4.78 and all the measurements are bounded within a range of ∓ 50 ppb.

(3)   *Frequency accuracy measurement of A2, TAF-FLL*

Figure 6.73 is the measurement of TAF-FLL (configuration A2, Fig. 6.69) under the settings of N = 8. There are two curves in the display shown on the right. The blue one is from TAF-FLL while the red one from a MMCM. The input is a 19.44 MHz signal, used for both the TAF-FLL and MMCM. Both loops are set at N = 8. As seen, the output from MMCM has a slightly better precision in term of lock accuracy. The result from TAF-FLL output is bounded within a range of ∓ 0.025 ppb. Figure 6.74 shows the measurement result for the fractional case of N · n = 6.0078125. The fraction is n = p/q = 1/128. The frequency error is bounded at ∓ 2 ppb. For this fractional setting, we cannot compare TAF-FLL with MMCM since the MMCM is unable to achieve this ratio. Its finest increment is only 0.125 (p/q = 1/8).

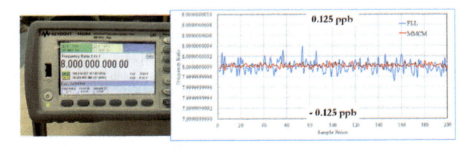

Fig. 6.73 Frequency accuracy of A2, N = 8

6.6 Frequency Lock on Time-Average-Frequency: All Digital TAF-FLL

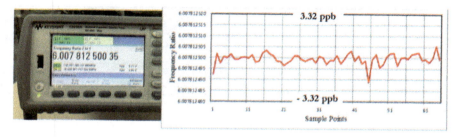

Fig. 6.74 Frequency accuracy of A2, N.n = 6.0078125

(4) *Frequency accuracy measurement of A3, TAF-FLL made from fractional NCO*

In the cases of Figs. 6.73 and 6.74, the TAF-DPS DCO is designed under the condition of $\Delta = 1.25$ ns. Thus, the T_A and T_B dissimilarity $\delta_{AB} = |T_A - T_B| = 1.25$ ns. For configuration A3, Δ is reduced to 0.7 ns. Thus, the dissimilarity δ_{ab} is diminished. This smaller δ_{AB} also leads to a better frequency accuracy. Figure 6.75 shows the measurement result. The ratio is set as N = 4. The blue and red curves are for TAF-FLL and MMCM, respectively. As seen, the frequency accuracy of A3 is better than that of A2. Now, the frequency accuracies of the TAF-FLL and MMCM are at about the same level.

(5) *Experiment on frequency step change*

The issue of frequency step change has been studied on this TAF-FLL. Figure 6.76 shows the frequency trend captured when the input has a step change from 19.318 to 35.417 MHz. The ratio of multiply is set as four. On the left is the case of TAF-FLL while MMCM on the right. As seen, for TAF-FLL, the output steadily climbs from the starting frequency to the destination. The frequencies are realized in TAF fashion. The desired ratio of N = 4 is accurately achieved after the lock. This fact cannot be validated from this figure with high accuracy. However, Fig. 6.75 provides the evidence at accuracy of about 0.02 ppb (The Time-Average-Frequency T_{TAF} is exactly four times of input frequency). As a comparison, the same frequency step change is applied to the MMCM. Its output behavior is displayed on the right-hand side of Fig. 6.76. Significant ringing with large overshot is observed in the transition

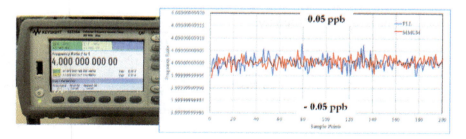

Fig. 6.75 Frequency accuracy of A3, N.n = 4

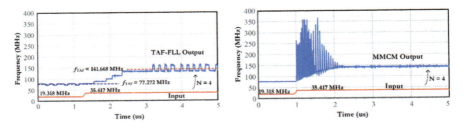

Fig. 6.76 Frequency trend captured when input having a step change: TAF-FLL (left) and MMCM (right)

region. The reason is due to the fact that MMCM is a third-order phase locked system while TAF-FLL is a first-order frequency locked system.

(6) *Experiment on frequency tracking capability*

Frequency tracking capability is another issue that has been studied. In this study, a frequency change pattern is created on the input signal as shown in the green curves in Fig. 6.77. The frequency of this signal changes from 6.25 to 20 MHz in 440.5 us, in an "inverse-linear" pattern (i.e., its period is changed linearly). This signal is fed into both the TAF-FLL and MMCM. The TAF-FLL and MMCM are both set as $N = 8$. The measured frequency trend from TAF-FLL output is plotted as the

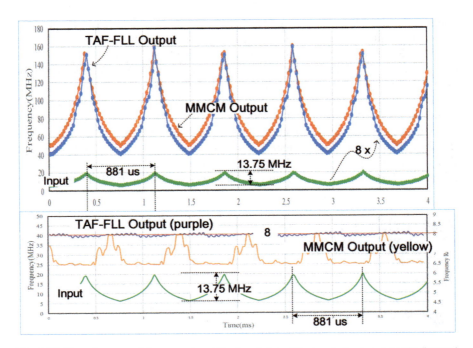

Fig. 6.77 Plot of measured frequency trend (top) and plot of the trend of measured ratio (bottom)

6.6 Frequency Lock on Time-Average-Frequency: All Digital TAF-FLL

blue curve while the MMCM output is red. In the bottom portion of this figure, the trend of measured ratio is also plotted. The purple is for TAF-FLL and the brown for MMCM. As evident from the plots, the purple curve stays with the setting of N = 8 reasonably well while the brown curve cannot follow it most of the times. This reveals the fact that our TAF-FLL has reasonably good frequency tracking capability (please refer to the curves on the right-hand side of Fig. 6.67). This is expected since our system is designed for frequency lock while MMCM is for phase lock.

(7)   *The issue of phase lock*

Figure 6.78 shows the waveforms captured from TAF-FLL operation. The yellow one is the input signal, the green one is the feedback signal (the DCO output passing the loop divider). As seen, the two signals' frequencies are equal (the TAF-FLL is in locked state now). However, their phases are not aligned. There are two reasons for this behavior. First, unlike conventional PFD, Alexander detector does not report the size of the phase difference (only its sign, the indication of early or late). Secondly, TAF-DPS DCO uses Time-Average-Frequency to match the input signal of conventional frequency. Thus, there is a phase movement whenever DCO changes the type of cycle that it currently uses. For those reasons, the phases of the input reference and DCO output are not aligned all the times, and this TAF-FLL loop structure is not claimed as phase locked loop.

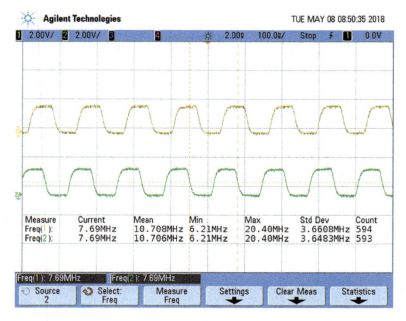

Fig. 6.78 Waveforms of the input signal (yellow) and TAF-FLL output (after the divider, green) when TAF-FLL reaches lock

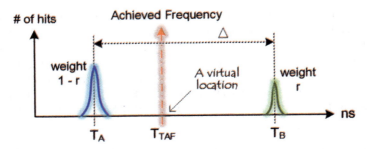

Fig. 6.79 The period distribution of TAF signal

(8) *Discussion on jitter*

As depicted in Fig. 6.79, the period distribution of TAF signal normally comprises of two groups corresponding to T_A and T_B, respectively. When fraction r = 0, the TAF signal retreats to signal of conventional frequency (T_B disappears). For this type of unconventional clock, when used in driving functional circuit, it is necessary to check its impact on the circuit operation. When a clock signal is used to drive digital circuit, the designer who is responsible for that circuit needs to pay attention on two things: setup and hold checks. Only by satisfying those two constraints, the digital circuit can be said "timing closed" and then becomes safe for operation.

When doing the task of timing closure, designer uses clock frequency (more precisely, clock period T) as the "setup constraint". It demands that all the logic-gate-operations between two sequential stages that have to be finished within one clock cycle must use a span of time that is less than T. Otherwise, the circuit will fail since there is no enough time for those logic gates to finish their actions. For clock signal of conventional frequency, the setup constraint is T. For TAF signal, the setup constraint is T_A (the short one). As long as the circuit is setup-constrained by using T_A, it does not matter where the T_B is (in other words, the size of Δ does not matter). This statement of "using shortest cycle for setup constraint" is true for all types of irregular clock signals including gapped clock, gated clock, spread spectrum clock, TAF clock and etc. For all those techniques, using the shortest cycle for setup constraint guaranties the safety.

One piece of good news is that, for hold check, it is irrelevant whether the driving clock is in conventional frequency or TAF since hold check only uses one clock edge. On another note, clock skew will have impact on hold-check. But clock skew is a rather different concept than jitter. It is not related to clock circuit design, but has much to do with the task of clock distribution. Please refer to Sect. 1.3 of Xiu (2012a) for more in-depth explanation on the issues of setup, hold, clock jitter, clock skew. Section 4.22 of Xiu (2012a) has specifically discussed TAF signal and setup constraint.

As depicted in Fig. 6.79, the T_A and T_B both have some kind of distributions associated with their respective groups. This type of distribution can be caused by the various innate noises associated with the semiconductor circuit, and also the mismatch occurred among the signals of the group of K phase-evenly-spaced signals.

6.6 Frequency Lock on Time-Average-Frequency: All Digital TAF-FLL   169

Namely, in the non-perfect world of real engineering practice, this group of K signals is not exactly "phase evenly spaced". This leads to the distributions around T_A and T_B (on top of the background noise). This mismatch induced noise is deterministic jitter. Although its magnitude is unknown beforehand, it is fixed once the layout is finished. For more information, please refer to Xiu et al. (2012) that discusses the TAF-DPS output's jitter characteristic intensively.

The above discussion on TAF-DPS related jitter is elucidated when TAF-DPS is standalone. When this TAF-DPS is used as a DCO in a frequency control loop, such as in TAF-FLL, the TAF-FLL output will behave the same as a standalone TAF-DPS in the view of jitter behavior. The TAF-FLL is a pure digital loop. The DCO is frequency controlled by a digital value, not analog voltage. The output from the frequency detector FD only affects the value of the frequency control word F. It only influences which integer value to be used for F. It doesn't have any impact on the electrical characteristic of the TAF-DPS output signal.

(9) *Resource usage*

Table 6.5 gives the numbers of LUT and FF used in A1, A2 and A3, respectively, all implemented on XC7K325T-2FFG900C FPGA chip. The whole structure of TAF-FLL is pure digital. The design is created from RTL code and is fully made of digital standard cells. Therefore, the design can be easily transported from process to process. Using a 130 nm TSMC digital process, the power consumption of TAF-DPS DCO with 5 integer- and 32 fractional-bits (i.e., the size of accumulator is 37 bits) is about 65 uW. This is the case for configuration A1 (Fig. 6.68). For configuration A2 (Fig. 6.69), the whole TAF-FLL loop uses about 35 uW when operating at N = 8 ($f_i = 20$ MHz, $f_o = 160$ MHz). The smaller power number is due to the smaller DCO used in A2 (it only has four integer bits with no fraction, 4 bits accumulator).

(10) *Distinguished from established architectures*

Table 6.6 lists the key difference between TAF-FLL and other established architectures, namely, the frequency concept used in the construction of the loop. Compared

Table 6.5 Resource usage

Architecture	Resource	Used	Utilization[1] (%)
A1	LUT	2784	1.37
	FF (Flip-Flop)	174	0.04
A2 integer	LUT	22	0.01
	FF	35	0.01
In A2 fraction	LUT	31	0.02
	FF	51	0.01
A3 integer	LUT	23	0.01
	FF	38	0.01
A3 fraction	LUT	32	0.02
	FF	54	0.01

[1] The numbers of total LUT and FF are 203800 and 407600, respectively

Table 6.6 The key difference between TAF-FLL and established architectures

	Oscillator's output (VCO/DCO)	Two signals compared at FD or PFD	FD or PFD output	Clock driving the loop
Traditional PLL and FLL	CF[1]	All are CF	Analog pulses, or digital value	CF
Bang-Bang type PLL and FLL	CF	All are CF	Digital value	CF
TAF-FLL	TAF[2]	One CF one TAF, or all are TAF	Digital value	TAF

[1] CF stands for conventional frequency
[2] TAF stands for time-average-frequency

to analog oscillator based FLL, whether it is CMOS ring oscillator (made of delay stages) or inductor-based oscillator (made of LC tank) or RC relaxation oscillator, TAF-FLL's DCO is a pure digital circuitry. This fact naturally leads to the possibility of a digital structure for making the whole frequency control loop. This can subsequently bring in many advantages offered by digital circuit such as PVT invariance, circuit scalability and process portability, low cost, low design complexity, among others. Compared to other digital-intensive frequency control loop (such as ADPLL or NCO based FLL), TAF-FLL uniquely utilizes Time-Average-Frequency (i.e., the TAF concept). The use of TAF is naturally in harmony with the working principle of bang-bang type detector. It makes the loop design simple and robust and, at the same time, preserves high precision on frequency lock.

Compared to the scenario of TAF-DPS being a standalone clock generator, the TAF-DPS circuit in TAF-FLL configuration A2 and A3 does not need, or can use less, circuitry for its fractional part in circuit implementation. This can increase its circuit speed and uses less resource while preserving its operation accuracy. Having said all those positive things, it is worth to mention that its applications is preferred to be in digital domain since TAF-DPS uses Time-Average-Frequency on its output. Therefore, TAF-FLL is mainly used in digital-oriented products, which is the dominant majority of modern electronic design. It is not targeted for generating carrier signal in wireless communication, nor for directly driving data converters (i.e., ADC and DAC). For those applications, analysis has to be carried out case by case.

Paradigm Shift:

In the emerging Time-Oriented paradigm, the classic problem of frequency lock can be investigated from the viewpoint of phase-lock-in-time-average. This new perspective opens possibility for new circuit architectures.

6.7 Random Number Generator Using Interplay of Frequency Sources as Entropy

Security has become one of the major concerns after the explosion of connected devices and the advent of cloud computing and Internet of Things (IoT). High entropy random number source is needed in many cryptographic applications since it is an essential component to provide secret key or token for the cryptographic algorithms used to guarantee information security (FIPS 140-2 1999). The quality of the random number has a significant impact on the vulnerability of those algorithms. It also plays an important role in the accuracy of Monte-Carlo and stochastic simulations. A series of random number with unpredictability, which can neither be observed nor controlled by attacker, is therefore highly desired. While the feature of non-recurrence could be fulfilled by almost all random number generators in a relatively easy manner, unpredictability is more difficult to assure. In decades of research and engineering, a variety of deterministic and hybrid random number generators have been proposed.

High unpredictability however must originate, in one way or another, from hardware. The hardware attributes can be physical noise, structure defect or manufacturing imperfection (Jun and Kocher 1999). True random number generator (TRNG) is usually referred to such devices whose unpredictability and randomness come from one or more such hardware attributes. Among TRNG systems, their structures typically consist of three key parts: entropy source, harvesting mechanism and post-processing unit. The entropy source is the place where randomness and unpredictability originate. Harvesting mechanism is the circuit that collects the entropy embedded in the noise source and converts it into random bitstream. Since most physical source is strong at unpredictability and somehow weak at randomness, a post-processing unit is often required to enhance the quality of the resultant random series' randomness.

Figure 6.80 depicts the general architecture of TRNGs. The entropy source can include a myriad of physical phenomena from nature. The magnitude of physics noises, which is usually small, is first amplified and conditioned and then fed into

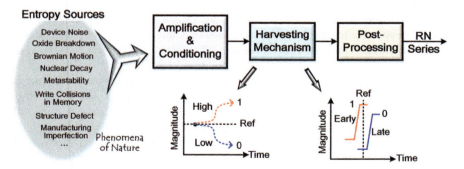

Fig. 6.80 The general architecture of TRNG

the harvesting circuit. The harvesting mechanism can be categorized into two types: voltage-domain (high/low) and time-domain (early/late), as illustrated in the figure. In the voltage-domain, the conventional method is to amplify noise directly with a high-gain and high-bandwidth amplifier followed by quantization. For example, resistor thermal noise is used in Petrie and Connelly (2000), oxide trap noise is utilized in Brederlow et al. (2006), and SiN device noise is employed in Matsumoto et al. (2008). Those designs are analog intensive. They require careful calibration of the amplifier and ADC to remove the bias in the resultant random number series. The extensive use of analog circuits also makes them less attractive in terms of system integration and technology portability. Another type of voltage-domain harvesting mechanism is to utilize metastability, which is provided by a pair of cross-coupled inverters (Mathew et al. 2012; Tokunaga et al. 2008). This method offers good operating frequency and power efficiency, but often requires large design efforts and run-time calibration to remove systematic and temporal mismatch in devices which are sensitive to structure defect and manufacturing imperfection. Those factors, if not taken care of, can result in bias in the resultant random series.

In time-domain harvesting, the principle is to utilize jitter from oscillator's output. It usually involves two ring oscillators (RO), called method of coupled oscillators. A slow RO is used to sample a jittery fast RO and the resultant data is the random bitstream (Bucci et al. 2003; Yang et al. 2014; Tang et al. 2014; Amaki et al. 2013). This method bears relatively low entropy and hence has low performance due to the limited amount of jitter in a single RO. To overcome this issue, using more oscillators (hundreds of them) is suggested in Sunar et al. (2007). In addition, this kind of design is vulnerable to PVT variation and power supply attacks since the quality of jitter is influenced by those factors greatly. To improve robustness, an edge racing even-stage RO is used in Yang et al. (2016), which employs a mechanism of frequency-collapse to resolve the ones and zeros. It however has the problem of low efficiency since it takes a long time for the circuit to reach the final state. Hence, the data rate could be impaired.

The metastability and coupled oscillators methods can be implemented in digital fashion. Thus, they have significant advantage over the noise amplification method which is analog extensive and hence has some obvious drawbacks. However, they still suffer the vulnerability problem from the variations of PVT (process, voltage and temperature). To alleviate those problems, in this section, we present a novel entropy-generation-mechanism which is based on interaction of multiple frequency sources of different and evolving frequencies. This plan is first proposed in Wei and Xiu (2020) and its two schemes are illustrated in Fig. 6.81.

There are two schemes in this plan of generating randomness from frequency interaction. The first one, shown in the top portion of Fig. 6.81, uses a method of frequency mixing. The signal waveforms from multiple frequency sources, each with its own unique frequency and initial phases, are XORed by a XOR tree. Due to the different frequencies and phases of those waveforms, the output waveform from the XOR tree is expected to bear an irregular shape. It is then sampled by another clock, resulting in a stream of random number. The second scheme, shown in the bottom of the figure, utilizes an approach of frequency-tracking. A free-run oscillator

6.7 Random Number Generator Using Interplay of Frequency Sources as Entropy

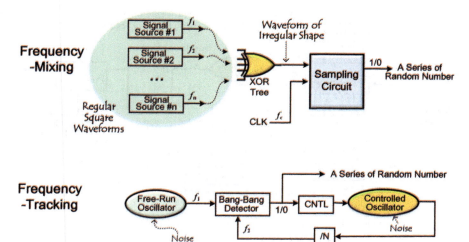

Fig. 6.81 Frequency-interaction for producing random number series: frequency-mixing (top) and frequency-tracking (bottom)

is frequency tracked by another oscillator through a frequency locked loop (please refer to Sect. 6.6). A bang-bang type detector is used as the frequency comparator. Its output is supposed to be a stream of random values since one oscillator is tracking the other in real time under various operating conditions.

Unlike the conventional mechanisms of providing entropy as discussed around Fig. 6.80, the source of entropy of the two methods discussed here is frequency interaction, which is inherently a mathematical operation but is influenced by many physically random and unpredictable factors. The three distinguished features of the frequency-interaction based architectures are pure digital structure, high programmability, and robustness against PVT variations. The effectiveness of those two schemes depends on the capability of generating frequency in a cost-efficient way. Ample supply of frequencies from a low-cost circuitry is the key for these two architectures. This requirement can be fulfilled by the strong frequency generation capability of the Time-Average-Frequency Direct Period Synthesis (TAF-DPS). The detail of these two structures will be discussed in the following text.

6.7.1 Frequency-Mixing TAF-DPS TRNG

Figure 6.82 shows the circuit block diagram of frequency-mixing TAF-DPS TRNG where MORO stands for multiple-output-ring-oscillator. There is a group of TAF-DPS circuits functioning as the signal sources of regular square waveforms. Each of them has its unique frequency, which is controlled by its respective frequency control word F_i. Their outputs are combined together by a XOR tree, whose output is an

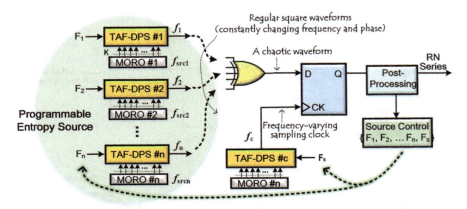

Fig. 6.82 Circuit block diagram of frequency-mixing TAF-DPS TRNG

irregular waveform resulted from the interaction of the waveforms. It is then sampled by a sampling clock, which can also be generated from a TAF-DPS or from any other type of clock generator. By inspecting this working mechanism, it is believed that the following factors will influence the shape of this resultant waveform; that is, the number of sources n, their frequencies f_i, the initial phases of those waveforms and the frequency of the sampling clock f_c. To improve the quality of randomness of the output, certain algorithm can be applied in the step of post-processing. For this issue, there are many algorithms that have been successfully used in the past, each is designed for specific physical entropy. They can be used in here for assisting the TAF-DPS TRNG. Two examples are available in Addabbo et al. (2006), Holcomb et al. (2009).

In the structure of Fig. 6.82, the components can be created by using either analog or digital method. For low-cost design, all the components can be made from digital circuit. The MORO, which is responsible for producing a plurality of K phase-evenly-spaced signals, can be generated from a ring oscillator made of K/2 differential delay stages, or from K cross-coupled NAND gates, or simply from a divider chain (e.g., Johnson counter). Those three approaches are shown in Fig. 6.83, each can be appropriate and effective for its respective application environment. For example, the cross-coupled NAND gates and divider methods can be implemented through pure digital fashion while the differential delay stages method has to be designed carefully using custom design style. In principle, each of the TAF-DPS in Fig. 6.82 shall have its own base time unit (i.e., MORO) so that the total amount of entropy can be maximized. To save area and power, however, all the TAF-DPS can share one MORO to trade area and power for amount-of-entropy.

Figure 6.84 illustrates the working principle of the TRNG. When a user presents its request to use the TRNG, a configuration step is performed to locate a set of control words $\{F_1, F_2, \ldots F_n, F_s,\}$ from a mapping table and assign it to this user. This approach of using a user-specific seed can enhance security level in sophisticated

6.7 Random Number Generator Using Interplay of Frequency Sources as Entropy

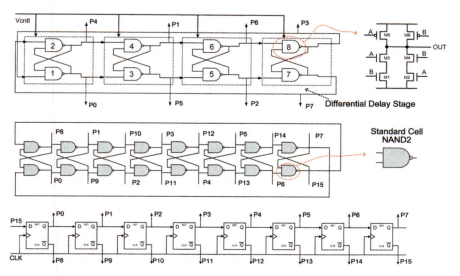

Fig. 6.83 Three MORO styles for generating a plurality of K phase-evenly-spaced signals

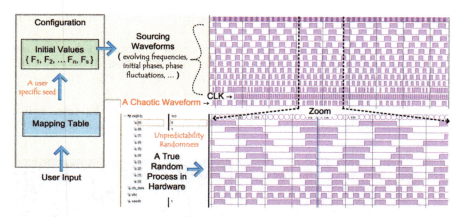

Fig. 6.84 Producing an ever-evolving chaotic waveform by mixing waveforms of varying frequencies as source of entropy

application scenarios. This set of control parameters is fed into the n TAF-DPS circuits which produce n square-wave pulse trains, each with its own frequency and initial phase. Moreover, the controlling F words are updated constantly based on the generated random number series through a feedback mechanism (Fig. 6.82), making all the TAF-DPS outputs' frequencies and phases constantly evolving. The chaotic waveform resulted from this hardware-based process bears the unpredictability and randomness preferred by TRNG. The total number of states in the mapping table (i.e., the seeds in Fig. 6.84) is theoretically $S = 2^{n \cdot m}$ where n is the number of pairs of {TAF-DPS, MORO} and m is the number of bits used for representing F (both the integer and fractional parts). In real case, some of the F settings may be unusable because the corresponding frequencies are too high to be realistic in the implementing technology.

Compared to other TRNGs, the main entropy in this TRNG is the interplay among electrical waveforms, which is fundamentally different from all the entropy-producing mechanisms discussed around Fig. 6.80. The frequencies and phases of those waveforms are constantly change, making it a time-variant entropy source. Consequently, the resulting chaotic waveform can continuously provide fresh entropy for a nonstop flow of random number (more precisely, it will take extremely long time for this chaotic waveform to complete one repeatable cycle if it can ever repeat itself). This TRNG architecture is robust against PVT and power supply variations since the entropy mainly originates from the interaction among the signal waveforms while the jitter from the ring oscillators (i.e., MORO) only contributes partially in a secondary role. Moreover, it is nontrivial to attack this circuit from outside since it is all digital. Furthermore, this architecture has high programmability since the shape of the resultant irregular waveform, and hence the entropy profile, can be easily altered by adjusting the TAF-DPS circuits' frequency control words in real time. The duty cycles of the TAF-DPS outputs can also be changed when needed. This can further improve its programmability. This high programmability makes it flexible for various applications.

This frequency-mixing TAF-DPS TRNG has been implemented on a 180 nm CMOS. The entire design is created in digital fashion and synthesized from an ASIC standard cell library. During the layout stage, all the circuits are collapsed into one flat netlist for automatic place and route. The die photo and layout are shown in Fig. 6.85. This style of implementation makes it feasible for a soft IP. In the top-level netlist, a TAF-FLL is also included on-chip for the testing of frequency-tracking TAF-DPS

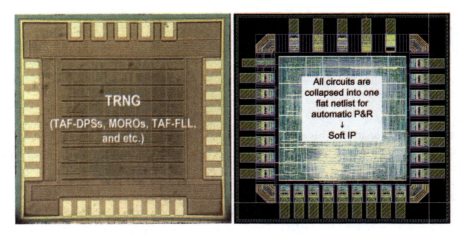

Fig. 6.85 Die microphotograph (left) and layout (right) of the TRNG testchip

6.7 Random Number Generator Using Interplay of Frequency Sources as Entropy

TRNG (will be discussed in next subsection). For the evaluation of frequency-mixing TAF-DPS TRNG, this TAF-FLL can function as an attacker capable of generating noise at a variety of frequencies.

There are 10 pairs of "TAF-DPS plus MORO" designed into the TRNG as programmable entropy source. Another pair is used for the sampling clock. For all the pairs of TAF-DPS plus MORO, a big value of K = 128 is used for more entropy and high programmability. To save power and area, the chip also includes another version where all the 11 TAF-DPSs share a common MORO. For all the TAF-DPSs, we use m = 16 bits for control word F, 8 bits for integer and 8 for fraction. The total number of states in the mapping table (i.e., the seeds in Fig. 6.84) is theoretically $2^{n \cdot m}$ = 2^{160}. In real case, about 10% of F settings are unusable since the corresponding frequencies are too high to be realistic in this technology.

In experiment, three different formations have been explored: (A) all 11 TAF-DPSs use their own MOROs and feedback is enabled, (B) same as (A) but feedback is disabled, (C) all 11 TAF-DPS circuits share a MORO without feedback. For each run in experiment, a set of 100 Mb data is collected for evaluation. For formation A, the resulting random series consistently passes all the 15 NIST 800-22 tests. For B, occasionally, some tests can fail. For C, as expected, the failure rate increases since all TAF-DPSs share the same MORO and the amount of entropy is largely reduced. However, with many $\{F_1, F_2, \ldots F_n, F_s\}$ settings, all the tests can still successfully pass.

The design is created on transistor of 1.8 v. In experiment, we have tested the circuit with supply voltage varying from 1.4 to 3.3 v. The Shannon entropy and average NIST pass rate versus supply is plotted in the top left of Fig. 6.86. The throughput and energy-efficiency versus supply is plotted in the top right. In the bottom right, a result of ACF test is presented on a data set of 1,048,576 bits, μ = 0 and $\sigma = 7.68 \times 10^{-8}$, 95% confidence level within 2σ of $\pm 1.54 \times 10^{-7}$. In the test chip, as said, there is an all-digital FLL implemented. It can generate a variety of frequencies inside the chip to inject noise around the TRNG. At 1.8 v, all the MOROs run roughly at 30 MHz. A noise injection test is carried out by making the FLL produce frequency from 10 to 300 MHz, in step of 10 MHz. Data is collected and Shannon entropy is calculated at each frequency point to check the robustness of this TRNG. The result is shown in bottom left. As expected, there is a drop around 30 MHz which is the MORO frequency. All values however are above 0.9999999.

The quality of the random bit streams is tested against the RNG standard tests listed in the NIST suite (NIST 2010, 2018). Long bitstream of length 100 Mb is generated and checked for all the tests suggested in NIST test suites. Figure 6.87 gives an exemplary result of NIST 800-22 and 800-90B tests and Fig. 6.88 is a speckle imaging of 1,048,576 bits of random number displayed in a 1024 × 1024 array. As mentioned, a highlight of this TRNG is its configurability. For different users or the same user but operating at different times, a dedicated seed can be assigned at the run time. This is not the deterministic seed used in PRNG but a hardware based initial condition whose evolution trace is unpredictable. This can furnish addition level of security. The benefit of configurability also lies in the fact that power and throughput can be conveniently programmed at user's desire. Figure 6.89 is the measured TAF-DPS

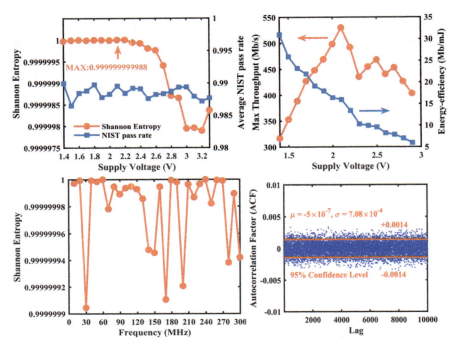

Fig. 6.86 Entropy and test pass rate versus supply (top left), data rate and efficiency versus supply (top right), entropy versus noise frequency (bottom left), autocorrelation function test (top right)

Test Name	P-value	Proportion
Frequency	0.779188	1
Block Frequency	0.574903	1
Cumulative Sums	0.419021	1
Runs	0.971699	0.99
Longest Run of 1s	0.779188	0.99
Rank	0.699313	1
FFT	0.897763	0.99
Non-overlapping Template	0.534146	0.99
Overlapping Template	0.514124	0.99
Universal	0.534146	1
Approximate Entropy	0.334538	0.99
Random Excursions	0.756476	0.99
Random Excursions Variant	0.484646	0.99
Serial	0.657933	0.98
Linear	0.987896	1

Test Name		Result	
excursion		905.039	PASS
Number of Directional Runs		796652	PASS
Length of Directional Runs		13	PASS
Number of Increases and Decreases		747833	PASS
Number of Runs Based on the Median		$4.9977*10^6$	PASS
Length of Runs Based on the Median		23	PASS
Average Collision Test Statistic		19.6679	PASS
Maximum Collision Test Statistic		72	PASS
Compression		$1.5581*10^6$	PASS
Periodicity Test Statistic	Periodicity(1)	245612	PASS
	Periodicity(2)	245600	PASS
	Periodicity(8)	245244	PASS
	Periodicity(16)	245540	PASS
	Periodicity(32)	245117	PASS
Covariance Test Statistic	Covariance(1)	$1.9976*10^7$	PASS
	Covariance(2)	$1.9978*10^7$	PASS
	Covariance(8)	$1.9974*10^7$	PASS
	Covariance(16)	$1.9981*10^7$	PASS
	Covariance(32)	$1.9974*10^7$	PASS
Chi-square Independence		PASS	
Chi-square Goodness-of-fit		PASS	
Length of the Longest Repeated Substring Test		PASS	
Restart Tests		PASS	

Fig. 6.87 NIST testing result: 800-22 (left) and 800-90B (right)

6.7 Random Number Generator Using Interplay of Frequency Sources as Entropy

Fig. 6.88 Speckle imaging of 1,048,576 bits displayed in a 1024 × 1024 array

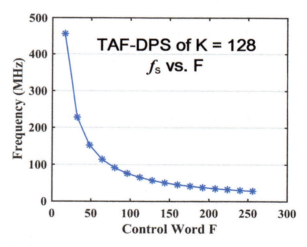

Fig. 6.89 Measured TAF-DPS output frequency versus its control word F

output frequency versus its control word F. The control words determine operating frequencies, and thus the power usage and data rate. This provides flexibility to user. In this testchip, the maximum throughput achieved is 530 Mb/s while the typical is 450 Mb/s. The energy efficiency achieved is 31 Mb/mJ (max.) and 19.1 Mb/mJ (typ.).

The amount of entropy in this TRNG is related to the number of TAF-DPS plus MORO pairs involved in the process. Intuitively, more pairs lead to more entropy. However, it can be understood that too many of them could make some of the signal-interplay effects be cancelled out if the frequency settings are not delicately balanced.

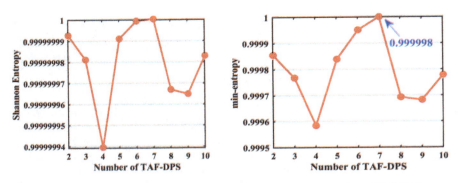

Fig. 6.90 Shannon entropy and Min-entropy (NIST 800-90B) versus number of TAF-DPSs

This leads to resource wasted. This is an open issue requiring rigorous mathematical study in future. In the experiment, we simply check this using the metric of Shannon entropy versus number of pairs. It is found that 7 pairs can give maximum Shannon entropy and Min-entropy (which is proposed in NIST 800-90B), as shown in Fig. 6.90.

In summary, compared to other TRNGs, the main entropy in this TRNG is the interplay among waveforms, which is fundamentally different from the mechanism of device noises. The frequencies of those waveforms are constantly changing, making it a time-variant entropy source. The resulting chaotic waveform can continuously provide fresh entropy for a nonstop flow of random numbers.

6.7.2 Frequency-Tracking TAF-DPS TRNG

Figure 6.91 is the circuit block diagram of frequency-mixing TAF-DPS TRNG. This is in principle a frequency locked loop. It is based on the TAF-FLL presented in Sect. 6.6. The output from a digital oscillator, TAF-DCO, is tracking in real time the frequency of the oscillator attached on the input side. This free-run oscillator at the

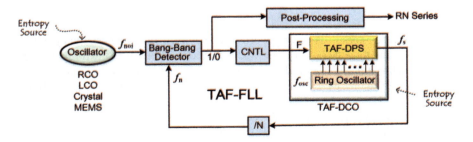

Fig. 6.91 Circuit block diagram of frequency-tracking TAF-DPS TRNG

6.7 Random Number Generator Using Interplay of Frequency Sources as Entropy

input can be RO (ring oscillator), LCO (LC-tank based oscillator), crystal oscillator or MEMS based oscillator, each having its unique noise profile. The two frequencies of f_n and f_{noi} are compared by the bang-bang detector. Its output, the status of lead or lag which corresponds to 1 or 0 respectively, is a stream of zero and one.

The operation of this TRNG depends on the unique characteristics of TAF-FLL. Firstly, the oscillator TAF-DCO is controlled by a digital value, not analog voltage. This enables the implementation of the entire structure in digital fashion. More importantly, this loop works on the principle of TAF (not the conventional frequency). Therefore, at any given moment, the two instant frequencies most likely will not aligned. They are only matched in time-average sense (the matching accuracy is good under TAF definition, can reach deep sub-ppb (ppb: parts per billion) range (Xiu et al. 2019)). Therefore, in most of the times, the detector output is not at an equilibrium state. The use of TAF positively contributes to the level of entropy, which is beneficial to the realization of strong randomness and high unpredictability. For post-processing, many algorithms can be applied here. In this work, the same chaos-based post-processing module employed in previous case is used, just for illustration purpose.

This frequency-tracking TAF-DPS TRNG has also been implemented in the 180 nm CMOS. The tunable parameters are the type of input frequency source, the input frequency f_{noi}, and the divider ratio N (which determines the DCO frequency f_s). With some settings, it is found that all the NIST tests can pass. Under the setting of $f_{noi} = 13.75$ MHz, N = 8 ($f_s = 110$ MHz), the power consumption is measured as 0.59 mW, giving an energy-efficiency of 23.3 Mb/mJ. The NIST 800-22 test report is shown in the left of Fig. 6.92 while the ACF test result is at the right. Currently, the ring oscillator (i.e., the MORO) and the TAF-DPS in Fig. 6.91 are designed with

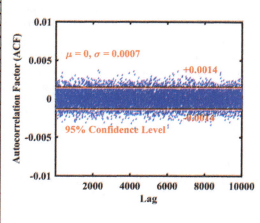

Test Name	P-value	Proportion
Frequency	0.334538	1
Block Frequency	0.401199	1
Cumulative Sums	0.275709	0.98
Runs	0.616305	0.99
Longest Run of 1s	0.657933	1
Rank	0.304126	1
FFT	0.304126	0.99
Non-overlapping Template	0.595549	0.99
Overlapping Template	0.554420	
Universal	0.275709	1
Approximate Entropy	0.678686	1
Random Excursions	0.834308	0.99
Random Excursions Variant	0.275709	1
Serial	0.637119	0.99
Linear	0.236810	0.99

Fig. 6.92 Frequency-tracking TAF-DPS TRNG: NIST testing result: 800–22 (left) and autocorrelation function test (right)

K = 128. It is believed that much smaller K (e.g., 32) can be used. This can significantly reduce the power consumption and area.

6.7.3 Some Open Issues for Further Investigation

A novel frequency-interaction based entropy-generation-mechanism is presented in this section for generating true random number. Compared to the noise amplification and harvesting method, this approach can be implemented in pure digital fashion which leads to advantages of low-cost, low-power and process-portable. Moreover, compared to the method of using metastability as source of entropy, this frequency-interaction approach is more robust against environmental variations, manufacturing imperfection, and malicious attacks. Furthermore, high programmability makes it friendly for a variety of applications.

The work presented in here is only a preliminary effort of employing frequency interaction as the entropy-generation-mechanism for producing randomness and unpredictability. There are many open issues that needs to be addressed to make this approach more effective and applicable. In particular, the large number of tuning parameters in the two architectures all can have impact on the quality of the final random bitstream. Hence, they need to be scrutinized more carefully. Several open issues are listed below for inviting interested researchers to look into them. They are issues requiring in-depth mathematic analysis, quantitatively if possible or qualitatively when difficult. The goal of this analysis is for creating a general guideline in generating as much entropy as possible.

For frequency-mixing TAF-DPS TRNG (Fig. 6.82), it is helpful to gain understanding on the following.

- The relation between the number of frequency sources and the amount of entropy. In other words, how many TAF-DPS sources are needed to get an acceptable result?
- The optimal relationship among the frequency values of those sources, and the determination of optimal sampling frequency.
- the development of post-processing algorithms

In frequency-tracking TAF-DPS TRNG (Fig. 6.91), there are also several topics deserving further investigation.

- the impact of noise source type (e.g., crystal, MEMS, RO, LCO) on the spectrum content of the random number series
- the impact of the deterministic jitter caused by the mismatch in DCO circuit layout on the spectrum content of the random number series
- the impact of the value of divider N on the spectrum content of the random number series.
- the optimal TAF-FLL operating frequency range for producing maximum entropy
- the development of post-processing algorithms

Paradigm Shift:

In the Time-Oriented paradigm, the classic problem of generating noise entropy of random nature can be investigated in the dimension of time. Entropy can be found in time. This new perspective opens possibility for new circuit architectures.

6.8 Inscribing Temporal Encryption on Spatial MPV Imprints for PUF

With the growing popularity of IoT and 5G, the number of connected devices is forecasted to reach ~40 billion by 2025. With so many interconnected endpoints all over the world, the leakage of device identity could lead to the leakage of physical location and movement, which can be exploited for grievous intelligent terrorism and criminal activity. Security hence has become a crucial dimension in the design of those devices. Its aim is to prevent attacks on our daily activities that rely on the exchange of confidential data, such as those related to finance, purchases, banking, health care, government and etc. Security services of confidentiality, integrity, authentication, nonrepudiation, and digital signature are commonly required to preserve the integrity of data. Nowadays security problem is especially exacerbated by ubiquity of IoT devices (large amount, heterogeneous, everywhere) and complicated user-case scenarios. In recent years, a trend-shift from software security to hardware security is increasingly strengthened as seen in a wide range of applications, from systems demanding a stringent level of security to tightly constrained systems where the power penalty of traditional software-based security methods cannot be afforded. In this increasingly hardware-central environment, PUF (Physical Unclonable Function) has emerged as a foundational IP in the implementation of hardware security.

The current consensus on the term PUF is that it describes a diversity of circuit topologies developed to extract the parametric mismatches from fabrication process of silicon devices for security purposes, such as device authentication and secret key generation. The variation caused by those uncontrollable deviations in the manufacturing process is referred to as MPV (manufacture process variation). It is unique from one chip to another, even all manufactured from the same mask. The imprint of this type of variation is permanent. The complex and random nature of this variation makes PUFs practically very hard to be cloned. Further, the sensitivity of the variation imprint to physical probing renders PUFs to be tamper-resistant against invasive attacks.

In general, process variations introduced during various IC fabrication stages (ion implantation, chemical mechanical polishing, chemical vapor deposition, subwavelength lithography, gas flow, thermal processes, spin processes, microscopic processes, photo processes and etc.) lead to parameter variations (threshold voltage, dielectric thickness, channel length and width, doping concentrations and etc.). Those variations can influence circuit's electrical characteristics (propagation delay, leakage current, voltage drop, aging rate, temperature performance and etc.) that can be observed by user. The validity of PUF is built on those observable characteristics (Chang et al. 2017; Alioto 2019; Herder et al. 2014).

Operation-wise, PUF is a circuit block that responds to an input (challenge) with a repeatable output (response), where the input–output mapping is unpredictable and unknown to external observers. When PUF is used in applications, the responses are not stored but recreated on-the-fly as determined by the chip-specific random variations. This requires the chip to be powered on so that responses can be delivered and pre-described properties can be satisfied. The quality of a PUF is assessed through a number of metrics such as uniqueness, stability, randomness, sensitivity to variations, area and energy efficiency, and etc.

Structure-wise, PUFs are essentially made up of circuits that ideally only magnify certain targeted variations while rejecting the effect of contributions from all other variations. A variety of materials, systems and technologies have been considered as source of uncontrollable randomness for PUFs. Among them, CMOS technology has already been exploited extensively. Non-idealities such as variations in physical dimension, random dopant fluctuation and line-edge roughness that are unique to each die have been used to encode secret information. In operation, the variations of MPV are first reflected as the mismatched analog parameters (delay, power, voltage, current, etc.). Those mismatched parameters are then converted into digital format as the response.

The popular PUF architectures can be classified into two major styles: memory-based and delay-based. Delay PUFs are based on the variations of delays that occurred on wires and logic cells. The effect of such delay variation is distinguished and utilized as useful output either through small random frequency differences among ring oscillators (RO-PUF) or by introducing a race condition between two paths (arbiter PUF, or A-PUF). For memory-based PUF, every cell has its own preferred state every time the memory is powered, resulting from the random differences in threshold voltages. This randomness is expressed in the startup values of "uninitialized" memory. It is unique to a particular memory and hence a particular chip.

There are extensive PUF related studies reported due to its popularity as a security IP. One of the early examples of classic A-PUF circuits is presented in Lim et al. (2005). It is a popular subject in PUF research (Lin et al. 2012; Govindaraj et al. 2020; Zhang and Shen 2021). A-PUF has a large challenge-response pair (CRP) space, making it a good candidate for strong PUF. However, its response might be predicted using a systematic approach, such as machine learning. Regarding RO-PUF, two methods can be adopted for identification: relative frequency relationship and absolute frequency value. For example, frequency comparison between difference rings is utilized to construct PUF response in Liu et al. (2017b). In Barbareschi et al. (2018), it uses absolute frequency value as PUF identification. Although having decent reliability in general, RO-PUF suffers from the problem of instability caused by environment variation, especially temperature. Various techniques have been proposed to mitigate this problem, such as those exampled in Yu et al. (2012), Cao et al. (2015),

Azhar et al. (2018). In Yu et al. (2012), a design using configurable ring oscillators and an orthogonal re-initialization scheme is proposed to improve RO-PUF repeatability. In Cao et al. (2015), a hybrid ring oscillator structure made of regular inverter-ring and negative temperature coefficient inverter-ring is used to mitigate the variation of frequency with temperature. In Azhar et al. (2018), instead of the frequency of ring oscillator, authors suggest a scheme of using duty-cycle for comparison, also for the purpose of improving reliability.

Memory-based PUF is also very popular due to its simple structure and high bit density. However, the CRP space of this type of PUFs is limited, making them weak PUFs. A variety of memory structures can be used to explore PUF. For instant, in Chellappa and Clark (2016), an example of SRAM-based PUF is presented. A DRAM-based PUF is described in Tehranipoor et al. (2017) while a resistive RAM-based PUF is discussed in Kim et al. (2018). In Ben et al. (2019), a spintronics memory-based PUF is proposed.

Besides the two major classes of memory-based and delay-based, other circuit techniques have also been explored. They can be roughly put into the category of analog PUF. Several examples are presented here. In Wan et al. (2015), He et al. (2018), switched-capacitor circuits are used for designing PUF. In Danesh et al. (2020), a symmetrical pair of DACs and VCOs is employed as the base and the resultant mismatch is used for producing PUF output. In Wang et al. (2018), a current mirror structure is employed as the base for PUF. A current array is used in Zhuang et al. (2020). While CMOS technology has now succeeded in many security solutions, the search is still on for PUFs that are secure, cheap, small, and energy-efficient. In Rahman et al. (2017), the authors have discussed the technologies for security beyond CMOS. Three options of phase-change memory, graphene, and carbon nanotubes have been investigated in this front. Furthermore, several emerging variants of PUFs developed in recent years include FinFET PUF, eNVM based PUF, quantum-secure PUF, sensor PUF and etc. (Chang et al. 2017; Alioto 2019; Herder et al. 2014; Kim et al. 2018; Lim et al. 2020).

From the extensive literature research presented above, an observation can be made that all current PUF architectures collect physical entropy from chip and produce response in a bitwise fashion. In other words, within the structure of PUT response, there is no intentionally designated temporal relationship among the bits. It is our belief that, once a temporal interconnection is established among the bits, the security level can be enhanced to next level. This new class of PUF can be generally termed as TeS-PUF for expressing the spirit of "Temporally encoding Spatial imprint". In its implementation, one possible approach of incorporating the temporal footprint into bits is through synthesizing an electrical pulse train. The nonideality occurred during the process of making the waveform of this pulse train can be assembled into a PUF response bitstream. In this section, a new perspective for devising PUFs that follow this principle will be discussed. The conceptual difference between this TeS-PUF and all other PUFs is illustrated in Fig. 6.93. The distinguished feature of TeS-PUF is the introduction of temporal action, in addition to the conventional spatial measure, into the construction of PUF.

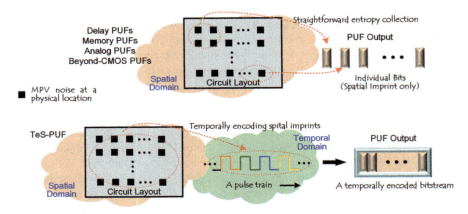

Fig. 6.93 The conceptual difference between conventional PUFs (top) and TeS-PUF (bottom)

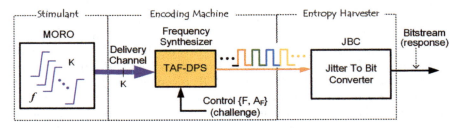

Fig. 6.94 A architecture for implementing TeS-PUF

In Fig. 6.94, a model for implementing TeS-DPS is depicted. It consists of three major blocks: a stimulant generator, an encoding machine and an entropy harvester. The stimulant generator is responsible for generating a group of K temporally ordered signals, which can be used later to inscribe a temporal footmark into the response bitstream. In its simplest form, this generator can be a multi-output ring oscillator (MORO). The encoding machine is materialized by a pulse train synthesizer, namely the emerging frequency synthesizer of Time-Average-Frequency Direct Period Synthesis (TAF-DPS). A pulse train can be generated from TAF-DPS with a designated frequency. Its physical configuration is configured by a pair of parameters F and A_F, where F is the frequency (period) word and A_F is the starting place. The entropy harvester is realized by a jitter to bit converter (JBC). During the processing of making the pulse train, manufacture nonideality will play a role and the electrical characteristic of the pulses hence can vary from chip to chip even the control parameters are all the same. When compared to an ideal reference pulse train (will be discussed later), the physical imprint can be extracted and be converted into a bitstream. This extraction of physical imprint, different from all other PUFs, is influenced by the temporal order inscribed inside the pulse train that is controlled by the pair of $\{F, A_F\}$. The CRP generation will be a two-step process. Originally, the control {F, A$_F$} is provided by user and an intermediate bitstream can be generated at the JBC output from the first run. For the second time, the bitstream from the first run is formatted as a new pair of {F, A$_F$} and be fed into the PUF again. As a result, the temporal relation among the bits of PUF output bitstream is no long related to user input but solely determined by hardware (the MPV nonideality). Working in this way, not only the values of all individual bits are MPV generated but also their temporal interconnections. Robustness against attacks is enhanced and a higher level of security is therefore expected.

This inclusion of temporal encryption upon spatial imprints is unique in the field of PUF design and thus is the key point of this novel perspective. It opens possibilities for a variety of new PUF architectures. The key points of discussion in this Sect. 6.8 will be carried out as follow.

1. A novel perspective of incorporating temporal encryption upon spatial imprints is presented.
2. A prototype on FPGA is presented for proof-of-concept of the principle.
3. From this new perspective, several topics for future research are discussed. Options for new architectures are explored.

6.8.1 Brief Review on the Characteristics of TAF-DPS Operation

A. *TAF-DPS As Pulse Train Synthesizer*

The core of TeS-PUF is the mechanism of establishing a temporal order among the bits in PUF response. One way for accomplishing this is by building a pulse train where a group of pulses are consecutively concatenated. This pulse train then can be the base for creating the PUF response. The emerging frequency synthesizer TAF-DPS, which is based on the Time-Average-Frequency (TAF) concept proposed in 2008 (Xiu 2008b), is an effective tool for this task. Fundamentally different from the popular phase-locked loop (PLL) based frequency synthesizers, TAF-DPS directly builds each of its pulses by waveform construction (that is the reason behind its name of direct period synthesis). Figure 4.8 depicts the working principle of TAF-DPS. From a base unit Δ, as explained in Sect. 4.6, the synthesizer first creates two types of cycles T_A and T_B. When synthesizing a particular frequency f_s (period T_{TAF}), it uses T_A and T_B in an interwoven fashion. Output period (frequency) is calculated as $T_{TAF} = 1/f_s = F \cdot \Delta$ where $F = I + r$ is the frequency control word. Fraction r controls the occurrence probability of T_A and T_B.

A particular circuit technique to implement TAF-DPS is the Flying-Adder frequency synthesis architecture, first developed around late 1990s. Selected examples on its development can be found in Mair and Xiu (2000), Xiu and You (2002, 2003, 2005), Xiu (2009) and Xiu et al. (2013). Over the years, it has been used in various commercial products for on-chip clock frequency generation and many other purposes. Its circuit block diagram is shown in the left of Fig. 6.95. A plurality of K

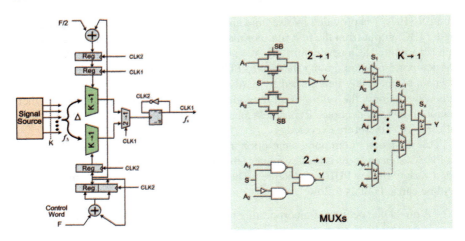

Fig. 6.95 Flying-Adder circuit (left) and its key component MUX (right)

phase-evenly-spaced signals of frequency f_Δ, generated from a signal source, is fed into the synthesizer circuit. The base time unit Δ, is the time span between any two adjacent signals of said plurality of signals. Flying-Adder's output frequency f_s is controlled by the frequency word F, as shown in (6.1.2). In principle, Flying-Adder synthesizer is an edge selector and combiner. At any particular moment, it selects one signal, among those K signals, and passes it to the output. Over time, starting from a place controlled by an initial address, selecting signals by a predetermined schedule enables a pulse train of desired frequency to be created.

From the perspective of each individually-synthesized pulse, the core operation of creating this pulse is the selection of signals, among the K signals, for functioning as the rising and falling edges of the resulting pulse. For this task, the key circuit components are the two K $\rightarrow$ 1 MUXs, whose selection addresses are controlled by the accumulator at the bottom (for the low-portion of the pulse) and the adder at the top (for the high-portion). Structure-wise, K $\rightarrow$ 1 MUX is typically made of multiple levels of 2 $\rightarrow$ 1 MUXs as shown in the right-hand side of Fig. 6.95. The 2 $\rightarrow$ 1 MUX is a foundational component in the field of circuit design. It can be constructed using transmission gates by starting from transistors, as shown in the figure. It can also be assembled using logic cells from an ASIC standard cell library, as depicted.

B. *Creating Temporal Order from MORO*

The waveforms of the K phase-evenly-spaced signals are depicted in the top of Fig. 6.96. As said, they can be generated from a MORO. In circuit design, the K signals, $\varphi_0, \varphi_1, \varphi_2, \varphi_3, \ldots, \varphi_{K-1}$, can be created in several ways for different levels of quality. Refer back to Fig. 6.83, in the top of that figure, a ring oscillator of K = 8 outputs is created from four differential delay stages. The design is crafted from transistors with careful attention paid on balancing the eight outputs (Xiu et al. 2013). In the middle illustration of that figure, an oscillator with 16 outputs is made from 16 cross-couple NAND2 gates. Those gates are directly instantiated from an

6.8 Inscribing Temporal Encryption on Spatial MPV Imprints for PUF

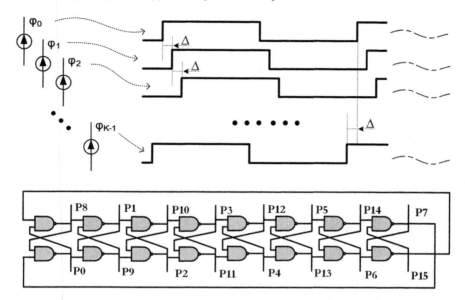

Fig. 6.96 The K stimulating signals (top) and a MORO example of K = 16 (bottom)

ASIC standard cell library. The placement and routing of those cells can be done automatically by tool (Xiu et al. 2019). In the bottom of the figure, an even simpler scheme is presented. A group of 16 signals is generated from a frequency divider (e.g., Johnson counter), which is made of 8 serially-connected flip-flops clocked by a signal CLK (Xiu and Chen 2017). Those options are available to user so that trade-off can to be made in various application scenarios. In the bottom of Fig. 6.96, the case of 16 cross-couple NAND2 gates is shown.

$$T_s(F, j) = t(\varphi_{j+F}) - t(\varphi_j) = \sum_{i=j}^{j+F} \Delta_{i,i+1} \quad (6.8.1)$$

A temporal order can be created by using those signals to make a pulse train. Starting from an initial place A_F (picking a particular signal from the group of K), selecting signals over time by a predetermined schedule enables a pulse train of desired frequency to be created. The control word F determines the length-in-time of each pulse (i.e., the number of Δ), as expressed in (6.8.1) where T_s is the length of this pulse and j is its starting address (A_F is the starting address of the entire pulse train, or the j value of the first pulse). Here, $\Delta_{i,i+1}$ is the time span between signal i and $i + 1$ since, in real world, the time span between any two adjacent signals most likely will not be equal to the theoretical value $\Delta = T_\Delta/K = 1/(K \cdot f_\Delta)$ due to manufacture nonideality.

After a pulse train is created by using the configuration $\{F, A_F\}$, a temporal order is established. This point can be further elucidated with the help of a tool called

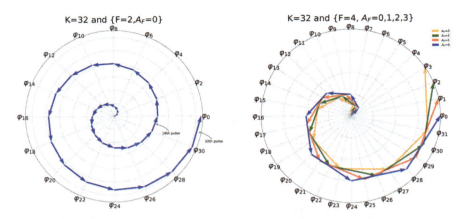

Fig. 6.97 Flight trajectory of {F = 2, A_F = 0} (left) and four trajectories of F = 4, each with its own A_F (right)

space–time flight trajectory. Figure 6.97 left shows a flight trajectory with setting of {F = 2, A_F = 0} on a TAF-DPS circuit of K = 32. In this graph, the spatial information (for example the physical locations of the K NAND gates in Fig. 6.96) are identified at the θ-axis in a polar coordinate system. In this case, a K = 32 MORO is represented and the locations of the 32 MORO internal elements are labeled along the θ-axis (assuming that its circuit layout is settled). Starting from A_F = 0 (the location of φ_0), the flight advances continuously. After every F (=2) Δs, a pulse is created (represented by a segment in the trajectory). After 16 pulses, the flight returns to the φ_0 and a full trajectory is completed. In Fig. 6.97 left, two full trajectories are shown with 32 segments (a pulse train consisting of 32 pulses). Three points can be said about this trajectory.

- A unique temporal order has been established among the segments of the trajectory.
- This trajectory has two facets of space and time. In one hand, it represents a physical path structure when the MORO circuit layout is settled. On the other hand, it has an attribute of time-of-flight (TOF) since it takes time for electrical signals to travel from places to places.
- Each segment in this trajectory corresponds to a distinct electrical path due to its peculiar pattern of visiting the signals φ_j and φ_{j+F}, leading to a distinct electrical delay (please refer to Eq. (6.8.1)).

Those three points are the foundation for building TeS-PUF. In Fig. 6.97 right, we show four such trajectories with same F but different A_F. All the four bear its own unique temporal orders and travels along its unique physical paths. And thus, they possess different electrical characteristics that can lead to different PUF output bitstreams.

C. Periodicity in TAF-DPS Operation

From $T_{TAF} = F \cdot \Delta$, each pulse (each segment on trajectory) is made of F number of Δ. Mathematically, creating such a pulse is an action of counting (the concatenation of multiple Δs). From Fig. 6.95 left, it is seen that this is accomplished by the use of a fixed-size accumulator. There is therefore an inherent periodicity in the advance of MUXs' addresses (the accumulator's output). From work in Xiu et al. (2012), the characteristics of TAF-DPS operation can be summarized as follows.

1. For a TAF-DPS working on K input signals ($K = 2^n$, n is an integer), the control word F can be chosen from the range of [2, 2 K].
2. Each of the K signals has a temporal index for representing its position in temporal domain and, once circuit layout is settled, a physical mark for its position in spatial domain.
3. Each MUX address corresponds to a signal bearing both a temporal index and a physical mark. The A_F can be chosen from the range of [0, K − 1].
4. The process of synthesizing a pulse starts from a value j. The theoretical value of its period is $T_{TAF} = F \cdot \Delta$. In reality, the actual period is "the sum of all the $\Delta_{i,i+1}$ in between". The difference between theoretical and actual values is jitter.
5. The MUXs' address-advance pattern is periodical due to the modulo operation of the accumulator. Its periodicity Ω is expressed in (6.8.2) where I = int(F). The pattern will repeat itself after Ω pulses (Ω segments in trajectory).

$$\Omega = K/\gcd(I, K) \quad (6.8.2)$$

Flight trajectory is a helpful tool that can visualize the TAF-DPS operation and reveal the inherent periodicity buried inside the address movement. Figure 6.98 presents a few more cases of F = 8, 10, 42 and 54. It is interesting that the trajectories

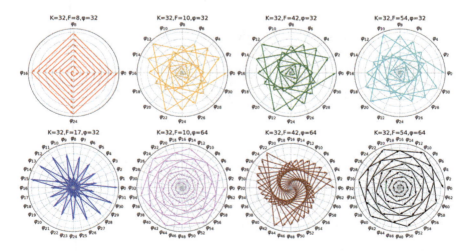

Fig. 6.98 Flight trajectories of F = 8, 10, 42 and 54

of F = 10 and F = 42 look identical (the 2nd and 3rd graphs on the first row, yellow and green). This is due to the reason that those two values have a difference of 32, which is a full MORO cycle of 32Δ. If the trajectories are plotted using a larger θ-scale of 64 (TAF-DPS circuit can virtually use 2 K = 64 identifiable φ_i), the two trajectories will be different as evidenced in the 2nd and 3rd graphs on the second row (purple and brown). It is also interesting to point out that the trajectories of F = 10 and F = 54 are mirror images about the θ-axis. This is because that the two values are symmetric about K = 32. This symmetry is invariant to θ-axis scale as shown in the 2nd and 4th columns of both rows (i.e., the symmetry is preserved in both the 32 and 64 scales).

This periodicity can be useful or detrimental to PUF. In one hand, the periodicity in time domain can show its face as recognizable pattern in frequency domain. This fact can be used to devise new PUF architectures (for example, the patterns in frequency spectrum can be used as PUF response). On the other hand, the periodicity can be a weak point under attack (in this case, however, some techniques could be employed to break the periodicity).

6.8.2 TeS-PUF Architecture and Its Circuit

A. Entropy Harvester

In Sect. 6.8.1, the discussion on TAF-DPS has cleared the path for constructing the first two blocks of simulant and encoding machine (please refer to Fig. 6.94). The last major block, entropy harvester, will be discussed here. A simulation model for TeS-PUF has been created as illustrated in Fig. 6.99. The entropy harvester, or JBC, is materialized by a hypothetical waveform comparator. The synthesized pulse and a

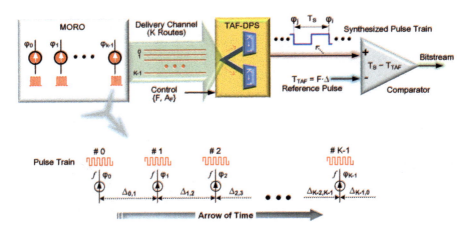

Fig. 6.99 TeS-PUF simulation model

6.8 Inscribing Temporal Encryption on Spatial MPV Imprints for PUF

reference pulse are fed into the comparator. The reference pulse possesses a period (frequency) of the theoretical value $T_{TAF} = F \cdot \Delta$ while the synthesized one has a value of T_s as expressed in (6.8.1). The difference between T_s and the theoretical value T_{TAF} is clock jitter (the so-called period jitter). From one synthesized pulse, one bit can be produced by comparing T_s and T_{TAF}. The value of $T_s - T_{TAF}$, which can be either positive or negative, determines the value of this PUF response bit (zero or one). Meanwhile, a temporal order is established among the bits since the pulses are created consecutively as a train.

B. Sources of MPV Noise

The manufacture induced jitter resulted from this waveform synthesis process is incorporated in this pulse train and is subsequently converted into a bitstream as PUF response. This type of jitter is caused by the nonideality in manufacture. Hence, it is deterministic. When TAF-DPS is used for building PUF (not frequency synthesizer as it is originally invented for), the circuit is designed and layouted in such a way that this type of deterministic jitter is intentionally amplified and it overwhelms the entire jitter spectrum. The generated bitstream therefore bears the imprint of the chip.

The circuit represented by Figs. 6.94 and 6.99 can induce two types of MPV noises: spatial and temporal. Figure 6.100 is the illustration of their causes. From the discussion on Fig. 6.95, the TAF-DPS circuit can be viewed as a fictitious switch made of a group of controlled channels. The electrical edges from the MORO are selectively passed to the output for making the synthesized waveform. From the MORO to the final output port, the physical route that each stimulant travels can be modeled as an electrical path with a delay of σ_i, as depicted in the top of Fig. 6.100. Theoretically, all the paths share a common value σ since their design and layout are fully symmetrical. However, due to manufacture imperfection, each path is expected to have its own delay of σ_i. This is the cause of spatial MPV noise.

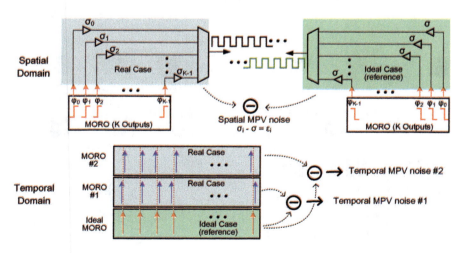

Fig. 6.100 The spatial MPV (top) and temporal MPV (bottom)

194   6 Old World and New Insight: Solving Problem with a Gestalt Switch

In the bottom part of Fig. 6.100, the cause of temporal imprint is illustrated. It happens inside the MORO. Ideally, all the K stimulants (please refer to Fig. 6.96) shall be evenly spaced in a full MORO oscillation cycle. Due to manufacture imperfection, this however most likely will not be the case. The deviations of their phases from the theoretical locations are temporal MPV noise. It is called temporal noise because it is always there and can be directly observed if an appropriate instrument is used (such as an oscilloscope). In contrast, the spatial MPV noise will not "come into being" unless an electrical signal travels through the physical path and produces an electrical delay. Both the spatial and temporal noises will influence the resultant PUF response bits, in the values of the bits or order-of-sequencing among the bits.

C. *Simulation*

Using the model shown in Fig. 6.99, a Simulink model is created in Matlab for studying the behavior of TeS-PUF. To make the waveform of a to-be-synthesized pulse, two of the K stimulants are selected to form the first and second rising edges (to be precise, another one is needed for the falling edge in between). The selected signals are marked as φi for the first edge and φj for the second. In this process, F and A_F are the control variables. The time span between signals φi and φj is T_s. Based on the analysis carried out around Fig. 6.100, MPV noises are added into the blocks of MORO, delivery channel and TAF-DPS circuit as independent noise sources. It is assumed that those noises follow normal distribution. Their mean values are zero while standard deviations are adjustable for representing various mismatch scenarios. It is worth mentioning that the delivery channel can be a major playground for amplifying the effect of spatial MPV noise due to the large number of wires and the large length of those wires. After adding those noises, the actual time span between any two adjacent stimulants (including spatial and temporal MPV noises) is no longer the universal $\Delta = T/K$ but a unique value of $\Delta_{i,i+1}$.

Figure 6.101 shows a simulation result from a TeS-PUF of K = 32 with setting of {F = 11, A_F = 0}. The theoretical period value is $T_{TAF} = F \cdot \Delta = 11\Delta$. In the

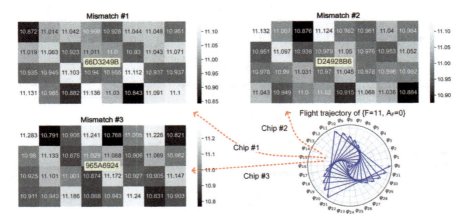

Fig. 6.101 Distribution from three manufacture scenarios

simulation, there are three mismatch scenarios simulated (three different seeds for the distributions, representing three different fabrications of the same design). For each case, a consecutive 32 pulses are synthesized. The 32 actual values of period are displayed in a 4 × 8 Gy map with the number labeled on each element. After the comparator, the output bitstreams are 66D3249B, D24928B6 and 965A6924, respectively. The flight trajectory shown on the bottom right is the space–time regulation configured by {F, A_F}, which can be used as the base for making PUF challenge. This simulation is just for illustration purpose. It shows that the structural imprint of the chip can be harvested and discriminated by the TeS-PUF.

The information displayed in Fig. 6.101 is MPV signature in time domain. MPV can also leave a mark in frequency domain. TAF-DPS, although invented primarily as a frequency synthesizer, can also be used as a tool for reflecting structure-related information to frequency spectrum. For TeS-PUF, as said, the circuit is intentionally implemented in such a way that it provides plenty of room for MPV to occur. Each configuration of MPV, when govern by a pair of {F, A_F}, can produce its unique spectrum signature. Figure 6.102 includes four spectrum plots for F = 11. The simulations are carried out under the setting of $\Delta = 1$ ns. Hence, $f_s = 1/T_{TAF} = 1/(11\Delta)$ = 90.9 MHz. The plot on the top-left is the spectrum of an ideal case (no mismatch). The plots on other three graphs correspond to three different mismatch scenarios. As discussed in Sect. 6.8.1, periodicity Ω reflects a regular action on the movement of address. From spectrum domain, this action is a frequency modulation on the main tone. Hence, the effect of MPV mismatch will appear as spurious tones spaced at f/Ω (please refer to Chap. 5 of Xiu (2012a)). As shown in all the three plots, there

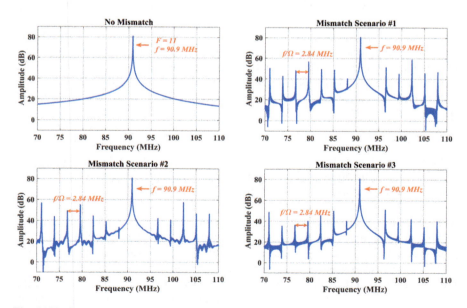

Fig. 6.102 Spectrum signatures of F = 11

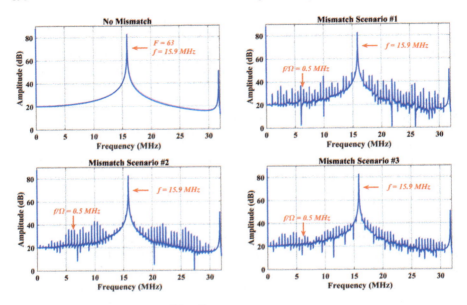

Fig. 6.103 Spectrum signatures of F = 63

are spurs spaced at 2.84 MHz ($\Omega = 32/\gcd(11, 32) = 32 \rightarrow f/\Omega = 90.9/32 = 2.84$). Although all the spurs sitting in the same positions horizontally, their magnitudes are different. Those magnitudes are the signatures of the MPV noises, each uniquely associated with a chip.

Figure 6.103 presents another case of F = 63. In this simulation, the mismatch configurations are the same as that of Fig. 6.102. Using same calculation, main tone locates at $f = 1/(63\Delta) = 15.9$ MHz and spurs are spaced at $f/\Omega = 15.9/32 = 0.5$ MHz. The point of this simulation is that, for the same MPV noise, the spectrum signature is difference if the control F (challenge) is changed. This is a necessary condition for spectrum-domain PUF.

D. *The Circuit*

Figure 6.94 is the generic model for TeS-PUF which shows its principle while Fig. 6.99 is a simulation model further elucidating its mechanism. They however cannot be directly implemented on chip, mainly due to the difficulty in constructing the JBC. There is no easy way to realize an ideal reference pulse train completely noise-free (this is actually impossible since, fundamentally, time in electronics is embodied through voltage transition). To circumvent this obstruction, a solution is presented in Fig. 6.104. Here, another TAF-DPS is employed and its output is treated as the "ideal reference pulse train". The beauty of this solution lies in the fact that it does not contradict the spirit of PUF at all. As shown, the two TAF-DPSs share the same control $\{F, A_F\}$. In principle, the two resultant pulse trains shall have exactly the same physical configuration and electrical characteristic. Due to manufacture nonideality, however, this will not be the case and the twos are different. When the

6.8 Inscribing Temporal Encryption on Spatial MPV Imprints for PUF

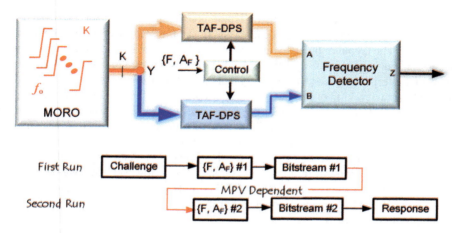

Fig. 6.104 Circuit implementation of TeS-PUF

essence of PUF is concerned, as long as structure-induced information is extracted, it does not care whether the reference is ideal or not.

In operation, as illustrated in the bottom part of Fig. 6.104, the procedure for creating a pair of CRP is carried out in two steps. In first run, a challenge is made from $\{F, A_F\}$. It produces a bitstream after being fed into the TeS-PUF. This bitstream is subsequently formatted into a new $\{F, A_F\}$. This new one is fed into the TeS-PUF at the second run. The output from the second run is the TeS-PUF response. As can be appreciated, the new $\{F, A_F\}$ is hardware generated. It is MPV dependent. It cannot be observed. This is fundamentally different from the post-processing algorithms that are often found in other PUFs. For this reason, the relation between the challenge and the response is not traceable by logical or mathematical reasoning, however sophisticated that might be.

By this configuration, a timing race is established between the two synthesizers. The comparator in Fig. 6.104 is realized by a frequency detector that detects slow-fast relation between the two pulse trains and generates one bit for each pair of pulses. This mechanism of timing race is more or less similar to that used in A-PUF. However, different from A-PUF where only two edges are compared in the fashion of one-time-operation, two pulse trains are compared in the case of TeS-PUF. This scenario resembles the continuous frequency comparison in FLL (frequency-locked loop) (please refer to Sect. 6.6) or phase comparison in PLL. By this token, a frequency detector is used here to evaluate the two pulse trains and generate a bitstream. The bang-bang binary detector commonly used in CDR (clock data recovery) circuit is borrowed. The most well-known bang-bang detector is the Alexander detector (Alexander 1975). It uses four flip-flops to do "three-point sampling" for comparing the edges. It can generate three output states based on the edges' relation: early, late or no-transition. In TeS-PUF, the no-transition state will never occur since we are comparing two clock-like signals (there is always transition on every cycle).

In implementation, to amplify manufacture mismatch, the transistors' sizes in the TAF-DPS and MORO circuits can be made small and their layouts can be made more spread-out. The layout of the delivery channel can be made narrower and longer than necessary (but being kept symmetrical) so that there is more chance for MPV noises to occur. It is especially worth mentioning that the inclusion of a MORO is significant. It introduces a new kind of imprint, the temporal imprint (please refer to the drawing at the lower part of Fig. 6.100), into the design of PUF. More significantly, it is the fuse of a sequencing engine that inscribes temporal order into a sequence of bits. Only by incorporating such a device that has an inborn sequencing capability, the temporal order can be brought into being in the TeS-PUF response. As a ring oscillator, the MORO circuit is prone to frequency injection attacks. However, viewing from the perspective of sequencing engine, the MORO outputs are just fuse to fire the engine. Neither the MORO's absolute frequency nor the noise on its outputs can impact the TeS-PUF operation in any harmful way.

6.8.3 Proof-Of-Concept on FPGA

The proposed TAF-DPS-PUF can be packaged as either a hardware or a software IP. It can be designed into a large ASIC as an on-chip security IP for SoC applications using custom design method as the rest of chip. On another front, due to its full digital nature, TeS-PUF can be conveniently marketed as a software IP. This is especially interesting when it is used as a security IP for FPGA based designs since TAF-DPS circuit can be readily reconfigured on the fly. In this section, a basic version of TeS-PUF is implemented on an FPGA platform for proof-of-concept. A group of 20 Kintex UltraScale + chips, from Xilinx, have been used for validation.

A. *Uniqueness and HD Distribution*

$$U = \frac{2}{m(m+1)} \sum_{i=1}^{m-1} \sum_{j=i+1}^{m} \frac{HD(R_i, R_j)}{n} \times 100\% \qquad (6.8.3)$$

PUF uniqueness can be described by using the average Hamming distance (HD) as expressed in (6.8.3), where m is the number of PUF instances, n is the number of bits of response, $HD(R_i, R_j)$ is the Hamming distance between two responses of R_i and R_j collected from two different PUF instances. All the responses are excited using the same challenge. In this experiment, 100 of K = 32 TeS-PUF designs are implemented on 20 different FPGA chips with 5 TeS-PUF instances on each chip. For each integer value of F in [2, 33], $A_F = 0$, a bitstream of length K = 32 is generated as the PUF output. A bitstream of 1024 bits is created by concatenating those 32 outputs. For the 100 instances, their respective bitstreams are used to calculate the inter-PUF Hamming distance by (5) where m = 100 and n = 1024. There are 4950

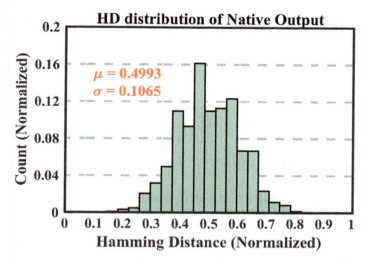

Fig. 6.105 Inter-PUF HD from 100 instances implemented on 20 chips

pairs of 1024-bit responses. The histogram of the 4950 HD values is illustrated in Fig. 6.105. The mean μ and standard deviation σ of the best-fit Gaussian distribution to the histogram are calculated as 0.4993 and 0.1065, respectively. The uniqueness of these responses is 0.4993.

B. *Reliability*

$$\text{BER} = 1 - \text{Realiability} = \frac{1}{k}\sum_{j=1}^{k}\frac{\text{HD}(R_n, R_j)}{m} \times 100\% \qquad (6.8.4)$$

Reliability concerns the variation of PUF response under different working conditions. Two major factors are voltage and temperature. Reliability is defined through bit error rate (BER) as expressed in (6.8.4), where R_n is an m-bit response produced by a PUF instance under normal operating condition. The same challenge is then applied k times to the same PUF instance to obtain the responses R_j. The average, calculated from (6.8.4), is used as the BER. The operation can be performed under different voltage or temperature. On the left of Fig. 6.106, data for reliability versus supply voltage is displayed. The voltage is swept from 0.65 to 1.05 V with a step of 0.05 V. The nominal voltage is 0.85 v. At each voltage, k = 100 tests are repeated on the same PUF instance using same challenge. The average BER sensitivities against supply voltage are calculated as 1.6395 and 1.606%/V at the respective two sides. The plot at the right-hand side of Fig. 6.106 shows the similar test on temperature change. The temperature range is from −15 to 75 °C. R_n is the response obtained at the nominal condition of room temperature 25 °C. The BER sensitivities against

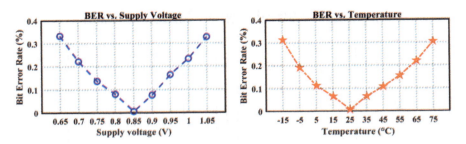

Fig. 6.106 Reliability: supply voltage (left) and temperature (right)

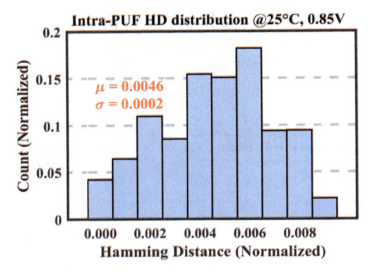

Fig. 6.107 Intra-PUF HD distribution at normal condition

temperature are 0.008 and 0.006%/°C at the respective two sides. Under nominal condition of 0.85 V and 25 °C, the intra-PUF HD is displayed in Fig. 6.107. The mean value μ and standard deviation σ of intra-PUF distribution are 0.0046 and 0.0002, respectively. The separation of intra- and inter-HD is reported as μintra/μinter = 109.

C. *Autocorrelation*

$$\text{ACF}_k = \frac{\sum_{i=1}^{N-k}(r_i - \bar{r})(r_{i+k} - \bar{r})}{\sum_{i=1}^{N}(r_i - \bar{r})^2} \tag{6.8.5}$$

6.8 Inscribing Temporal Encryption on Spatial MPV Imprints for PUF

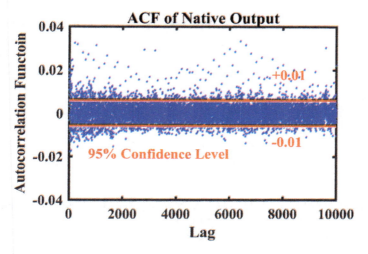

Fig. 6.108 ACF correlogram

ACF (autocorrelation function), as expressed in (6.8.5), is a well-known operation for determining the degree of independence among bits in a bitstream. For PUF design, ACF test result can indicate its capability of resisting autocorrelation attack. In this experiment, the ACF test is performed on the data of 102,400 bits collected from the 100 TeS-PUF instances implemented on the 20 chips. The resulting correlograms (autocorrelation plot) are shown in Fig. 6.108. The ACF result of native output is displayed. The maximum of lag for illustration is chosen as 10,000. It is observed that, within 95% confidence bound of a Gaussian distribution, the ACF value fluctuates between -0.01 and 0.01.

D. *On-the-Fly Reconfigurability*

The TeS-PUF can be easily reconfigured on FPGA when situation arises. One of the most convincing reasons for doing this is to enhance security level of a system after the PUF has been used in the system for a significant period of time. This goal can be easily fulfilled in our case. Figure 6.109 includes three responses generated from

Fig. 6.109 Speckle diagrams from three circuit sizes of $K = 16$, 32 and 64

Table 6.7 Resource usage

Circuit size K = 2^n	LUTs	Flip-flops	Bit/cost[a]
8 (n = 3)	48	34	0.09
16 (n = 4)	73	46	0.134
32 (n = 5)	125	62	0.171
64 (n = 6)	171	86	0.249
128 (n = 7)	284	126	0.312
256 (n = 8)	499	198	0.367
512 (n = 9)	931	334	0.405

[a] Cost: number of (LUT + Flip-Flops), Bit: K

three circuit sizes of K = 16, 32 and 64, respectively, shown as speckle diagrams from challenges generated by using F = 9.5 for all the three cases. The change of configuration is conveniently done by a straightforward procedure of HDL reprogramming and remapping to FPGA. As can be appreciated, the difficulty of reverse-engineering the response is significantly increased.

E. *Resource Usage*

The 2nd and 3rd columns in Table 6.7 are the numbers of LUT and Flip-flop used when TeS-PUF is implemented on Kintex UltraScale FPGA, for several configurations of K = 8, 16, ..., 512. The numbers in this table correspond to the circuit depicted in Fig. 6.95. The 4th column of Bits/Cost is the number of response bits (i.e., K) divided by the sum of 2nd and 3rd columns. For weak PUF applications, smaller K is suggested (e.g., K ≤ 32). For strong PUF applications where a large CRP space is required, larger K can be used.

6.8.4 Directions for Future Research

In previous discussion, we have presented a new perspective of temporally encrypting spatial imprints through the temporal regulation performed in a frequency synthesizer of TAF-DPS. A prototype TeS-PUF has been implemented on FPGA for validating this principle. This novel perspective opens new possibilities for devising PUF circuit. Due to the freshness of this idea and the space limitation of this work, many issues have not been touched. In current subsection, we discuss several directions that can potentially lead to new PUF architectures, all enabled by the spirit of this perspective. Due to space limitation and the nature of issues-to-be-resolved-in-future, they are discussed only briefly. Interested researchers are invited to looking into those issues.

A. *TeS-A-PUF for Strong PUF*

The circuit discussed in Sects. 6.8.2 and 6.8.3 is the basic version of the proposed class of TeS-PUF. The configuration of Fig. 6.104 is regarded as weak PUF since its entropy space is limited even through the entropy space can be significant when

6.8 Inscribing Temporal Encryption on Spatial MPV Imprints for PUF

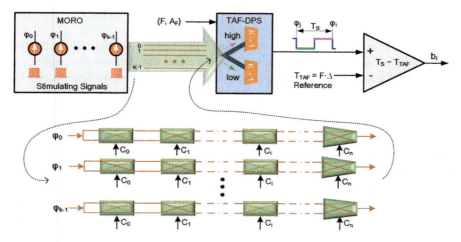

Fig. 6.110 The configuration of TeS-A-PUF

K takes a large number (for instance, 512). To enlarge entropy space and make it a strong PUF, the idea of A-PUF can be merged into TeS-PUF. Figure 6.110 presents one of such ideas. The delivery channel between the MORO and TAF-DPS contains K physical routes. For being a PUF, those routes can be intentionally made long and narrow so that MPV-induced mismatch can have more chance to occur. To go one step further, as shown in Fig. 6.110, those routes can be made in A-PUF fashion. The architecture resulted from this combination is called TeS-A-PUF. It can dramatically enlarge the entropy space and potentially make it a strong PUF. Several important issues, however, still need to be worked out, such as the analysis of CRP space.

B. *Collecting both Spatial and Temporal Imprints*

Referring back to Fig. 6.100, the TeS-PUF contains two types of structure imprints: spatial and temporal. Different from the spatial imprint that is resulted from geometrical mismatch (which is the case for almost all the existing PUFs), the temporal imprint is the temporal deviation on the phases of the stimulants from their theoretical location in time. Although, by tracing to the ultimate physical cause, those temporal deviations are caused by geometrical mismatch (inside the MORO) as well, it distinguishes itself from the spatial imprint by the fact that it is active (already there) while the other is passive (requiring a signal's involvement to show its face). However, the circuit of Fig. 6.104 is not able to harvest this temporal imprint. This is due to the reason that this temporal MPV noise originated from the MORO is a common factor to the two TAF-DPSs and thus is canceled by the symmetry in circuits (before the Y point, all are common factor to the two synthesizers). In other words, using current circuit, the effect of this temporal imprint cannot be sensed at the PUF output. Currently, the MORO only functions as the fuse of a sequencing engine. In future, we want its temporal noise contribute to the PUF response as well. The challenge here is to find a better way in creating this "reference pulse train" (Fig. 6.99).

C. *Introducing Chaos into TeS-PUF*

The temporal regulation in TAF-DPS is controlled by the {F, A_F}. Normally, the space–time trajectory is well-behaved and can be predicated (however, only when the {F, A_F} are known). To improve security level, chaos can be introduced into TeS-PUF by using the fraction part of the F. When fractional number is involved, each time the waveform of a pulse is constructed, the advance of the address will have two possibilities of I and I + 1 (instead of just one choice of I). The control of their appearances is determined by fraction r (please refer to Sect. 6.8.1). If r is randomly selected, chaos can be introduced into the system. A feedback mechanism can be employed and, each time, the fraction r is generated in real-time from the current PUF output which is MPV dependent. After multiple times of iteration, a chaotic space–time trajectory can be formed. As a result, the PUF response is harder to be decoded or predicted. More serious work is required to materialize this plan. The goal is to thwart reverse engineering attacks.

D. *Spectrum Signature*

Currently, the MPV imprint is first reflected into time domain and then converted into bitstream. An alternative is the spectrum domain. The spectrum signatures displayed in Figs. 6.102 and 6.103 are uniquely associated with the fabrication of the chip. It thus provides another path for exploration. It is worth mentioning that this is only possible with TeS-PUF since, in all other PUFs, the response consists of only individual bits (there is no temporal relation among the bits and thus no spectrum). In-depth exploration is required to materialize this idea.

Paradigm Shift:

In the emerging Time-Oriented paradigm, a particular problem of collecting noise entropy from chip for security purpose can be inspected from a new angle. A new kind of entropy is recognized by the interaction of space and time.

6.9 TAF-DPS as a PWM Signal Generator

Pulse width modulation (PWM) is a method of controlling the average amount of "information" delivered by an electrical signal through an operation of chopping the signal into discrete parts. The "information" specified can be electrical charge, electrical voltage, time duration, digital message, energy (power) and etc. The very early use of PWM is for control servomechanisms (i.e., servo control), an automatic device that uses error-sensing negative feedback to correct the action of a mechanism. Later, PWM becomes very popular in the application of power delivery where the amount of power delivered to a load suffers much less loss when compared to the linear power delivery by resistive means. PWM is also often used in voltage regulators. By switching voltage to the load with the appropriate duty cycle, the output will approximate a voltage at the desired level. In telecommunications, PWM is a

6.9 TAF-DPS as a PWM Signal Generator

form of signal modulation where the widths of the pulses correspond to specific data values encoded at one end and decoded at the other. PWM is also found in audio application for achieving certain audio effects and audio amplification. Moreover, PWM has played an important role in soft-blinking LED indicator, which is a less disturbing solution than "hard-blinking" on/off indicator. In recent years, PWM has been used in stochastic computing as stochastic number generator.

As its name suggests, the media in PWM is an electrical pulse of square waveform. This pulse has two distinguished states of high and low. The ratio between the time durations of those two states is the intended information. This ratio is reflected as a numerical value called duty cycle, symbolized by the width of the pulse (the time duration of "high" state). The traditional way of generating PWM signal is by using a sawtooth or triangle waveform and a comparator. The sawtooth or triangle waveform, called modulation waveform, is compared against a reference signal. When the value of the reference signal is higher (or lower) than that of the modulation waveform, the PWM signal is in the high state, otherwise it is in the low state. In its implementation, the comparison between the modulation and reference signals can be carried out in a more sophisticated fashion, such as delta modulation and delta-sigma modulation.

With the advance of digital technology, nowadays, PWM signal is often generated by using digital circuits. A simple digital counter driven by a clock signal can do the trick. At every clock tick, the content of the counter incrementally grows. When the counter value reaches a predetermined reference value, the PWM output changes state from high to low (or low to high). It is reset at the end of every period of the PWM operation. This technique is appropriately referred to as time proportioning, for indicating the fact that certain proportion of a fixed cycle time is spent on the high state.

This principle of time proportioning can be easily realized by the TAF-DPS technology. As a circuit level tool for direct waveform construction, TAF-DPS is naturally suitable for being a PWM signal generator. Compared to the regular digital counter method, the key advantage of TAF-DPS PWM modulator is finer time resolution that can subsequently lead to higher performance when controlling the load attached at its output. Moreover, it has high flexibility in generating a variety of waveforms of various timing characteristics. In particular, it can generate three types of PWM waveforms, Type-I of varying frequency, Type-II of fixed-period-varying-duty-cycle and Type-III of fixed-pulse-length-varying-period.

In Fig. 6.111, the general architecture of TAF-DPS PWM is illustrated. The profile for a desired PWM signal is first generated by a profile generator and then be converted into frequency control word, which is fed into the TAF-DPS circuit. It subsequently produces the corresponding PWM signals. In Fig. 6.112, the three types of waveforms are displayed. Those waveforms are captured from the output of a real chip.

As an example, we will use electrical voltage regulator for illustrating the advantage of TAF-DPS PWM modulator. DC-to-DC converter is an electronic circuit or electromechanical device that converts a source of direct current (DC) from one voltage level to another. It is a type of electric power converter. A very popular type

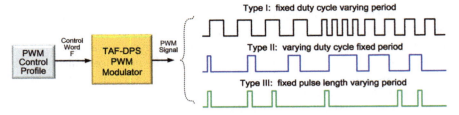

Fig. 6.111 The principle of time proportioning is realized by TAF-DPS in generating three types of PWM waveforms: varying frequency (top), fixed-period-varying-duty-cycle (middle) and fixed-pulse-length-varying-period (bottom)

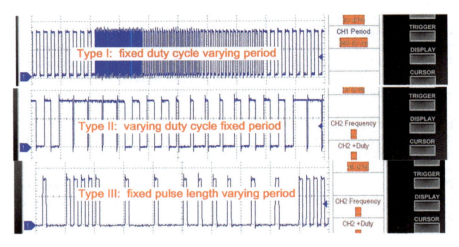

Fig. 6.112 Three types of TAF-DPS PWM waveforms captured from a real chip

is switch-mode DC-to-DC converter. One of the key parts of a switched-mode DC-to-DC converter, whether the energy storage element is in the fashion of magnetic field (inductor, transformer) or electric field (capacitor), is the FET switch which operates at certain frequency. The ON/OFF of this switch is controlled by a PWM signal.

For modern design in the environment of advanced digital technology, this switch control can be realized by a digital circuit. The TAF-DPS circuit is ideal for this case since, as shown in Fig. 6.111, it is a very powerful PWM. Listed below are some of the features enabled by the TAF-DPS in controlling a DC-to-DC converter.

- The output frequency controlling the FET switch can be arbitrarily set (arbitrary frequency generation) which is good for designing various conversion efficiency.
- More, the output frequency from TAF-DPS can be quickly switched in a quantifiable fashion, which can be an enabler for innovative solutions in the designing of DC-to-DC converter.

- The duty-cycle of the TAF-DPS output pulses can be programmable, easily through software. This is good for programming the input/output voltage ratio.
- In practice, the operating frequency of DC-to-DC converters is low, usually below hundred MHz. At this frequency range, TAF-DPS circuit can be easily realized with very low cost.

One of the drawbacks in using switch-mode DC-to-DC converter is the electromagnetic interference (EMI) generated by the converter and radiating to the environment. In other words, the high frequency switching associated with a PWM pulse train presents a EMI problem to the system that uses the DC-to-DC converter as its power supply. In Sect. 6.4, we have presented a spread spectrum clocking (SSCG) technique to reduce the clock EM radiation. This technique can be readily applied here in DC-to-DC converter design.

In short, TAF-DPS is ideal for being used in the design of DC-to-DC converter, both for innovative architecture-level design and for EMI reduction. Furthermore, DC-to-DC converter is just an example. In electronic system design, PWM is a widely used technique which plays crucial role in many applications. It is therefore expected that TAF-DPS will find many new uses as a powerful PWM.

Gestalt Switch:

The resolution of the two long-lasting problems (arbitrary frequency generation and instantaneous frequency switching) enables new treatment on old problems. In this section, the traditional problem of pulse width modulation is simplified by using a new tool.

6.10 A Case of Signal Processing in Time Dimension: Deterministic Stochastic Computing

As intensively discussed in previous chapters, one of the major driving forces for rapid economic growth in modern epoch is the Moore's Law, which has governed the development of microelectronics for more than seven decades. The success of Moore's Law relies heavily on Boolean logic as its core mathematic foundation. From an abstraction point of view, Boolean functionality is first defined in a deterministic logical layer. It is later translated into a physical layer that uses electrical voltage to convey information. In practice, the depiction associated with Boolean operations is precise in the light of fact that the voltages are interpreted as the exact logic values. As Moore's Law progresses into deep submicron realm, this practice however becomes ever more costly in emerging technologies. All forms of noise and uncertainty occurred in the physical layer have to be compensated by using complex and energy-hungry circuits with large design margins. In other words, the cost of using voltage to faithfully represent logic values and carry out Boolean operations is ever increasing. Consequently, a variety of new researches are emerging to find different approaches for computation in the era of beyond Moore's Law (Xiu 2019a). One of the directions is to handle device uncertainty in a more efficient way. Stochastic Computation (SC), which exploits probability theory to deal with variations, is a promising technique in this direction.

In SC domain, instead of being deterministically represented, numbers and signals are encoded in a probabilistic form (Gaines 1967a). The history of SC starts from the 1950s when Von Neumann proposed the initial ideas concerning probabilistic and error-tolerant electronic design (Neumann 1952). This work has inspired a great deal of subsequent researches. In the mid-1960s, influenced by the developments in both analog and digital computers at that time, SC was explored concurrently in the U.K. (Gaines 1967b, 1969) and the U.S. (Poppelbaum 1976 and Ribeiro 1967). Several general-purpose stochastic computers ever actually implemented were built around that time. From the development of those SC machines, numerous shortcomings of the technology have been uncovered. Among them, it is observed that "short sequences are untrustworthy" and that a major drawback of SC is low bandwidth and therefore low computational speed (Poppelbaum 1976).

On the positive side, it is generally agreed that, compared to the deterministic binary computing, SC has two distinct advantages of low hardware complexity and fault-tolerant computing. Although the stochastic representation of a number is usually much lengthier than conventional binary radix, complex operations however can be performed with remarkably simple logic (Qian et al. 2011). For instance, a single AND gate can perform multiplication with unipolar representation [0, 1] and a single XNOR for multiplication in bipolar representation [-1, 1]. A multiplexer (MUX) can realize the functionality of scaled addition and subtraction. Besides simple and compact logic, SC offers the advantage of tolerant to soft transient errors (such as bitflips and supply voltage ringing). In a noisy environment, the phenomena of bitflip and supply ringing can affect all the bits with equal probability. Within the conventional binary system, the high-order bits represent a larger magnitude. Consequently, faults in these bits can produce larger errors. In contrast, in SC domain, all the bits are equally weighted. Hence, a single bitflip has a much smaller impact on accuracy (Alaghi et al. 2018).

In its circuit implementation, SC offers a third advantage of parallelism (massively parallel processing). This is attractive for many hardware-demanding applications, including digital image processing (Li et al. 2014; Alaghi et al. 2013), neural networks (NNs) (Morro et al. 2018; Brown and Card 2001a, b; Ardakani et al. 2017; Liu et al. 2019, 2020), soft polynomial solving and filtering (Liu and Parhi 2016; Saraf et al. 2014), as well as modern error-correcting coding and decoding (Gutnik and Chandrakasan 2000; Tehrani et al. 2006). Furthermore, again in its circuit implementation, SC provides the possibility to create system of scalable or progressive precision where shortened computation can already provide an early estimate of a target value. This concept allows the trading of precision for energy at run-time, a feature that can be well exploited in emerging ultra-low energy applications (Alaghi and Hayes 2014).

In SC implementation, the key building block is the stochastic number generator (SNG) that produces a random number series of designated possibility of logic one (or zero). The SNG by itself is a large block compared to the other SC elements,

6.10 A Case of Signal Processing in Time Dimension: Deterministic ...

occupying most of the area of the whole design (Najafi et al. 2019). For example, in the cases of Ichihara et al. (2014), Mohajer et al. (2019), SNG circuits consume around 80% or even 90% of the total area. The most commonly used approach for designing SNG is by comparing the output of a true random number generator (TRNG) or pseudo random number generator (PRNG) to a fixed number of designated value. From the comparison, a series of bits are generated wherein the probability of one's occurrence approximates that given value, as shown in the left of Fig. 6.113. The output of TRNG or PRNG is required to be uniformly distributed (equal probability of one and zero for each bit). The construction of such high quality TRNGs or PRNGs are nontrivial and area-hungry. Moreover, its operations for SNG requires many running cycles, consuming significant power. There is hence a strong motivation to improve (Salehi 2020) or eliminate the SNG (Temenos and Sotiriadis 2021; Najafi et al. 2017).

The idea proposed in Najafi et al. (2017) is especially interesting in its way of representing value. It encodes value in the duty cycle of a periodical signal, the PWM (pulse width modulation) signal. Computation is carried out through logic operation between two such PWM signals. It is different from other conventional SC by the fact that information (i.e., the value) is encoded directly in time. This idea can be traced back to old time when analog circuit design technique is predominant for sophisticated signal processing (Toral et al. 2000). The idea proposed in Najafi et al. (2017) still heavily depends on analog PWM techniques and thus is not well suited for advanced digital technologies in modern era. Due to this circuit implementation issue, although utilizing time as information, it is however not developed into a general-purpose platform for time domain SC.

In Najafi et al. (2019), a new concept of deterministic stochastic computing is proposed. The most important observation from this work is that randomness is not an absolute requirement for stochastic computing. Instead of representing value by random bitstream, value can be expressed by using the so-called unary bitstream where all the 1's are grouped together at the beginning or end. Similar to pseudo-random bitstream generated by PRNG, unary bitstream is a deterministic bitstream. It

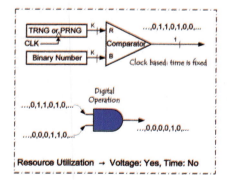
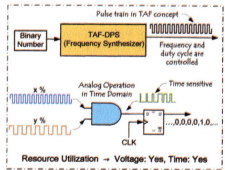

Fig. 6.113 The principal difference between conventional time-insensitive SC (left) and new time-sensitive SC (right)

however has a much cleaner format: all 1's followed by all 0's. This opens a variety of new possibilities for efficient circuit implementation. More significantly, it is elucidated in Najafi et al. (2019) that, when properly structured, this deterministic SC can be completely accurate, free of random fluctuation.

The example in Najafi et al. (2017) and the framework of Najafi et al. (2019) has inspired us to develop a generic circuit platform for exercising stochastic computing in the dimension of time. To fulfill this desire, a fundamental support from circuit level is a necessity. In the past several decades, most of the circuit technologies are developed around the idea of manipulating voltage, through the means of either analog or digital. As Moore's Law moves into deep submicron realm, signal processing in voltage domain becomes ever more error-prone and costly. Now, "It is time to use time", as advocated in Xiu (2019a). Stochastic computing is one of such promising lands where manipulation in time can be a more efficient option for processing signal. In particular, for the current task, we need a digital-friendly generator for producing PWM-like signal. More precisely, we demand a frequency synthesizer that can generate many frequencies and can change its output frequency quickly after a command is received. This frequency synthesizer has to be digital oriented in its implementation so that it can be merged into the picture of large-scale integration predominating the modern era. The TAF-DPS described in Sect. 4.6, with the features of "arbitrary frequency generation" and "instantaneous frequency switching", achievable simultaneously for a given design, is a circuit-level enabler for system-level innovations. This innovative tool from circuit level can be squarely suited for exploiting potential in the dimension of time for stochastic operation.

The proposal of time domain SC is depicted in the right-hand side of Fig. 6.113. Compared to the conventional time-insensitive SC, the proposed stochastic operation is time-sensitive. In other words, for an electrical pulse train, time is used as information carrier while voltage only plays an assisting role of distinguishing states. As shown, using the concept of TAF, TAF-DPS is able to generate an electrical pulse train as the signal whose frequency and duty cycle can be precisely controlled. When two of such signals interact through a logic operator, a time sensitive waveform is resulted where the proportion of time between the two different states of high and low is the intended value of calculation. As can be appreciated from this depiction, the efficiency on resource utilization is improved since both time and voltage are actively involved in the signal processing while time is only passively used as an indexer in the conventional SC. Another significant benefit of this new approach is that the computation accuracy can reach to 100% as long as enough time is given to the circuit for operating. Meanwhile, with the same hardware, the computation accuracy is dynamically adjustable and progressive precision can be achieved by simply programming the runtime. The details will be discussed in the following subsections.

6.10.1 The Scheme of Using TAF-DPS for Stochastic Operation

In the top portion of Fig. 6.114, the scheme for conventional stochastic operation is illustrated. A SNG is made of a register containing the input value, a TRNG for producing a uniformly distributed random number series, and a comparator for generating the stochastic bitstream. For a given stochastic operation, a logic operator is fed with two inputs generated by two SNGs. After that, computation result is produced cycle by cycle. In the bottom portion, the proposed time-domain scheme is depicted. As seen, the input value for operation is formatted as the control words {F, D} for TAF-DPS. After that, the intended value is embodied as an electrical pulse train created by the TAF-DPS. For a stochastic operation, two of such pulse trains (i.e., the waveforms) are interacted inside the particular operator and produce another waveform, most likely in an irregular shape. The resulting waveform is then sampled by a clock to generate computation result.

By inspecting those two schemes, it is immediately clear that the most distinguished difference between the two lies on the way of how they treat time. In the conventional approach, the operation is clock cycle based. Within each cycle, the operation is insensitive to time. In other words, time does not play any part in the operation, only voltage does. Time is only reflected as an indexing tool. In the proposed

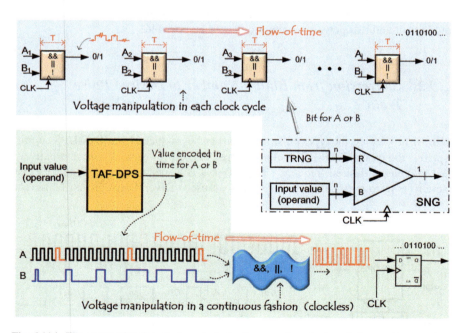

Fig. 6.114 The conventional cycle-based stochastic operation (top) and the continuous time-domain stochastic operation (bottom)

Table 6.8 The resolution of two independent variables: voltage and time

	Resolution and range	
	Analog design	Digital design
Voltage	Resolution: noise floor Range: VDD to VSS	Resolution: high and low Range: VDD to VSS
Time	Resolution: infinitely small (limited by Heisenberg limit) Range: infinitely large	Resolution: clock period Range: infinite large

scheme, however, the interaction of the two waveforms is continuously carried out. Every moment in the dimension of time counts. Voltage in this case is used as a tool for indexing the states of high and low. Hence, at a more profound level of understanding, the roles of voltage and time are switched.

In Table 6.8, the two independent variables at the foundation of electronic signal processing are listed and compared. For conventional SC, it falls into the category of voltage-domain digital design. For the proposed scheme, it is in the new realm of doing signal processing in the time-domain analog design. In this new approach, unlike the old one where only voltage is actively used for processing signal, both voltage and time are utilized. Operation-wise, from a circuit design perspective, voltage is used the same way as before. The benefit comes from the gain-in-potential from the dimension of time since time is the added new resource. Therefore, resource utilization rate is improved. As a result, theoretically, more compact circuit could result and less power consumption could be possible.

6.10.2 Conversion from Binary Number to Desired Pulse Train

Figure 6.115 depicts the types pulse train that can be produced from TAF-DPS circuit. Type 0 is the typical TAF-DPS output where pulse type T_A and T_B are alternatively used to create a desired frequency T_{TAF}. Type I is a special case of Type 0 in that fraction r is set to zero (i.e., only pulse type T_A is used). Several waveforms of different

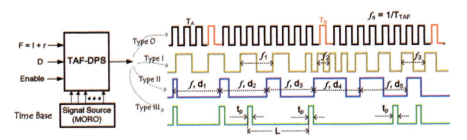

Fig. 6.115 The types of pulse train that can be produced from TAF-DPS

frequencies are displayed to show the feature that frequency can be instantly changed. Type II is a pulse train having different duty cycles. As depicted, the duty cycle can be adjusted cycle by cycle. Type III is the case where a fixed-length pulse is created in a user-specified time duration. This powerful device of manipulating electrical pulses is our tool for encoding digital value, through F and D, in the dimension of time.

For a TAF-DPS circuit, its output period is $T_{TAF} = F \cdot \Delta$ where control word $F = I + r$ and I is an integer. The fraction r can be expressed as $r = p/q$, p and q are integers and $\gcd(p, q) = 1$. In general, the waveform of TAF signal is not simply periodical as in a signal of conventional frequency since T_A and T_B are not equal in their lengths. But it is still periodical in a more complex fashion. The fundamental period T_{FD} (fundamental frequency $f_{FD} = 1/T_{FD}$) is defined as the minimum time span wherein the waveform of a TAF signal can repeat itself. From the study in Sect. 3.5 of Xiu (2012a), the length of T_{FD} can be expressed in (6.10.1).

$$T_{FD} = q \cdot T_{TAF} = (q - p) \cdot T_A + p \cdot T_B = (q \cdot I + p) \cdot \Delta \qquad (6.10.1)$$

In time-domain SC, we use the portion of time where the pulses stay at high voltage (i.e., logic high) over a specified total time to represent the number used for stochastic operation. The duty cycle D_{FD} of the fundamental period T_{FD} is defined for this purpose as $D_{FD} = T_{HI\_FD}/T_{FD}$, where T_{HI_FD} is the time-span-of-logic-high within one T_{FD}. For a TAF signal that use two types of cycles T_A and T_B in its construction, the duty cycles of T_A and T_B can also be controlled. Their logic-high portions can be expressed in (6.10.2), where ζ is an integer in the range of $[1, (I-1)]$.

$$T_{HI\_A} = T_{HI\_B} = \zeta \cdot \Delta, \quad \zeta \in [1, (I - 1)] \qquad (6.10.2)$$

$$T_{HI\_FD} = (q - p) \cdot \zeta \cdot \Delta + p \cdot \zeta \cdot \Delta = q \cdot \zeta \cdot \Delta \qquad (6.10.3)$$

$$D_{FD} = \frac{T_{HI\_FD}}{T_{FD}} = \frac{q \cdot \zeta \cdot \Delta}{(q \cdot I + p) \cdot \Delta} = \frac{q \cdot \zeta}{(q \cdot I + p)} = \frac{\zeta}{(I + r)} \qquad (6.10.4)$$

$$\frac{1}{I+1} < \frac{q}{q \cdot I + p} \leq D_{FD} \leq \frac{q \cdot (I-1)}{q \cdot I + p} \leq \frac{I-1}{I} \qquad (6.10.5)$$

Then, considering a T_{FD} which is made of $(q - p)$ T_A and p T_B (please refer to the Eq. (6.10.1)), its time duration on logic high can be derived in (6.10.3) and its duty cycle in (6.10.4). The range of D_{FD}, namely the value that can be used for stochastic operation, is expressed in (6.10.5). Its resolution also depends on the control parameters used in TAF-DPS operation. It can be influenced by the group {I, r, ζ}, as indicated in (6.10.4). In principle, the larger the I, the better the resolution. At the same time, the more bits we use to represent r, the better the resolution. Figure 6.116 left is the curve of resolution versus I (r is set at 0.01, or 7 bits for r). It can be used for reference in operation. As can be appreciated from (6.10.5), a

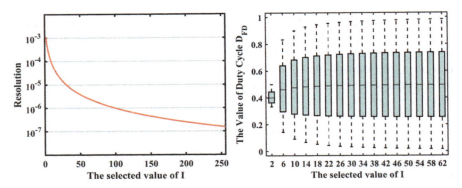

Fig. 6.116 The resolution versus I (left) and range versus I (right)

significant part of range [0, 1] can be covered. The regions at the two extremes, very close to 0 and 1, however are unreachable. The plot in the Fig. 6.116 right shows its dependence on I. For instance, when I = 62, the coverable range is [0.01587, 0.9839]. It is worth mentioning that, when r = 0 (i.e., p = 0), the TAF signal retires to signal of conventional frequency and the equation for D_{FD} is simplified from (6.10.4) to $D_{FD} = \zeta/I$. This is the special case that has been studied in (Najafi et al. 2017).

6.10.3 Deterministic Bitstream and Stochastic Operation

For implementing the time-domain SC, we have several options. Type I waveform has regular shape and thus is trivial and not used. Type II and III might be useful for some special cases. Type 0 is the general case that we will focus our attention on.

- Type II: conventional frequency is used (only pulses of type T_A are used, or $r \equiv 0$); a number of value x (0 < x < 1) is represented through duty cycle (please see Type II waveform in Fig. 6.115).
- Type III: pulse of fixed-length is used; value of x (0 < x < 1) is represented through the ratio of t_p/L (please see Type III waveform in Fig. 6.115). This low-density pulse train can have some special applications.
- Type 0: TAF concept is used (i.e., $r \neq 0$); value of x (0 < x < 1) is represented through the use of {I, r, D}. The value is NOT representable by a single cycle (pulse). It takes a full fundamental period T_{FD} to accurately represent the value; q pulses are required in this case (please see Type 0 waveform in Fig. 6.115).

For the purpose of illustration, a cycle in the TAF-DPS output waveform can be viewed as made from a consecutive series of I or (I + 1) base unit Δ. Within each Δ, the waveform can take voltage level high or low. If the waveform of this cycle is sampled by a clock of period Δ, it will result in a bitstream of length I or (I + 1) bits. This bitstream will be in unary format since 1's and 0's are grouped together.

6.10 A Case of Signal Processing in Time Dimension: Deterministic ...

Figure 6.117 shows the composition of a full cycle of fundamental period T_{FD}. It is the minimal repeatable portion of the Type 0 waveform. Visually, it can be viewed as made from $(q \cdot I + p)$ units of Δ, although in real circuit operation the waveform is not exactly created in this way but in a fashion of edge selection and combination. With the help of this viewpoint, it is immediately clear that this waveform corresponds to a bitstream if each Δ is treated as a bit. This is the extended case of unary bitstream introduced in Najafi et al. (2019) and it is a deterministic bitstream.

This powerful device of manipulating electrical pulses is our tool for encoding digital value, through F and D, in the dimension of time and then converting it into deterministic unary bitstream. Table 6.9 provides several examples of TAF-DPS setting and its corresponding bitstream. Each of those settings uniquely corresponds to a signal of frequency $f = 1/T_{TAF} = F \cdot \Delta$, and to a value of $x = D_{FD}$. The waveform of T_{FD} is repeatable and so is this T_{FD} bitstream. A parameter Iqp is defined as Iqp $= q \cdot I + p$, which represents the length of this T_{FD} bitstream.

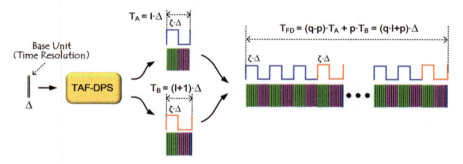

Fig. 6.117 The composition of a full cycle of fundamental period T_{FD}

Table 6.9 Examples of TAF-DPS setting and T_{FD} bitstream

Setting	T_{FD} Bitstream
$F = I + r = 4.5$, $\zeta = 2$, $p = 1$, $q = 2$, Iqp $= 9$ $\rightarrow x = D_{FD} = 0.4444444$	110,011,0**0**0*
$F = I + r = 4.125$, $\zeta = 2$, $p = 1$, $q = 8$, Iqp $= 33$ $\rightarrow x = D_{FD} = 0.4848485$	110,011,001,100,110,011,001,100,110,011,0**0**0
$F = I + r = 16.375$, $\zeta = 3$, $p = 3$, $q = 8$, Iqp $= 131$ $\rightarrow x = D_{FD} = 0.1832061$	11,100,000,000,000,001,110,000,000,000,000 111,000,000,000,000,0**0**1,110,000,000,000,000 111,000,000,000,000,011,100,000,000,000,000 111,000,000,000,000,011,100,000,000,000,000
$F = I + r = 16.375$, $\zeta = 14$, $p = 3$, $q = 8$, Iqp $= 131$ $\rightarrow x = D_{FD} = 0.8549618$	11,111,111,111,111,001,111,111,111,111,100 111,111,111,111,110,0**0**1,111,111,111,111,100 111,111,111,111,110,011,111,111,111,111,000 111,111,111,111,110,011,111,111,111,111,000

*The bold text **0** is the effect of fraction r

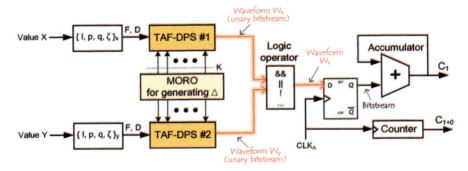

Fig. 6.118 Illustrative diagram for TAF-DPS based stochastic computing

A statement made in Najafi et al. (2019) is that two deterministic bitstreams can be used for performing stochastic operation as long as every bit in one bitstream sees every bit in the other bitstream (i.e., convolution). When this requirement is fulfilled, the result of stochastic operation is completely accurate. In this case, the issue of correlation between the two bitstream has no impact. This important conclusion can be used here in TAF-DPS based deterministic stochastic computing. Figure 6.118 is the illustrative diagram for performing stochastic computing on TAF-DPS circuit platform. Two operands X and Y, in decimal or binary format, are first converted into $\{I, p, q, \zeta\}$ as TAF-DPS frequency and duty-cycle control words F and D, which are in conventional binary format. Two waveforms W_x and W_y are correspondingly generated and fed into a logic operator for stochastic computing. Although the operation is actually carried out upon two waveforms as an analog action, it can be analyzed as a digital logic operation if W_x and W_y are treated as unary bitstreams. The resulting waveform, after sampled by a clock CLK_Δ of $f_\Delta = 1/\Delta$, leads to the bitstream that carries the information regarding computed value. The actual value can be obtained by the ratio of two counters C_1 and C_{1+0}, which are all in binary formats. From implementation perspective, this circuit platform is resource efficient since the unary bitstreams do not need to be created physically and stored in real hardware. The creation of operands and the logic operation occur simultaneously. Further, the computing result is naturally available when the stochastic information is converted back into the binary world. All those benefits are materialized by the deeper utilization of time resource.

One important technical issue is to determine the stop point for this continuous time domain operation. In other words, how many CLK_Δ cycles, denoted as N_{stop}, are required to reach the accurate result? or to allow every bit in bitstream W_x to see every bit in bitstream W_y? If Iqp_x and Iqp_y are relatively prime, the N_{stop} can be easily set as $N_{stop} = Iqp_x \cdot Iqp_y$. This is similar to the relatively prime method suggested in Najafi et al. (2019). From (6.10.4), for a user-specified value x, it is seen that we have the freedom to set four values for the group $\{I, q, p, \zeta\}$ to make x $= D_{FD} = q \cdot \zeta/(q \cdot I + p)$. For two operands x and y, we can impose a constraint that

6.10 A Case of Signal Processing in Time Dimension: Deterministic ...

$\gcd(\mathrm{Iqp_x}, \mathrm{Iqp_y}) = 1$. Therefore, the guideline for using the TAF-DPS platform can be summaries as follow.

For an operation between two operands x and y, set values for $\{\mathrm{I, q, p, \zeta}\}_x$ and $\{\mathrm{I, q, p, \zeta}\}_y$ so that:

- $x = q_x \cdot \zeta_x / \mathrm{Iqp_x}$ and $y = q_y \cdot \zeta_y / \mathrm{Iqp_y}$
- $\gcd(\mathrm{Iqp_x}, \mathrm{Iqp_y}) = 1$
- making $\mathrm{Iqp_x}$ and $\mathrm{Iqp_y}$ as small as possible, $N_{stop} = \mathrm{Iqp_x} \cdot \mathrm{Iqp_y}$.

An example can be helpful to illustrate the operation. If we have $x = 0.4444444$ and $y = 0.1832061$ (the two values in the 1st row and 3rd rows of Table 6.9), a multiplication can be performed by using a AND gate for the logic operator in Fig. 6.118. The expected result is $z = x \cdot y = 0.08142485$. Figure 6.119 presents the waveforms of W_x, W_y and W_{xy} in three zoom scales. The top two rows are waveforms of W_x and W_y. The 3rd row is the waveform after AND operator (multiplication). The time-domain operation can be appreciated from those waveforms. The corresponding bitstreams are also labeled in the figure for viewing the stochastic operation in the traditional way. In the left of Fig. 6.120, the plot of ratio C_1/C_{1+0} versus time (in unit of CLK_Δ cycle) is presented. In this example, $N_{stop} = \mathrm{Iqpx} \cdot \mathrm{Iqpy} = 9 \cdot 131 = 1179$. As seen, at points of multiple of N_{stop}, the C_1/C_{1+0} ratio is the theoretical result of 0.081425 (the calculated error is exactly zero). In operation, we can stop the circuit when counter C_{1+0} reaches this number of N_{stop} and get the completely accurate result.

In the circuit platform of Fig. 6.118, if the logic operator is implemented as a 2 → 1 MUX, scaled addition can be carried out handily. The MUX selection signal

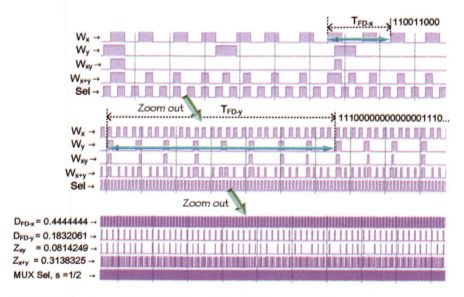

Fig. 6.119 Waveforms of W_x, W_y, W_{xy}, and W_{x+y} in three zoom scales

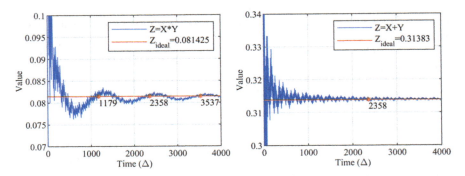

Fig. 6.120 Results of multiplication (left) and scaled addition (right) as progressively expressed in C_1/C_{1+0}

can be a 50% duty cycle clock of period (frequency) $T = 1/f = s \cdot \Delta$, where s is the scaling factor. This clock signal can be easily generated from another TAF-DPS based on the same MORO. The N_{stop} in this case shall be set as $N_{stop} = s \cdot Iqp_x \cdot Iqp_y$. When s = 2, the operation produces $z = (x + y)/2$. At points of multiple of N_{stop}, the computation results are accurate. On the right-hand side of Fig. 6.120, the result of s = 2 scaled addition of x = 0.4444444 and y = 0.1832061 is plotted. As see, at $N_{stop} = 2 \cdot Iqp_x \cdot Iqp_y = 2358$, we get the accurate addition result. The waveform of addition W_{x+y} is shown at the 4th row of Fig. 6.119 while the MUX selection is at the 5th row.

Multi-stage stochastic computing is important. The platform shown in Fig. 6.118 can be expanded to do multi-level operation. The generic architecture is depicted in Fig. 6.121. We can use a three-level multiplication between four values as an example: $z = [(x_1 \cdot x_2) \cdot x_3] \cdot x_4$. This multiplication is accomplished by configuring the structure in Fig. 6.121 to three stages. The four values are listed below.

- $F_1 = 2 + 1/2, \zeta = 1, \rightarrow x_1 = 0.4, Iqp_1 = 5,$
- $F_2 = 3 + 1/4, \zeta = 2, \rightarrow x_2 = 0.615384615, Iqp_2 = 13$

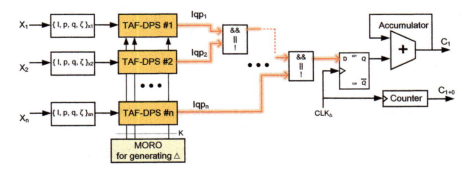

Fig. 6.121 TAF-DPS based multi-stage stochastic operation

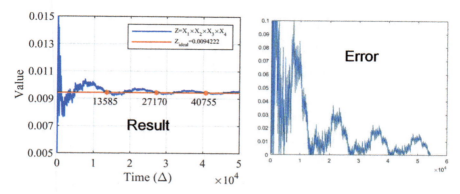

Fig. 6.122 Result of three-stage multiplication (left) and error (right)

- $F_3 = 4 + 3/4$, $\zeta = 1$, $\rightarrow$ $x_3 = 0.210526315$, $\text{Iqp}_3 = 19$
- $F_4 = 5 + 1/2$, $\zeta = 1$, $\rightarrow$ $x_4 = 0.181818181$, $\text{Iqp}_4 = 11$
- $N_{\text{stop}} = \text{Iqp}_1 \cdot \text{Iqp}_2 \cdot \text{Iqp}_3 \cdot \text{Iqp}_4 = 13{,}585$.

Figure 6.122 shows the result of this three-stage multiplication. As expected, at points of multiple of N_{stop}, the computation results are accurate. As shown in the left, the errors are zero at those points. It is worth mentioning that, unlike the quantization error experienced in conventional multi-level SC, there is no loss in resolution in this operation. This is own to the analog-like time-domain operation (the left column of 2nd row in Table 6.8).

From Figs. 6.118 and 6.121, it can be understood that the biggest advantage of time-domain SC is that the bitstreams representing the operands do not need to be created and stored. There is no need of TRNG or PRNG. Similar to conventional SC, the computation accuracy is progressive improved (Figs. 6.120 and 6.122). However, unlike conventional SC where computation result is randomly fluctuating, completely accurate result can be reached if the operation is carried out long enough to the point of N_{stop}, which can be predicted beforehand. The tradeoff lies between accuracy and latency (and the energy consumed). In conventional SC, the length of the sequence (bitstream) for operand depends on the hardware. Thus, computation accuracy is constrained by hardware. It cannot be changed once the design is fixed. In TAF-DPS circuit platform, the length of the unary bitstream is flexible. The resolution in expressing operand can be set by user (please refer to Fig. 6.116). And therefore, computation accuracy can be programmed at runtime without altering the hardware. The circuit sizes of TAF-DPS and MORO are small, compared to TRNG and PRNG.

The scheme of using TAF-DPS as a generic circuit platform for deterministic stochastic computing has been tested on Kintex UltraScale FPGA. Table 6.10 includes four cases of a circuit implemented in this FPGA. The TAF-DPS circuit uses the configuration of K = 16 and F[20:0], where F[20:16] for integer and F[15:0] for fraction. Table 6.11 reports the resource used.

Table 6.10 Cases studied on FPGA

	Case 1	Case 2	Case 3	Case 4
Operands x and y	0.4 0.44444444	0.22222222 0.30769231	0.34782609 0.36090226	0.14545455 0.34782609
$z = x \cdot y$	Expected: 0.17777778	Expected: 0.068376068	Expected: 0.12553122	Expected: 0.05059289
F values*	$F_x = 2.5$ $F_y = 2.25$	$F_x = 4.5$ $F_y = 3.25$	$F_x = 5.75$ $F_y = 8.3125$	$F_x = 13.75$ $F_y = 8.625$
$\{I, q, p, \zeta\}$	$\{2, 1, 2, 1\}_x$ $\{2, 1, 4, 1\}_y$	$\{4, 1, 2, 1\}_x$ $\{3, 1, 4, 1\}_y$	$\{5, 3, 4, 2\}_x$ $\{8, 5, 16, 3\}_y$	$\{13, 3, 4, 2\}_x$ $\{8, 5, 8, 3\}_y$
Iqp	$Iqp_x = 5$ $Iqp_y = 9$	$Iqp_x = 9$ $Iqp_y = 13$	$Iqp_x = 23$ $Iqp_y = 133$	$Iqp_x = 55$ $Iqp_y = 69$
N_{stop}	45	117	3059	3795
Values of C_1 and C_{1+0}	8 (HEX) 2D (HEX)	8 (HEX) 75 (HEX)	180 (HEX) BF3 (HEX)	C0 (HEX) ED3 (HEX)
Experiment result $z = C_1/C_{1+0}$	FPGA: 0.17777778	FPGA: 0.068376068	FPGA: 0.12553122	FPGA: 0.05059289

Table 6.11 Resource usage for the cases in Table 6.10

	LUT	FF	CARRY8	F7 MUX	F8 MUX
MORO	1	16	0	0	0
TAF-DPS (2)	68	66	6	8	2
Accumulator	5	16	0	0	0
Counter	6	19	0	0	0

Paradigm Shift:

In the emerging Time-Oriented paradigm, instead of voltage, the task of signal processing can be performed in the dimension of time. For example, stochastic computing can be carried out by using time as the processing medium.

6.11 Reducing FIFO Memory Size and Smoothing Data Flow

In Sect. 5.3, it has been discussed that one of the ubiquitous issues in VLSI circuit design is the data transport from one place to another. It can happen between small circuit blocks, between large on-chip systems and even between chips. Oftentimes, the transmitting and receiving parties are driven by their own clocks and the two clocks are at different rate (frequencies). For successful data transfer, a storage device must be inserted between the two parties for accommodating the differences on data-flow-rate and on clock frequencies, to prevent data loss, as illustrated in Fig. 5.8. In practice, the storage is embodied in the form of FIFO (first in first out) memory.

6.11 Reducing FIFO Memory Size and Smoothing Data Flow

This problem of FIFO design is so omnipresent in circuit design that almost every sizeable chip project comes across this issue and almost every digital design engineer has encountered it in one way or another. In most SoC (system-on-chip) design, a large portion of silicon area is dedicated to various FIFO memories. Therefore, a slight improvement on the efficiency of this data transport problem can have a wide-reaching impact on the field of chip design, for the purpose of saving area and power.

From a functional perspective, FIFO is used for the purpose of data synchronization (i.e., the hi-fidelity transportation of information from one place to another). Structural-wise, a FIFO device primarily consists of a set of read and write pointers, storage and control logic. Storage may be static random-access memory (SRAM), flip-flops, latches or any other suitable form of storage. For FIFOs of non-trivial size, a dual-port SRAM is usually used, where one port is dedicated to writing and the other to reading. From clocking perspective, synchronous FIFO is a FIFO where the same clock is used for both reading and writing while asynchronous FIFO uses different clocks for reading and writing.

Implementation-wise, FIFO is often implemented as a circular queue with two pointers and a set of status flags. Examples of flags include: full, empty, almost full, and almost empty. In the left of Fig. 6.123, the block diagram of a typical FIFO device is shown. At the center is the circular buffer with two pointers of write and read attached to it. At each side of the FIFO, there is a corresponding clock domain wherein the data is handled by its relevant clock and communicate to the FIFO. The clock for write (CLKW @ f_w) and the clock for read (CLKR @ f_r) can be the same signal (synchronous FIFO) or two different signals (asynchronous FIFO).

In the design of any FIFO, the primary goal is the minimization of the size of the storage (i.e., reducing the size of the circular buffer as much as possible). Besides this demand, there is another subtle issue concerning the continuousness of the data flow. In real operation, one of the characteristics of incoming data (attached to the data input port of the FIFO) is that there is no guaranty of valid data at every clock cycle. In other words, for some cycles of CLKW, there can be no valid data. This data validity at the input side is indicated by a flag (i.e., the "data valid flag" shown on the figure). At the output side, therefore, a possibility exists that the outgoing data flow might be a broken stream (for example, there can be no data for a particular CLKR cycle when the FIFO is empty). In most applications, however, it is preferred that

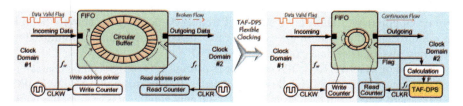

Fig. 6.123 FIFO implementation: traditional fashion (left) and TAF-DPS clocking version (right)

the outgoing data flow is a continuous one (i.e., there is valid data on every CLKR cycle).

In short, the current situation of FIFO design can be described as follow.

- For the two clocks respectively driving the write and read ports, they can be the same signal or two signals having different frequencies. In both cases, the clock frequencies are fixed (not dynamically adjustable in real-time).
- The incoming data flow to FIFO can be broken. Or, there is no guaranty of valid data on every CLKW clock cycle.
- In all the cases, the primary goal in FIFO design is the minimization of its storage size.
- In some cases, another goal is to smooth the outgoing data flow: make its flow continuous and its rate as close to that of incoming flow as possible.

The two design goals listed above are hard to be achieved when the clock frequency is rigidly fixed. In synchronous FIFO case, the outgoing data flow simply cannot be continuous when the incoming data flow is broken. This is due to the reason that both sides are working on the same rate (driven by the same clock). For asynchronous FIFO, this is not an easy task neither since the rate difference between f_w and f_r needs to accommodate the pattern of "valid data", which is dynamically varying and usually unknown at design time. For the size of the FIFO storage, it primarily depends on the rate difference between f_w and f_r. Besides rate difference, the pattern of "valid data" can significantly influence the size as well. Thus, when clock rate is rigid, the size of storage is hard to be reduced. The larger the rate difference between f_w and f_r is, the more irregular the pattern of "valid data" is, the larger the storage size needs to be.

The TAF-DPS provides an elegant solution to this problem. Its features of arbitrary frequency generation and instantaneous frequency switching make TAF-DPS output a very flexible clock signal. This flexibility in frequency can accommodate the variation on data pattern and be used for the purpose of reducing storage size. As shown in the right-hand side of Fig. 6.123, the TAF-DPS clock generator is employed to drive the outgoing flow of the FIFO. Depending on the emptiness/fullness status of the FIFO, the TAF-DPS output frequency can be adjusted dynamically to slow down and speed up the outgoing data flow rate under the constraint of having valid data on every CLKR cycle. During this process, the adjustment on the rate f_r can be deliberately calculated to make the circular buffer small. When fraction is used in TAF-DPS control word, as discussed in previous sections, the frequency resolution can reach ppb range. This ample supply of frequency can be great help to the task of minimizing buffer size. In the extreme case, the buffer size can be minimized to hold only two units of data (Yang and Haider 2012).

The use of adaptive TAF-DPS clock for assisting FIFO design can be characteristically called TAF-DPS FIFO design approach. The advantage of TAF-DPS FIFO can be appreciated from two directions. In some applications, it is extremely important to reduce the size of the FIFO storage. In others, it is crucial to keep the outgoing bitstream continuous, as smoothly as possible from a data-flow perspective. Next, we will use two examples to illustrate the points.

6.11 Reducing FIFO Memory Size and Smoothing Data Flow

In our first example, we study the issue of reducing storage size. Referring to the left-hand drawing in Fig. 6.123, we assume that there is 100,000 bits of data to be transported from clock domain #1 to #2, whose respective clock frequencies f_w and f_r are 50 and 40 MHz. Under this assumption, the size of storage must be at least $100,000 * (50-40)/50 = 20,000$ bits. Using any circular buffer smaller than this size will lead to data loss. If we use TAF-DPS FIFO design strategy where the read clock's frequency can be dynamically adjusted, this storage size can be significantly reduced. For instance, a strategy can be applied that sets the f_r of CLKR to 66.7 MHz if the amount of data inside FIFO is currently more than 100 bits. When current amount of data inside FIFO is less than 50 bits, the TAF-DPS output (i.e., the CLKR) can be slowed down to 40 MHz. Designing clock signal in this way, the size of storage can be controlled within 200 bits, which is 100 times smaller than the original size.

The corresponding simulation result is displayed in Fig. 6.124. On the left-hand side, the trends of amount of data inside FIFO are shown. The blue curve is for conventional FIFO while the red is for TAF-DPS FIFO. As seen, in conventional case, the FIFO will be completely full after 2 ms of operation. At this time, if the incoming bitstream continues, some data will be lost. On the other hand, the amount of data held inside the storage is never more than 100 bits in the case of TAF-DPS FIFO. This is more clearly shown in the right-hand side of the figure. As seen, the CLKR jumps between 40 MHz and 66.7 MHz to rein the amount of data inside the FIFO.

Our second example focuses on the task of keeping the FIFO outgoing stream as smooth as possible, which is crucial for certain applications. The incoming data stream to FIFO is still driven by CLKW of $f_w = 50$ MHz. However, the data flow is broken. In other words, there exist some clock cycles that does not have associated data. The situation is indicated by a "data valid flag" as shown in the top graph of Fig. 6.125. Although the incoming data stream is broken from time to time, we want the FIFO outgoing stream continuous and the data rate variation as smooth as possible. Figure 6.125 shows the simulation of such scenario. In the top graph, the data valid flag is shown. When its level is "0", there are no data in the corresponding

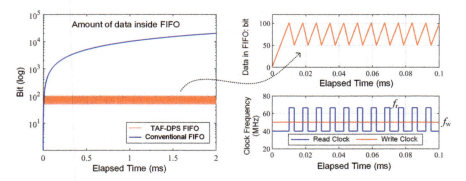

Fig. 6.124 TAF-DPS FIFO for reducing storage size

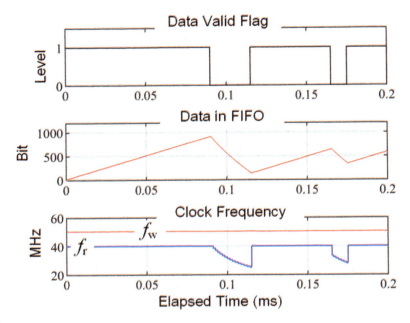

Fig. 6.125 TAF-DPS FIFO for smoothing data flow

clock cycles. In the bottom graph of Fig. 6.125, the f_w and f_r of write and read clocks are shown. As seen, the f_r is continuously adjusted when the incoming data stream is broken. The adjustment is based on the status of data amount inside FIFO, which is indicated in the middle graph. As a result, the outgoing stream is continuous since the FIFO is never empty. It is always kept at an appropriated level. Further, the change in f_r has been made as smooth as possible.

At circuit level, Fig. 6.126 shows the TAF-DPS's capability of dynamical frequency adjustment that qualifies it as an adaptive clock. The waveforms are obtained from a SPICE simulation at transistor level. In the top graph, two traces are displayed. On the top is the TAF-DPS control signal and at the bottom is the corresponding output clock waveform. As seen, the output clock waveform (and its frequency) faithfully follows the pattern of the control. In the middle and bottom graphs, the detailed views on frequency-switching are presented, from fast-to-slow and from slow-to-fast, respectively. The frequency-switching is smoothly done and the waveform transition is seamlessly (glitch free), which is an indispensable requirement for successful circuit operation. It is worth mentioning that, although the rate of clock is varying, the data flow it drives is continuous (i.e., having valid data at every cycle). This continuity in the flow of data can be important for many applications.

Gestalt Switch:

The resolution of the two long-lasting problems (arbitrary frequency generation and instantaneous frequency switching) enables new treatment on old problems. In this section, the classical problem of FIFO design is optimized by using a new tool.

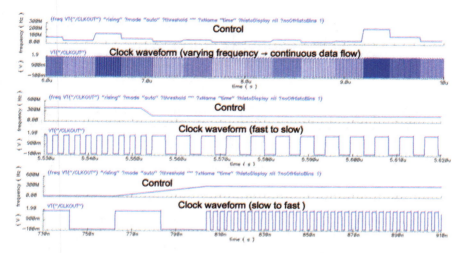

Fig. 6.126 TAF-DPS adaptive clock: control and waveform

6.12 Clock Data Recovery by Explicitly Using Time-Average-Frequency

In all our previous discussions regarding data transportation between two places, we have assumed that the clock on the receiver works independently of the clock in the transmitter. In other words, the clock on the receiving side has no knowledge of the clock on the transmitting side. This is one of the reasons for the very existence of the special circuit device called FIFO, discussed in Sect. 6.11. If the receiving side can somehow extract the clock frequency information from the incoming data, the receiver clock can be then correspondingly created and the data transportation process be facilitated. This idea of extracting information about transmitting clock from the transmitted data stream is called clock data recovery (CDR).

In the field of electronic system design, the task of data transportation is formally termed as data transmission. In its early day of development, parallel communication approach is widely used. It is a method of conveying multiple bits simultaneously by using multiple channels in parallel. Comparing to serial communication which conveys only a single bit at a time, parallel communication has higher communication capacity. But it requires additional channels for signals other than regular data, such as a clock signal to pace the flow of data, a signal to control the direction of data flow, and some handshaking signals for coordinating the transmitter and receiver. A well-known example of parallel communication standard is PCI (Peripheral Component Interconnect). In recent years, the decreasing cost and better performance of integrated circuits has led to the use of serial communication over parallel one. The most famous example is the USB (Universal Serial Bus). In serial link design, one of the keys is the CDR circuit since, unlike in parallel link, the clock signal is not transmitted but has to be extracted from the transmitted data. In this regard, serial

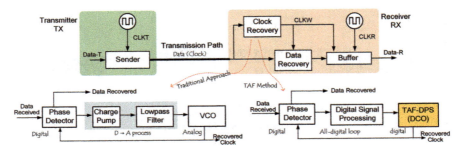

Fig. 6.127 Clockless data transmission (top) and traditional CDR (bottom left) and TAF based CDR (bottom right)

link is a scheme of clockless data transmission. It has become extremely popular in modern electronics, especially for high data rate inter-chip communication.

In the top portion of Fig. 6.127, the principle of clockless data transmission is illustrated where the key functional blocks are included. As seen, the clock information is embedded in the data transmitted. The advantage of this approach lies in the fact that it eliminates the skew problem between the multiple communication channels, which is a serious challenge in the design of parallel communication (getting worse with the increase of the data rate). On the receiving side, the embedded clock signal needs to be extracted by using the method of CDR. Inside the receiver, clock CLKW is generated from the CDR and, at the same time, used to latch the incoming data. Before sending the received data to the downstream processing unit which is driven by clock CLKR, a buffer might be inserted in between to accommodate the slight frequency difference between CLKW and CLKR. This difference could be caused by the action of removing some auxiliary bits, which is presented in the incoming data stream for assisting the establishment of the link.

$$T_{txavg} = T_{rxavg} = a_0 \cdot T_0 + a_1 \cdot T_1 + a_2 \cdot T_2 + \ldots + a_n \cdot T_n, \quad \sum a_i = 1 \tag{6.12.1}$$

On the bottom left of Fig. 6.127, the details of CDR are shown. As shown, it is a negative feedback control system. The goal of this feedback loop is to satisfy Eq. (6.12.1), where T_{txavg} (=$1/f_{txavg}$) is the transmitters TX's average clock period (frequency) and T_{rxavg} (=$1/f_{rxavg}$) is the receiver RX's average clock period (frequency). $T_0, T_1, T_2, \ldots T_n$ are the periods of individual clock cycles in RX; $a_0, a_1, \ldots a_n$ are the possibilities of occurrence of the corresponding cycles. It is important to point out that, during the process of clock recovery, the frequency matching between TX and RX is achieved in a long-term average sense (i.e., over multiple cycles). From each individual cycle's perspective, it is meaningless to discuss frequency matching (since the very concept of frequency is established in a long-term frame of one second). The requirement imposed on CDR circuit is for the average clock rates of the two sides being matched over multiple cycles so that, within each

6.12 Clock Data Recovery by Explicitly Using Time-Average-Frequency

cycle, the incoming data can be reliably latched by the RX using the extracted and "frequency-matched-to-TX" clock signal.

In the construction of this feedback loop, binary phase detector is often used because of its speed advantage (the high data rate makes it difficult for linear detector to work reliably). The output from this binary detector is in digital fashion. However, the local oscillator is a VCO, which is controlled by an analog voltage. Therefore, a D $\rightarrow$ A (digital to analog conversion) process is required between the detector and the VCO (the blocks of charge pump, analog filter and etc.). This will result in high cost (area and power) and, more significantly, slow response. For a typical CDR of this type, to satisfy Eq. (6.12.1), n probably needs to be a large number due to the slow response of this VCO-based loop. This will have negative impact on the performance of CDR (e.g., frequency tracking capability and jitter tolerance).

In the bottom-right drawing of Fig. 6.127, the architecture of Time-Average-Frequency CDR (TAF-CDR) is depicted (Xiu 2015e, f). In this approach, the enabling block is the TAF-DPS based digital controlled oscillator (DCO) which takes a digital value as its frequency control variable. TAF-CDR takes advantage the digital nature of detector output and DCO control. Consequently, the entire loop can be constructed using all digital elements. Due to the fast response speed of this DCO (two RX cycles), the loop latency M, which is the time elapse (in number of RX cycles) between the moment-of-detecting and the moment-of-action, can be designed small (i.e., the loop can respond very fast to the input change). Further, the value of M can be precisely determined, resulting in a quantifiable loop response speed. This TAF-CDR loop is similar to the one used in TAF-FLL (for more details, please refer to Sect. 6.6). One of the key benefits of TAF-CDR is the elimination of the D $\rightarrow$ A process required in conventional CDR. Furthermore, since the loop has fast response, the frequency tracking capability and jitter tolerance can be greatly improved.

$$T_{txavg} \approx T_{rxavg} = \sum_{i=1}^{M} a_i T_i, \quad \sum_{i=1}^{M} a_i = 1 \qquad (6.12.2)$$

Instead of using many frequencies (periods) to match the two sides as done in traditional CDR approach, as indicated in Eq. (6.12.1), the TAF-DPS based DCO only generates several discrete frequencies (periods) and uses them to achieve the frequency matching. For example, three unique cycles (i.e., three frequencies or periods) can be produced to serve three specific purposes of hold, speed-up and slow-down, respectively. In a relatively short time frame of M cycles, the loop dynamically matches TX and RX frequencies as shown in (6.12.2). Although the matching on frequency is only an approximation in short term, it is good to the degree that the TX data can be reliably captured by the RX. From the perspective of each individual data unit, the setup and hold margin of the sampling cell inside the phase detector is guaranteed by honoring Eq. (6.12.2). In long term, the Eq. (6.12.1) is satisfied precisely just as in the case of conventional CDR. As a result, no data is lost or created from nowhere during this data transmission process.

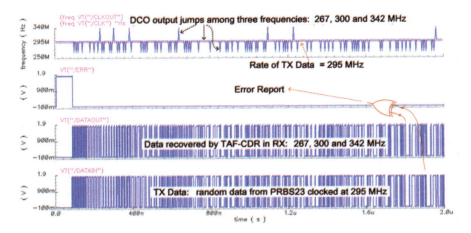

Fig. 6.128 A simulation of a TAF-CDR operating around 300 MHz by using M = 3

Figure 6.128 shows a transistor level simulation result for TAF-CDR. The TX data rate is at about 300 Mbps. The design is implemented in a 180 nm CMOS process. The loop latency is designed as M = 3. In this particular simulation, the incoming data (from TX) is driven by a clock intentionally set at 295 MHz (which is 0.16% slower than the nominal frequency 300 MHz). The DCO in the RX is designed to generate three discrete frequencies: 300 (hold), 342 (speed up) and 267 MHz (slow down). These three frequencies are produced by a TAF-DPS using the configuration of 8Δ, 7Δ and 9Δ, respectively. The base time unit Δ = 417 ps is generated from an 8-output ring oscillator running at 300 MHz.

The operation can be explained according to Eqs. (6.12.1) and (6.12.2). In this case, there is only one data rate of 295 Mbps in the TX side which is determined by the TX clock. The data rate in the RX block must be made equal to this rate (otherwise, a FIFO will be needed). The RX circuit operates in a fashion that its frequency is constantly changing among the three aforementioned values. The selection of the value is made in every M = 3 cycles, based on the result of calculation performed by the loop. As shown in the top trace of DCO instantaneous frequency over time, the loop dynamically adjusts DCO output frequency once in every three cycles. At any particular moment, it selects one among the three choices of 267, 300 and 342 MHz. The decision is made in real time based on the relationship between the edges of incoming data and RX clock. The trace at bottom is the TX data which is generated from a 23 bits PRBS (Pseudorandom Binary Sequence) encoder driven by the TX clock (=295 MHz in this case). The third trace from the top is the recovered data (outputted from the detector in RX). The second trace from the top is the output from an error detector which is made of the corresponding 23 bits PRBS decoder. The inherent logics of the PRBS encoder and decoder are matched. The error detector is fed by the recovered data and it is driven by the recovered clock. It will report error (logic high) if the recovered data is not agreed with the TX data.

6.12 Clock Data Recovery by Explicitly Using Time-Average-Frequency

The simulation shows that the incoming data is correctly latched by the receiver. It also implicitly reveals the fact that the clock information is rightfully recovered or extracted from the incoming data. Further, it is important to emphasize that this TAF-CDR circuit has correctly recovered the transmitted data under a rough condition of incoming data rate (295 MHz) being 16,666 ppm away from its nominal 300 MHz. The top graph in Fig. 6.128 illustrates that the 295 MHz of TX is averagely matched by the DCO output frequency in RX. As expected, more samples of slow-down-frequency (267 MHz) are used than that of speed-up-frequency (342 MHz). This is due to the fact that incoming data is slower than the nominal value of 300 MHz. The majority of the samples however are at the 300 MHz (which is the hold state). Please also see Sect. 6.8 of Xiu (2012a) for more explanation on the principle of TAF-CDR.

As shown in the bottom right of Fig. 6.127, the architecture of TAF-CDR is all digital. This can lead to elegant and extremely low-cost implementation in some cases. One such example is an ultra-low power and low-cost CDR function (<1 uW) for a low data rate application of 2 Mbps. The low-cost is achieved by building a TAF-CDR circuit entirely from standard cells. This design is implemented in a 65 nm CMOS process. A group of 16 phase-evenly-spaced signals is generated from a ring of 8 D-type flip-flops driven by a 16 MHz reference signal. The resulting 16 outputs are in 2 MHz frequency and are fed to the TAF-DPS DCO, which is also made from digital standard cells. The control block is made of digital standard cells as well. The base unit $\Delta = 31.25$ ns is generated from the 16 signals of 2 MHz ($\Delta = 500$ ns/16). The DCO is responsible for generating three types of cycles: 15Δ (468.75 ns), 16Δ (500 ns) and 17Δ (531.25 ns). As explained previously, each of these cycles is used in the appropriate moment so that the incoming data flow can be tracked properly. In this design, the entire operation chain of detect-calculate-adjust performed in the TAF-CDR loop (please refer to the bottom right drawing of Fig. 6.127) can be finished within one cycle since the clock rate is low (~2 Mbps, bit-time is around 500 ns). Thus, the loop latency can be designed as small as $M = 1$ (in contrast, $M = 3$ in previous simulation case since bit time is 3.3 ns and it operates in a slower process of 180 nm). As a result of this minimum latency of $M = 1$, this TAF-CDR circuit has a very fast response speed. This leads to a high tolerance for frequency error (and jitter) since the loop can respond very quickly to correct any error.

Figure 6.129 is the test configuration used in the lab to evaluate this TAF-CDR. An Agilent E4437B signal generator is used to provide the test bitstream (i.e., the TX data). Its internal BER (Bit Error Rate) tester is used to check the received bitstream. As shown in the drawing on the left, a 2 Mbps bitstream of PRBS data is sent from the signal generator to the TX port of the TAF-CDR circuit. The recovered data from the TAF-CDR is sent back through its RX port to the BER tester in signal generator for validation. The 16 MHz clock, which is used to generate the 16 phase-evenly-spaced signals needed for the DCO, is provided by a function generator. The frequency of this "16 MHz" can be adjusted by changing the setting of the function generator. By this configuration, the frequency tolerance of this CDR system can be studied.

Figure 6.130 shows some of the experimental result. The snapshot in the right-hand side is the BER test result. The input data is a 2 Mbps bitstream of PN9. The number of bits used is 4294967295 ($=2^{32} - 1$). The BER check result is zero. The

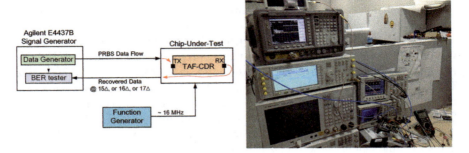

Fig. 6.129 The test configuration of TAF-CDR

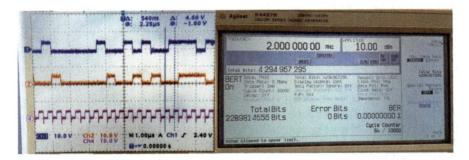

Fig. 6.130 Experimental result: waveforms (left) and BER result (right)

screen capture on the left are the waveforms of incoming TX data (top), recovered RX data (middle) and recovered clock (bottom). As seen, the recovered data follows the incoming data faithfully. It is however delayed by one RX cycle because of the latency of $M = 1$. From the waveform of recovered clock, if inspected closely, it can be seen that there are three types of cycles of different lengths. They are the slow-down, hold and speed-up cycles. For example, the cycle bounded by the two markers is measured as 540 ns, which is the slow-down cycle (the nominal value is 500 ns). This fact is more clearly revealed in Fig. 6.131.

The oscilloscope screen capture in the left side is obtained in single-run mode. The bottom waveform is the incoming TX bitstream of pattern "0001". The top waveform is the recovered data which shows a pattern of "1110" (there is a level of inversion added in the circuit). The waveform in the middle is the recovered clock by the TAF-CDR. If viewed closely, it can be seen that all the TX bits are at 500 ns length. But the recovered data and clock have several different lengths. This fact is easier to be viewed in the snapshot presented on the right-hand side, which is obtained in the continuously triggered mode. As seen, the TX data is clean while the RX data and RX clock are "jittery". This "jittery" is due to the use of different cycle lengths in TAF-DPS output. It is interesting to point out that, in contrast to

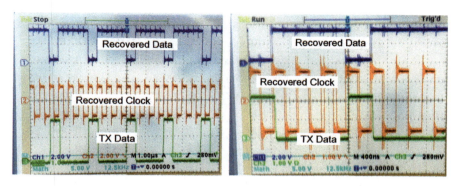

Fig. 6.131 The illustration of three types of cycles used in TAF-CDR: single run (left) and continuously triggered (right)

conventional wisdom, "jittery" clock is not necessarily a bad thing (please refer to Sect. 3.6 of Xiu (2015a) for more discussion on "jittery" clock).

Thanks to the small latency of $M = 1$, this TAF-CDR circuit has a high tolerance for frequency error. The DCO's nominal frequency (the frequency in the hold state) is designed to be equal to the nominal rate of the TX bitstream (2 MHz in this case). When the incoming data rate deviates from its nominal value, the TAF-CDR needs to track it as best as it could. This tracking capability is tested in the lab by using the function generator as an adjustable frequency source. During the test, the frequency of the 16 MHz reference is adjusted (this is equivalent to the adjusting of the incoming data rate) and the BER is correspondingly checked. Tables 6.12 and 6.13 are the results of using bitstream of PN9 and PN15 for test, respectively. In all the cases, the number of bits used is 2147483647 ($=2^{31} - 1$). The frequency has been adjusted from both directions: frequency+ and frequency−. From those tables, it is

Table 6.12 BER at various degrees of frequency error using data stream of PN9

Error	Frequency+ (MHz)	BER+	Frequency− (MHz)	BER−
1 ppm	16.000016	0	15.999984	0
10 ppm	16.00016	0	15.99984	0
100 ppm	16.0016	0.0000001	15.9984	0.0000002
500 ppm	16.008	0.000001	15.992	0.000001
1000 ppm	16.016	0.00002	15.984	0.00003
5000 ppm	16.08	0.000135	15.92	0.00009
1%	16.16	0.0002	15.84	0.0001
2%	16.32	0.0009	15.68	0.001
3%	16.48	0.0025	15.52	0.0034
4%	16.64	0.0085	15.36	0.0093
5%	16.8	0.012	15.2	0.0115

Table 6.13 BER at various degrees of frequency error using data stream of PN15

Error	Frequency+ (MHz)	BER+	Frequency− (MHz)	BER−
1 ppm	16.000016	0.00000001	15.999984	0.00000002
10 ppm	16.00016	0.0000007	15.99984	0.00000065
100 ppm	16.0016	0.0000087	15.9984	0.00003
500 ppm	16.008	0.00008	15.992	0.00008
1000 ppm	16.016	0.0007	15.984	0.00062
5000 ppm	16.08	0.0012	15.92	0.0009
1%	16.16	0.0019	15.84	0.0012
2%	16.32	0.0028	15.68	0.0019
3%	16.48	0.0054	15.52	0.0047
4%	16.64	0.0174	15.36	0.0185
5%	16.8	0.022	15.2	0.031

seen that this TAF-CDR can tolerate large frequency error. It is worth mentioning that still better result can be expected if PN5 bitstream is used in our test. In PN5 stream, 8b/10b encoding is used and consequently, the longest no-transition stream is 5 bits. In PN9 and PN15, longer no-transition period is possible and this will make the work of CDR harder. The BER result from PN9 shows a better tolerance of frequency error than that the case of PN15 since the no-transition period is longer in PN15.

Gestalt Switch:

The resolution of the two long-lasting problems (arbitrary frequency generation and instantaneous frequency switching) leads to a new direction for the investigation of an existing problem. In this section, the classical problem of clock data recovery is studied from a new perspective.

6.13 TAF-DPS for Dynamic Frequency Scaling

Dynamic frequency scaling (DFS), also known as CPU throttling, is a power management technique in computer architecture. Depending on actual need, the clock frequency of a processor can be automatically adjusted "on the fly" to conserve power and reduce the amount of heat generated by the chip. DFS can help preserve battery on mobile devices and reduce cooling cost on mainstream computing. In practice, DFS almost always appears in conjunction with dynamic voltage scaling since lower voltage can be used when clock of lower frequency is driving the circuit, still yielding correct result. The combined effort is known as dynamic voltage and frequency scaling (DVFS).

6.13 TAF-DPS for Dynamic Frequency Scaling

TAF-DPS, as a clock signal generator, is an ideal tool for DFS due to its features of arbitrary frequency generation and instantaneous frequency switching. A hypothetical example can be used here to illustrate the point. In this imaginal example, our goal is to keep the chip's temperature as steady as possible at a target value. It is well-known that power consumption can be estimated by $P = C \cdot V^2 \cdot A \cdot f$ where C is the capacitance being switched per clock cycle, V is voltage, A is the activity factor indicating the average number of switching events in the chip and f is the clock frequency. We further assume that power consumption and chip temperature has a simple linear relation expressed as $T_e = T_s + (P_c - P_d) \cdot \Delta t/C_m$, where T_e and T_s are ending and starting temperatures, respectively, P_c is current power consumption corresponding to current operating frequency f and P_d is the power dissipated as the heat, Δt is the time difference between start and end, C_m is a coefficient representing the power needed for a temperature increase of 1 °C. Under this postulation, the operation of TAF-DPS can be correspondingly controlled and its output frequency profile can be tuned toward the goal of keeping temperature steady. A simulation for illustrating the point is displayed in Fig. 6.132. As seen, by dynamically adjusting clock frequency, the chip operating temperature can be reined near the target of 40 °C. This is a hypothetical illustration that can be easily tailored in real situation. In real application, on-chip temperature monitor, which is readily available in today's chips, can be used for monitoring the temperature continuously. Its output can be used to guide the TAF-DPS operation. During this process, specific algorithms can be designed, conveniently at software level, to fulfill user's any need.

When applying DFS technique in real application, an issue that is crucial at circuit level is the requirement that the normal work flow must not be interrupted while clock frequency being adjusted. In other words, we want the system to continually operate even when its driving clock is making significant frequency change (for whatever purpose). TAF-DPS is ideal for meeting this challenge since it can switch its output frequency seamlessly, free of any glitch as evidenced in the simulation of Fig. 6.126. To demonstrate this feature for real, we create a circuit of multiplication-then-division

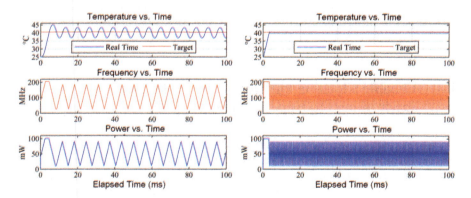

Fig. 6.132 Dynamic frequency scaling for keeping temperature steady: Cm = 0.01 (left) and Cm = 0.001 (right)

as depicted in Fig. 6.15. A 8-bit PRBS generator produces a random sequence whose last five bits $D_1(5)$ are fed into a multiplier. The $D_1(5)$ is then multiplied with a 5-bit constant $D_2(5)$. The output $D_3(10)$ is fed into a division circuit whose output is XORed with the original $D_1(5)$. The circuit is clocked by CLK. When it works correctly, the circuit should always output a logic "0". From a circuit operation perspective, this test circuit is a representative of all digital circuits regardless of their functionalities. We want to use this figurative circuit to show the effectiveness of TAF-DPS clock signal in a general sense.

This test circuit is mapped into a Kintex-7 FPGA. To test this feature of uninterrupted operation under dynamic frequency scaling, the setup constraint for the circuit is set to 120 MHz. The TAF-DPS control word F is switched in a pattern of $32 \rightarrow 4 \rightarrow 32$, in step of one. For each F value, a time frame of 50 cycles is allocated. This pattern is plotted as the brown curve in Fig. 6.16. The TAF-DPS output frequency is measured and displayed as the blue. As seen, the frequency sweeps from 20 to 160 MHz as F changes. This TAF-DPS output is used to drive the test circuit. The bottom potion of Fig. 6.16 is the screen capture of the test circuit's output waveforms. At the top is the waveform of TAF-DPS output. As seen, the waveform is continuously varying since its frequency is constantly changing. The red curve at the bottom is the test circuit output. It fails (output becomes "1") in the region where the clock frequency is high. This is due to the fact that the test circuit is constrained at 120 MHz and the setup constraint is violated when its driving clock is higher than that value. In all the other regions of lower clock frequency where the setup constraint is met, the circuit works correctly. To our best knowledge, this feature of seamlessly frequency switching has not been reported from other type of clock generators. This clocking style of no-interruption-to-system-operation is certainly preferred in driving processor and many other circuits (Xiu 2017a).

Gestalt Switch:

The resolution of the two long-standing problems (arbitrary frequency generation and instantaneous frequency switching) provides a new possibility for solving an old problem. In this section, the classical challenge of dynamical frequency scale is investigated by using a new tool.

6.14 Flexible Clocking for Resource-Constrained Environment in Edge and IoT

Loosely speaking, cloud computing is the on-demand availability of computation resources, especially data storage and computing power, to user without requiring active management from the user. The term is generally used to describe data centers available to many users over the Internet. Nowadays, large clouds often have functions distributed over multiple locations from central servers. Cloud computing relies on sharing of resources to achieve coherence and economies of scale. The availability

6.14 Flexible Clocking for Resource-Constrained Environment in Edge and IoT

of high-capacity networks, low-cost computers and storage devices as well as the widespread adoption of hardware virtualization, service-oriented architecture and autonomic and utility computing has led to the rapid growth in cloud computing.

On the other hand, against the background of cloud computing, there emerges a new trend of edge computing. Edge computing is a distributed computing paradigm that brings computation and data storage closer to the location where it is needed to improve response time and save bandwidth. It is specifically important in applications where real-time data processing is required. In comparison, cloud computing operates on big data while edge computing operates on "instant data" that is real-time data generated by sensors or users. For the purpose of processing real-time data, edge computing offers advantage in the aspects of privacy and security, scalability, reliability and more importantly, speed and efficiency.

The increase of IoT (Internet of Thing) devices at the edge of the network is producing a massive amount of data to be computed at data center, pushing network bandwidth requirements to the limit. Despite the improvement in network technology, data center designated for cloud computing cannot guarantee acceptable transfer rate and response time to most requests, which however could be a critical demand for many applications at the edge and for IoT. On another front, devices at the edge constantly consume data coming from the cloud, forcing people to build content delivery network to decentralize data and service provisioning. The aim of edge computing is to leverage physical proximity to the end user. It moves computation away from data centers towards the edge of the network, exploiting smart objects, mobile phones, or network gateways to perform tasks and provide services on behalf of the cloud. By moving services to the edge, it is possible to provide content caching, service delivery, persistent data storage, and IoT management. This shift in workload to edge can result in better response time and transfer rate. The partition of responsibility is illustrated in Fig. 6.133.

IoT has enjoyed a rapid growth in recent decades, from blood glucose, blood pressure, and oxygen saturation monitors in medical sector, to temperature and smoke detectors used in building automation, to e-locks used in building security, just to name a few. However, the lack of integration is still one of the major bottlenecks to the widespread of IoT adoption. The key factor in this regard is the crystal used as the component for timing reference in IoT devices. The problem is simply due to its large size and high-power consumption. There is a strong motivation for replacing crystal with something more integration-friendly and less power hungry, such as MEMS (Microelectromechanical system) and BAW (bulk acoustic-wave) resonators. From another viewpoint, the growing popularity of edge computing also calls for crystal-less solution. In its hardware implementation, edge computing faces the same challenge of integration due to the very fact that the devices sit at the "edge" between the real world and the computing world. The harsh environment of "edge" demands small size and low power consumption, in contrast to the luxury datacenter where cloud computing enjoys. This situation naturally calls for the elimination of the crystal, as the case in IoT.

As shown in Fig. 6.133, the three major tasks in edge and IoT are computing (information processing), sensing (environment assessment) and actuating (action

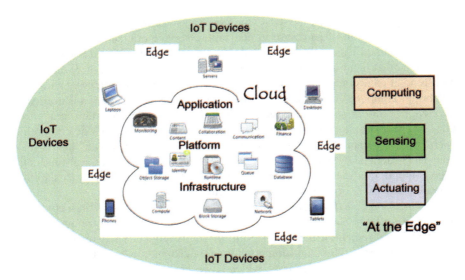

Fig. 6.133 Cloud computing, edge computing and IoT

upon physical world). The task of computing requires a reliable clock signal. The sensing and actuating demand a good capability on manipulating electrical pulse. By replacing crystal with more integration-friendly solutions (such as MEMS and BAW), the quality of frequency source is however suffered, especially frequency accuracy. The subject of clocking in this environment of large frequency variation is therefore a new challenge for edge computing and IoT. In this battle, the ideology of flexible clocking is the grand strategy, the TAF-DPS enabled architectures, described in previous sections of this chapter, are powerful weapons in our arsenal. Here, the term of flexible clocking is referred to the clock signal having the features of "arbitrary frequency generation" and "instantaneous frequency switching", achieved simultaneously for a given design. The TAF-DPS enabled flexible clocking ideology provides a fresh perspective for advancing in this new hardware design frontier. It can be readily used for assisting the various tasks at hand. This perspective is presented in Fig. 6.134.

In nodes for computing, the usage of TAF-DPS is illustrated in the top drawing of Fig. 6.134. Here, the working target is binary bits. There are three operations in this realm: bit manipulation for computation, bit transportation for communication and bit generation for security. The TAF-DPS FIFO, TAF-DPS CDR, TAF-DPS DFS, TAF-DPS SSCG, TAF-DPS RNG and TAF-DPS PUF can be used to handle the various problems caused by large frequency variation. For nodes targeting at sensing and actuating, the applicability of TAF-DPS lies in its power of manipulating electrical pulse. The plan is depicted in the bottom drawing. In this domain, the TAF-DPS is not used for generating clock signal but for creating various electrical waveforms. Those waveforms are used to, through the fundamentally tangible information-carrier of

6.14 Flexible Clocking for Resource-Constrained Environment in Edge and IoT

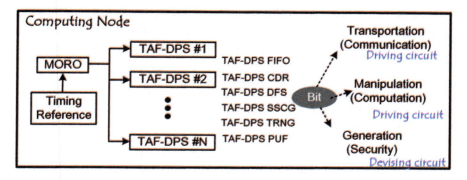

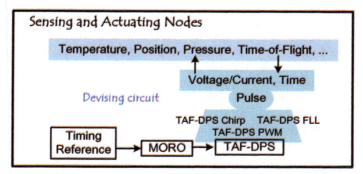

Fig. 6.134 TAF-DPS enabled architectures for computing node (top) and sensing & actuating node (bottom)

voltage/current and time, interact with a variety of sensors and actuators. The ultimate goal is to discern and control the change in natural phenomena of temperature, position, pressure, time-of-flight and etc. To fulfill those duties, the TAF-DPS PWM, TAF-DPS FLL and TAF-DPS chirp generator can be very useful tools. They are convenient and low-cost in implementation.

In Fig. 6.135, a more detailed picture is painted for illustrating the use of TAF-DPS. It is used to assist the baseband signal processing in an environment of large frequency

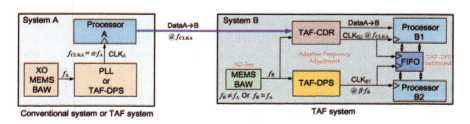

Fig. 6.135 Flexible clocking for system with large frequency variation: computation and data transportation

variation that is a very likely scenario when crystal timing reference is replaced by something more integration-friendly. One of the most noticeable drawbacks when crystal is removed from the system concerns the frequency accuracy of the time reference. With a relatively high-Q factor, crystal provides decent short-term noise performance and long-term frequency stability for most commercial applications. Generally speaking, crystal-less time reference suffers the problem of frequency error more severely than its crystal-based counterpart. This can cause many secondary problems. One of the issues is the data transportation between systems, as illustrated in Fig. 6.135.

There is requirement of data communication between two systems A and B, which is symbolled by the bitstream DataA → B. The two systems are asynchronous by the fact that they are all driven by their own clocks (since they are independent nodes in IoT or "edge"). We assume that the time reference on system A can be either XO (crystal) based or XO-less while system B is XO-less. Under this assumption, the reference frequency f_A for system A is most likely not equal to the reference frequency f_B for system B. The two frequencies can be completely out of sync $f_A \neq f_B$ when the two systems are heterogeneous, or $f_A \approx f_B$ when homogeneous. In both cases, there is nonignorable frequency mismatch between the twos.

The Processor A is driven by CLK_A with a frequency of $\alpha \cdot f_A$, where α is a frequency multiplying factor. The multiplication can be performed by a PLL or TAF-DPS. In system B, there can be multiple processors. Some of them have direct data communication with system A and others do not. We use processor B1 to represent those having communication with system A and B2 for those that do not. B1 and B2 communicate with each other through a FIFO. As discussed in Sect. 6.11, there is a need of large FIFO inserted in between A and B since they are frequency-heterogeneous. The larger the frequency mismatch is, the larger the size of FIFO needs to be. Large FIFO is certainly not preferred in the design of nodes for IoT and edge computing. In this situation, the TAF-DPS based flexible clocking can provide an elegant solution.

As shown in the Fig. 6.135, on the data receiving side (i.e., system B), the system can be driven by TAF-DPS generated clocks. The TAF-CDR discussed in Sect. 6.12 can be used to latch the incoming bitstream DataA → B. As a result, the large FIFO in between A and B can be eliminated. Moreover, the FIFO between processors B1 and B2 can be optimized by using the approach discussed in Sect. 6.11. In short, the system B is working in an environment where all the operating frequencies can be adaptively adjusted. This can lead to significant resource saving, which is critical to IoT and edge computing applications. The solution presented in Fig. 6.135 is applicable regardless the condition of system A. In other words, system A can work in the environment of conventional frequency or in Time-Average-Frequency.

Paradigm Shift:

In the Time-Oriented paradigm, the subject of circuit-clocking can be investigated from a new and larger perspective. It paves the way for meeting the challenges encountered in emerging hardware design frontier.

6.15 TAF-DPS Generator in Large SoC for Producing Clock Frequencies

In Sect. 6.14, the flexible clocking ideology enabled by TAF-DPS has been discussed for assisting signal processing amid crystal-less environment where a large degree of frequency variation might exist. In a more general case, TAF-DPS can be used as clock signal generator to drive on-chip processor (or any digital signal processing circuit) in large System-on-Chip (SoC) project, whether the timing origin comes from crystal or something else. The whole picture is illustrated in Fig. 6.136. On the left is the TAF-DPS circuit where its K-inputs come from a MORO (multiple output ring oscillator). A figurative TAF-DPS usage in a large SoC is depicted in the immediate right of the TAF-DPS circuit. Compared to the conventional solution of using PLL, the advantage of using TAF-DPS clock generator is the strength of ample supply of frequency and fast frequency switching. The techniques of TAF-DPS FIFO, TAF-DPS DFS and TAF-DPS SSCG, discussed in previous sections, are representative examples of utilizing such power. They are especially useful in the environment of large SoC for commercial products. This is due to the reason that, in large SoC projects, there are usually a large number of FIFO devices used in the design. Moreover, dynamic frequency scaling (DFS) is highly desired in large signal processing chip since the heat dissipation problem is very challenging there. Furthermore, spread spectrum clock generation (SSCG) is often a major concern for large SoC used in commercial products as well. In meeting those challenges, TAF-DPS is a powerful weapon.

In large SoC project, from pure implementation perspective, an important concern is the silicon area and power consumption that are budgeted for clock generators. In many cases, the power allocated for on-chip clock generators can be as high as one third of the total budget. The portion of area used for clock generators, when considered against the background of whole chip area, can also be significant. Therefore, there is strong motivation to reduce the amount of silicon area and the power usage of

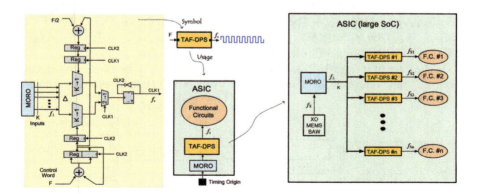

Fig. 6.136 Using TAF-DPS as clock generator for large System-on-Chip project

clock generators, especially for SoC used in commercial products where cost is very sensitive. TAF-DPS technology again can provide a decent solution to this specific problem. As shown in the right-hand side of Fig. 6.136.

In large SoC chip projects, there can be multiple clock generators required to support all the functions that will be performed by the whole chip. It is not unusual to see several, even tens of, PLLs being used in a single SoC project. Those PLLs occupy significant silicon area and consume lots of power. A solution to mitigate this problem is to use multiple TAF-DPS devices that are attached to a single MORO, as shown in Fig. 6.136. This solution, compared to the plan of using multiple independent PLLs, can save significant area since TAF-DPS circuit is very compact. By getting rid of the analog components (especially the large capacitors used in the PLL loop filter), the silicon area allocated for clock generated can be largely reduced. At the same time, the power consumption for clock generators can be seriously lessened as well. This solution has been applied in several price-sensitive consumer electronics products. Please see Xiu (2007), Xiu et al. (2013) for more details in this regard.

In large SoC projects, besides digital signal processing circuit (e.g., processor), mixed-signal components such as ADC (analog-to-digital converter) and DAC (digital-to-analog converter) are frequently used since the SoC has the need to interact with the outside world. When driving ADC, DAC and other mixed-signal circuits, the spectrum of clock signal is preferred to be clean. For this purpose, the TAF clock signal is not ideal since its spectrum contains spurious tones caused by the use of multiple types of cycles (e.g., the T_A and T_B, please see Chap. 5 of Xiu 2012a for more detail on spectrum analysis of TAF signal). In such scenario, the TAF-DPS is not recommended for being used as the clock generator to directly drive those mixed-signal components. It can however function as a fractional frequency divider for enhancing frequency resolution of integer-N PLL. This idea is delineated in Fig. 6.137 Xiu (2014a, d).

Thanks to the use of multiple signals as its input, the TAF-DPS circuit can resolve a time step that is finer than the cycle of the input signals. In other words, it can reach sub-cycle time resolution (Xiu 2015b). Using the structure shown in the left of Fig. 6.137 as example, the time span between any two adjacent VCO outputs is $\Delta = T_{VCO}/K = 1/(K \cdot f_{vco})$. The TAF-DPS output period thus is $T_S = 1/f_s = F \cdot \Delta = F/(K \cdot f_{vco})$. Or, $f_s = (K/F) \cdot f_{vco}$. When the PLL is in lock, we have $f_r = f_b = f_s/N = [K/(F \cdot N)] \cdot f_{vco}$. This leads to $f_{vco} = (F/K) \cdot N \cdot f_r$.

In the case of integer-N PLL where $f_{vco} = N \cdot f_r$, or $T_r = N \cdot T_{VCO}$, the signals from VCO and reference are compared once every N VCO cycles. For the case with

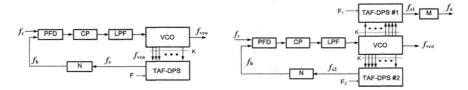

Fig. 6.137 TAF-DPS used as fractional frequency divider for enhancing frequency resolution

6.15 TAF-DPS Generator in Large SoC for Producing Clock Frequencies

TAF-DPS, the two signals are compared once every (F/K) · N cycles. Therefore, the time resolution can be smaller than that of integer-N when F < K. Consequently, there are more frequencies that can be generated from VCO. The resolution of the original integer-N PLL is improved. Furthermore, when a technique called "post divider fractional bits recovery (PDFR)" is applied, the resolution of this "integer-N plus TAF-DPS fractional divider" structure can be proven to be f_r/K. In other words, the original integer-N PLL's frequency resolution is improved by K times (Xiu et al. 2013). For more information on PDFR technique, please refer to Sect. 4.6 of Xiu (2012a). It is worth mentioning that, although the frequency granularity is improved, the design of loop bandwidth is not impacted since the reference frequency f_r is not changed.

The structure shown in the left of Fig. 6.137 can be further aided by another TAF-DPS for more power in frequency generation. This new structure is depicted in the right-hand side of the figure. Using some simple algebra, its output frequency can be derived as $f_o = [(F_2 \cdot N)/(F_1 \cdot M)] \cdot f_r$. This formular is similar to the one achieved from two cascaded integer-N PLLs, which is shown in Fig. 6.138. However, the structure depicted in Fig. 6.137 right is even more powerful when PDFR technique is used. In this structure, the [F_2, N] is a pair where the PDFR can be applied, as discussed before. At the same time, the PDFR can be applied on the pair of [F_1, M]. By this configuration, the frequency resolution can be further improved to the order of $O(f_r/(M \cdot K)^2)$ (Xiu et al. 2013).

The real purpose of using TAF-DPS circuit to improve integer-N PLL's frequency resolution is to generate clock signal in conventional frequency. This signal can then be used to drive on-chip ADC or DAC, which are often incorporated in large SoCs. As an example, we use a design case from a real commercial product to illustrate the point. One of the challenges for clock generator used in TV oriented SoC is the generation of audio frequencies. Table 6.14 lists all the required audio frequencies. The numbers inside the table are the audio frequencies in MHz. The difficulty of materializing this table comes from the requirement that all these frequencies have to be generated from a single reference frequency (e.g., a crystal) for cost concern. For this reason, in the past, two or three integer-N PLLs are cascaded together in producing those frequencies (fractional-N PLL could be used theoretically, but it does not provide a practical solution due to the fraction spurs which are hard to be eliminated completely). Thanks to the structure presented in Fig. 6.137 and the PDFR technique, the goal of generating all those frequencies can be achieved with

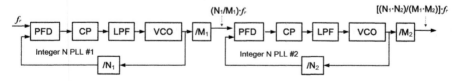

Fig. 6.138 Two cascaded integer-N PLLs

Table 6.14 Audio frequencies (in MHz)

	128[2]	256	384	512	768	1024
16[1]	2.0480	4.0960	6.1440	8.1920	12.2880	16.3840
22.5	2.8224	5.6448	8.4672	11.2896	16.9344	22.5792
24	3.0720	6.1440	9.2169	12.2880	18.4320	24.5760
32	4.0960	8.1920	12.2880	16.3840	24.5760	32.7680
44.1	5.6448	11.2896	16.9344	22.5792	33.8688	45.1584
48	6.1440	12.2880	18.4320	24.5760	36.8640	49.1520
64	8.1920	16.3840	24.5760	32.7680	49.1520	65.5360
88.2	11.2896	22.5792	33.8688	45.1584	67.7376	90.3168
96	12.2880	24.5760	36.8640	49.1520	73.7280	98.3040
128	16.3840	32.7680	49.1520	65.5630	98.3040	131.0720
176.4	22.5792	45.1584	67.7376	90.3168	135.4752	180.6336
192	24.5760	49.1520	73.7280	98.3040	147.4560	196.6080

[1] The first column is the sampling frequencies in KHz
[2] The first row is the oversample rate

just one component. Figures 6.139, 6.140 and 6.141 are the measured spectra of the six audio frequencies in the low right corner of the table (all the rest frequencies can be generated through division from those). The ratios of these six frequencies to 12 MHz (the reference) are all listed in the corresponding figures. For each frequency, the corresponding setting $\{N, F_1, F_2, M\}$ is also included in the figure.

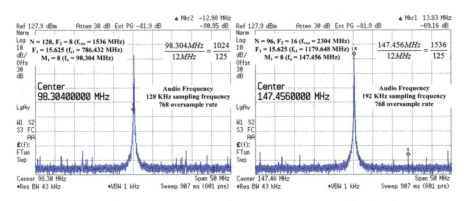

Fig. 6.139 The spectrum of two audio frequencies: 98.304 (left) and 147.456 MHz (right)

6.16 TAF-DPS for the Challenge of Global Clock Distribution

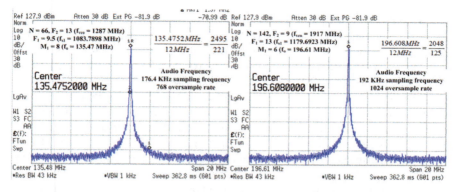

Fig. 6.140 The spectrum of two audio frequencies: 135.4752 (left) and 196.608 MHz (right)

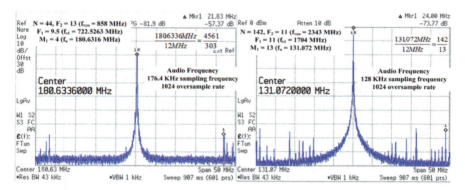

Fig. 6.141 The spectrum of two audio frequencies: 180.6336 (left) and 131.072 MHz (right)

Gestalt Switch:

The resolution of the two long-standing problems (arbitrary frequency generation and instantaneous frequency switching) enables new solution to old problems. In this section, the classical problem of improving clock generator's frequency resolution is handled with a new tool. The important issue of reducing the resource used by clock generator in large SoC is subsequently addressed.

6.16 TAF-DPS for the Challenge of Global Clock Distribution

Clock distribution is a critical task in the implementation stage of modern chip design. In recent years, the advance in CMOS technology has led to an exponential increase in chip complexity. The number of transistors on large chips has reached the level

244   6 Old World and New Insight: Solving Problem with a Gestalt Switch

of multi-billions. Structurally, modern SoC can be regarded as many on-chip micro-networks communicating to each other all the time. Clock signal is the key factor that makes this happen. From clocking perspective, chip architecture can be classified as Globally Asynchronous Locally Synchronous (GALS) and Globally Synchronous Locally Synchronous (GSLS). In GSLS approach, the clock signals driving all the on-chip modules must run at the same frequency. Among them, they also have to maintain fixed phase relationship. Those requirements require the distribution of a global clock signal. There are several considerations when distributing a clock signal globally: the skew caused by different distribution paths, the jitter accumulated along the individual distribution path, the silicon and metal resource required for routing the clock signal and the power used by the distribution network. Figure 6.142 illustrates the clock distribution methods commonly used in chip implementation practice.

In the top left of Fig. 6.142, conventional tree structures are used to distribute the clock signal. The distribution network can be constructed using the topology of branch tree, H-tree or X-tree. In the illustration of top right, a H-tree with active skew compensation is depicted. To alleviate skew problem, the delays at the ends of different branches are compared. The result is then used to drive delay lines so that the delays of the paths can be adjusted. Consequently, skew can be minimized. Clock mesh (clock grid) is also used for some designs, especially in high end microprocessor. In this method, a solid grid made of metals is constructed on-chip as shown in the bottom left of Fig. 6.142. Its purpose is to deliver the clock signal to all the spots on the chip. In practice, the tree and grid methods can be used together to achieve the

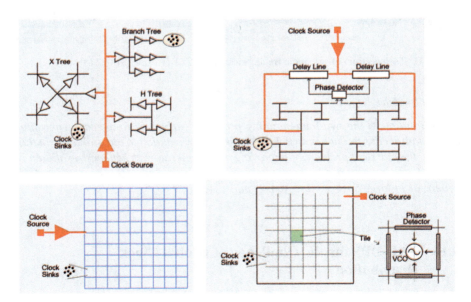

Fig. 6.142 Clock distribution methods: conventional trees (top left), delay/skew compensated H-tree (top right), clock mesh (bottom left) and distributed PLL array (bottom right)

6.16 TAF-DPS for the Challenge of Global Clock Distribution

goal of delivering a clock signal from a source to all the sinks across a large chip in a resource-efficient way (Chan et al. 2009). To minimize skew actively, a method of using distributed PLL array is proposed (Pratt et al. 1995; Gutnik and Chandrakasan 2000a, b; Zianbetov et al. 2013). In the drawing of bottom right, the entire chip is split into multiple small areas called tiles. Inside each tile, there is a local frequency generator (represented by the VCO symbol). Along the four boundaries of each tile, there are phase detectors used for comparing the delay differences between the local clock and its neighboring clocks. The result is used to drive the frequency generator and then to minimize the skew. In this approach, the array of distributed PLLs actively compensates the skew.

As semiconductor process technology advances, the tree structure faces difficult challenges. The circuit operating frequency becomes higher due to the reduction in transistor gate delay. The chip size becomes larger since more transistors needs to be packed into the chip. As a result, the global clock signal has to travel a longer distance. Moreover, both the gate and interconnect delay variations caused by PVT change become larger. Furthermore, the interconnect delay does not scale well with process advance. All these factors have made the problem of skew worse. Now, the skew has taken a larger percentage of clock period. Also, the variation of skew is harder to be controlled. To make it even worse, the delivery of clock signal crossing chip in high frequency requires large amount of metal resource (for shielding) and high consumption of energy (could be as high as 50% of the total power used by the chip). The distributed PLL array approach also has the problems of high resource usage and high power consumption. Further, it has additional stability problem due to the reason that multiple PLLs are required to lock to the same common reference.

The TAF-DPS clock generator introduced in Sect. 4.6 provides a new possibility for dealing with the clock distribution problem. As discussed in Sects. 6.2.2 and 6.3.9 and expressed in Eqs. (6.2.1) and (6.3.2), the TAF-DPS is able to generate frequency that is higher than its input. Furthermore, as explained in Sect. 4.25 of Xiu (2012a), TAF-DPS can control its output phase. In other words, the phase of its output has a deterministic relationship with that of the input reference and the position of the phase is adjustable as desired. This enables a clock-distribution method that can use low frequency in the global clock signal. When reaching destinations, the low frequency global clock can be frequency-boosted by local TAF-DPS to generate the function clock. This scheme is depicted in Fig. 6.143 (Xiu 2017c). It has several benefits as will be elucidated next.

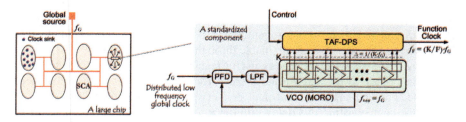

Fig. 6.143 TAF-DPS enabled clock distribution method of using low global frequency

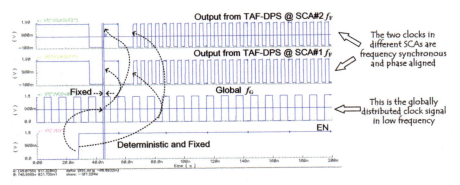

Fig. 6.144 The TAF-DPS clock sources in all SCAs are synchronous

As shown in the left drawing of Fig. 6.143, the whole chip is split into multiple synchronous clocking areas (SCA). Each SCA is small enough that conventional clock distribution method (such as tree-structure) can be efficiently used to deliver the clock signal within the area. All the SCAs are supposed to be synchronous to each other (i.e., having same frequency with fixed phase relationship among them). Within each SCA, there is one TAF-DPS clock generator designated in producing the needed frequencies for function clock serving local operations. Its structure is depicted in the right-hand side of the figure. There is a 1 × PLL used for generating the K outputs to support the TAF-DPS. The distributed low frequency global clock signal of f_G is used as its reference. The frequency of function clock f_F is generated as $f_F = (K/F) \cdot f_G$, where F is the frequency control word. As seen, when F < K, we can produce a function clock with higher frequency than f_G.

Figure 6.144 presents a transistor level simulation result to illustrate the scheme. The global clock signal is delivered to their respective PLLs for both the SCA#1 and SCA#2 in a relatively low frequency. In both PLLs, one of the K VCO outputs is phase aligned with the global clock. Therefore, the two VCO outputs are phase aligned to each other even though they are in different SCAs (i.e., the two VCOs are synchronous). Within each TAF-DPS, the circuit guarantees that its output has fixed and known phase relationship with that particular VCO output, as explained in Chap. 4 of Xiu (2012a). This feature is demonstrated here in the simulation of Fig. 6.144. As a result, the outputs of both TAF-DPS (the two function clocks associated with those two SCAs) are phase aligned, in addition to the fact of having same frequency. This ensures the synchronous operation of the GSLS system. As shown, this function clock has a higher frequency than that of the global clock (much higher frequency can be generated but it would be difficult for it being displayed clearly). Moreover, this frequency is programable through the control word F.

Figure 6.145 shows another very useful feature that TAF-DPS is able to adjust its output's phase. If the SCA#1 and SCA#2, for some reasons such as imbalanced delays accumulated in their global clock delivery paths, are required to have different phases for compensating this delay difference, one of the TAF-DPS can adjust its phase to take care of this requirement. This is because that, as explained in Sect. 4.25

6.16 TAF-DPS for the Challenge of Global Clock Distribution

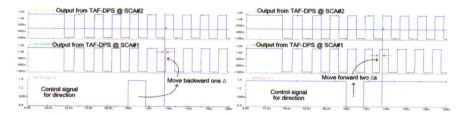

Fig. 6.145 The phase between TAF-DPS clock sources can be adjusted

of Xiu (2012a), the TAF-DPS can be viewed as a DLL (delay locked loop) with infinite delay adjustment capability. This scenario can happen when there is data communication between SCA#1 and SCA#2. In the simulation shown in the left-hand of Fig. 6.145, the output from TAF-DPS of SCA#1 is moved backward by one Δ. In the right-hand simulation, it is moved two Δs forward. This phase adjustment capability can help align data and clock appropriately so that the data can be captured more reliably.

In short, the features of TAF-DPS based low frequency global clock distribution method can be summarized as below.

- All the SCA function clocks can be made frequency synchronous (having same frequency), or they can be set for different frequencies if desired.
- All the SCA function clocks can be made phase aligned (having zero phase difference). This leads to zero skew.
- If desirable, each SCA clock's phase can be adjusted individually (useful skew between SCAs).

The advantages of this clock distribution method are summarized below.

- Low frequency can be used in global clock signal. This leads to significant reduction in routing resource and power consumption.
- Skew problem can be better controlled since global clock is in low frequency.
- Function clock is expected to have lower jitter since less jitter accumulation is experienced in the delivery path. Also, the PLL can filter out most of the high frequency jitters collected in the delivery path.
- This method scales well with technology since the performance of TAF-DPS improves with the advance of process.
- Overall, less electromagnetic radiation is generated since global clock is in low frequency.

Gestalt Switch:

A powerful new tool in circuit level enables novel solution in system level. A difficult challenge in the field of chip implementation, the clock distribution problem, is investigated from a new direction.

6.17 Frequency Issue in Clock Distribution Through Resonant Clock

The most distinct feature of clock network is its large capacitive loading presented to clock source. During operation, the clock source is responsible for the charge and discharge of this large capacitance. Therefore, an effective way of reducing power is to recycle the energy used by the charging and discharging of this large capacitance of clock network. This idea of globally distributing clock signal over a large chip is called "charge recovery clock distribution". This approach is realized by utilizing the principle of LC resonance, as illustrated in Fig. 6.146.

As shown, some amount of inductance is intentionally introduced into the chip's structure. The innate capacitance associated with the clock distribution network functions as the C of the LC resonator. During the process of charge and discharge, energy is stored and released periodically through this resonator. Ideally, 100% of the energy can be recycled and the electrical oscillation (the clock waveform) can be self-sustained. In practice, due to the parasitic resistance associated with the inductor and the clock sinks, some portion of the energy is lost as the generated heat. Hence, a compensation circuitry has to be incorporated on-chip to supply the energy required for sustaining the oscillation. It is expected that the energy needed in this approach is much lower than that consumed in conventional methods (e.g., those discussed in Sect. 6.16). This is due to the reason that, instead of $C \cdot V^2 \cdot f$ as in conventional case, the consumed power now is I^2R where R is the total parasitic resistance. This power is frequency independent and hence this approach is a good candidate for distributing clock in high frequency (~GHz range).

Figure 6.147 illustrates four methods reported in literature for delivering clock signal in the LC resonance fashion. In top left, the clock distribution network is divided into multiple clock sectors (Chan et al. 2003, 2004, 2009). The global clock signal is delivered to each sector using H-tree. Within each sector, the clock signal is further delivered to multiple (for example four) locations. In each of such location, there is a spiral inductor whose two ends are connected to the global clock network and a decoupling capacitor, respectively. The decoupling capacitors, the parasitic capacitance associated with the clock network and the inductors form a LC resonant circuit that is used to recycle the clock energy. In this structure, however, the actual

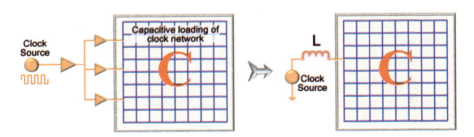

Fig. 6.146 The principle of recycling energy in clock distribution network

6.17 Frequency Issue in Clock Distribution Through Resonant Clock

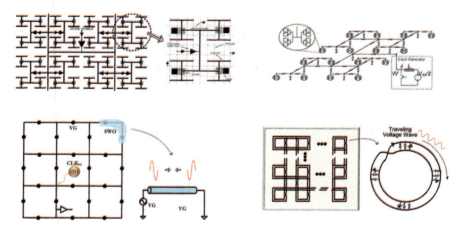

Fig. 6.147 Clock distribution using LC resonance: LC resonance at clock sector (top left), LC resonance at actual clock loading (top right), standing-wave LC clock distribution network (bottom left), traveling-wave clock generation and distribution network (bottom right)

clock sinks (flip-flops, latches, etc.) are driven by local clock buffers attached to a clock grid. This grid is connected to and driven by the global LC distribution network (see the drawing). Therefore, the large capacitance associated with the clock loadings is isolated from the LC network and is not participated in the LC resonance. This leads to a reduced efficiency in recycling the clock power.

In the scheme shown in top right of this figure, the clock LC distribution network reaches directly to the clock sinks (no clock buffer in the network as in previous case). Thus, all the parasitic capacitance takes part in the LC resonance. It results in a larger degree of power saving (Sathe et al. 2008). However, since the LC network outputs a sinusoidal waveform, the slow rising and falling clock edges limit the clocking speed. Further, due to the lack of isolation between the LC network and the clock elements, the data induced jitter will significantly degrade the quality of the clock signal. Further, the popular clock gating technique (for power saving) is hard to be implemented due to the elimination of the buffers in the delivery chain.

$$A\cos(\omega t - \beta z) + B\cos(\omega t + \beta z + \varphi)$$
$$= 2B\cos\left(\omega t + \frac{\varphi}{2}\right)\cos\left(\beta z + \frac{\varphi}{2}\right) + (A - B)\cos(\omega t - \beta z) \quad (6.17.1)$$

When two waves of same frequency travel in opposite directions and interact with each other, standing and traveling waves will be formed. Equation (6.17.1) reveals the result when two waves of $A\cos(\omega t - \beta z)$ and $B\cos(\omega t + \beta z + \varphi)$ travel in opposite direction within the same transmission line, where t and z are independent variables for time and location, respectively. In the right-hand side of the equation, the first term is the standing wave which has uniform phase in all the locations along the line. As seen, its amplitude is location dependent. The second term is the traveling wave whose phase is location dependent but its amplitude is constant. Since standing wave has the unique feature of uniform phase, it can potentially achieve zero skew when used for clock distribution.

In the bottom left of Fig. 6.147, a clock distribution network is constructed by a grid of metal structure with multiple virtual grounds (VG). The VGs are formed by shorting them together. Between two VGs, a standing wave oscillator (SWO) is formed since the electrical waves are reflected at the VGs. This leads to the interaction of waves, resulting in standing wave (the traveling wave is minimized by the fact that the two waves' amplitudes are almost same, or $A \approx B$ in (6.17.1)). A clock signal can be injected into the network to initial and maintain the oscillation. Clock buffers are attached to the metal structure to form the local clock distribution network (O'Mahony et al. 2003). In another design, the VGs are attached by inductors to alleviate the amplitude variation along the line (the amplitudes no longer go to zero at the ends) (Sasaki 2009).

In the bottom right of Fig. 6.147, it shows the configuration of using traveling wave to generate clock signal and to distribute clock signal across chip. The traveling wave is produced by connecting one end of the transmission line to the other so that a loop is formed (Wood et al. 2001). As shown, instead of tying the ends of the transmission line to virtual ground, the ends are cross connected (Möbius ring). In this way, no reflection is resulted and the standing wave is non-existing ($B = 0$ in (6.17.1)). Multiple antiparallel inverter pairs are added to the line to compensate energy lost and maintain rotation lock. The traveling wave has constant amplitude but different phases along the points on the path. After startup, the wave can travel in either direction (usually the direction of lowest lost) unless it is intentionally biased (directional coupling) (Zhang et al. 2009). The oscillation frequency can be slightly adjusted by adding tunable capacitors in between the line. This traveling wave can be distributed to the whole chip as global clock signal by connecting multiple such structures together, as illustrated in the drawing.

The LC resonance clock distribution methods described above have great potential in reducing the power consumption of clock network. They all, however, lack the frequency flexibility required by microprocessor operation. This is due to the fact that the oscillation frequency (for standing wave and traveling wave) or the optimal working frequency (for the two structures in the top row of Fig. 6.147 to achieve maximum power saving) is determined by the physical structure of the network. In other words, the frequency is constrained by the natural frequency of the LC resonator. In supporting LC resonance clock generation and distribution, as a circuit level innovation enabler, TAF-DPS can be used to enhance their frequency flexibility.

Out of the four LC resonance methods discussed in Fig. 6.147, standing wave oscillator (bottom left) and traveling wave oscillator (bottom right) are promising. The reason is that the quality of the clock signal (skew and accumulated jitter) can be better controlled since the global clock is distributed by chip-size distributed LC resonators. In both approaches, the clock generation (frequency synthesis) task can be accomplished by TAF-DPS. In Fig. 6.148, a scheme of using SWO and TAF-DPS for clock distribution and clock generation is illustrated. As described previously, all the locations along the SWO ring can provide clock signals of same frequency and

6.17 Frequency Issue in Clock Distribution Through Resonant Clock

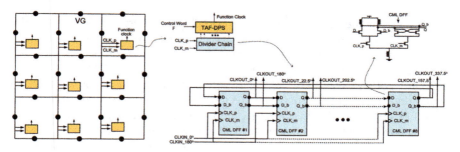

Fig. 6.148 Standing wave oscillator assisted by TAF-DPS for clock distribution and generation

same phase. Usually, this SWO ring can operate in high frequency (e.g., ~10 GHz in a 0.13 um process). The output is available as a differential pair. This high oscillating frequency provides an opportunity for generating multiple outputs through dividers. For example, a divider chain made of 8 CML (current mode logic) dividers can produce 16 evenly distributed phases from this pair of LC resonator output, as shown in the figure. This group of phases can be used to support the TAF-DPS operation. Each region shown in the figure can have a dedicated TAF-DPS. The task of frequency synthesis for functional clock is then accomplished by the TAF-DPS. As discussed in Sect. 6.16, all the TAF-DPS circuits' outputs can be made synchronous to each other since their inputs are synchronized by the SWOs, resulting in zero skew in an ideal case.

The traveling wave oscillator (RTWO) clock distribution depicted in the bottom right of Fig. 6.147 is further illustrated in Fig. 6.149. The LC transmission line is configured as Möbius ring. Within each ring, the signal's phases are uniformly distributed as shown in the left drawing. When tapped out, those phases naturally provide the input for TAF-DPS circuit. There is, however, a synchronization problem when these rings are connected together as ROA (rotary oscillator array). When a single ring is concerned, the wave traveling direction is the one with less impedance. It could be clockwise or counterclockwise. When connected together as shown in the

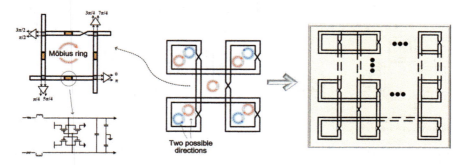

Fig. 6.149 RTWO structure (left), ROA (middle) and chip size clock distribution using ROA (right)

middle drawing, the neighboring rings can oscillate at different directions (nondeterministic). This leads to the difficulty for synchronizing the clocks around the chip because the same-phase-point of all the rings has to be tested beforehand.

In Teng et al. (2011), a solution is proposed by arranging RTWO rings in a different topology called ROA brick. The basic ROA brick structure is illustrated in the left drawing of Fig. 6.150. Compared to the structure in the middle drawing of Fig. 6.149, topology-wise, this one is its "mirror structure". The important feature of this structure is that the traveling wave in all the associated rings travels in the same direction, either all clockwise or all counterclockwise. This fact makes it possible to identify same-phase-point on all the rings, which is labeled in the drawing. Chipwise, the ROA bricks can be arranged in a fashion as shown in the right-hand drawing. Every four RTWO rings make a ROA brick and every ROA brick shares two RTWO rings with its neighbor. In this way, the same-phase-point can be established around the whole chip.

Starting from the same-phase-point, other locations can be tapped out to construct the multiple phases needed for TAF-DPS circuit. The sequential order is determined by their physical positions. With each region surrounded by the lines, a TAF-DPS can be created to generate the function clock. All the outputs from the TAF-DPSs will have the same frequency and are phase aligned, satisfying the requirement of clock distribution for synchronous operation.

The combination of RTWO and TAF-DPS can be a promising solution for the difficult challenge of clock signal distribution in modern large processing chip. The following list is the summary of its advantages.

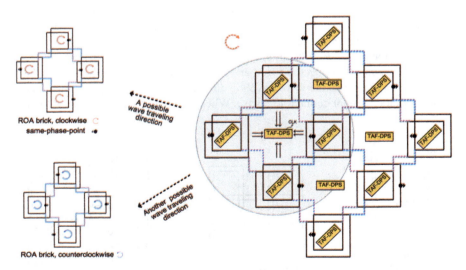

Fig. 6.150 ROA brick (left), chip size clock distribution using ROA and clock generation using TAF-DPS (right)

- Same-phase-point of the whole chip is identifiable by connecting RTWO rings in a ROA brick fashion.
- The sequential order of the multiple phases is established from the same-phase-point and the physical positions of the tap-out points.
- TAF-DPS circuit in each region operates from the phases generated from the particular RTWO ring. Its output is used as function clock for local operation.
- Frequency synthesis is accomplished by TAF-DPS circuit
- The TAF-DPS clocks of all the regions are synchronous.

Gestalt Switch:

A powerful new tool in circuit level enables novel solution in system level. A difficult challenge in the field of chip implementation, the issue of clock distribution through LC resonance, can be solved more efficiently through the assistance of the new tool.

6.18 TAF-DPS for Network-On-Chip GALS Strategy

With billions of transistors used in today's large chips, the advantage of uniprocessor architectures is diminishing due to many reasons, including clock-related issues such as its demand for high clock frequency and the global distribution of such clock signal. Multicore architecture is emerging as the prevailing scheme in both general-purpose and application-specific markets since it allows the distribution of computation load to multiple cores which can operate at their optimum speeds (clock frequency). Consequently, the focus in architecture design is shifted from computation to communication. As the core count increases, the need for a scalable on-chip communication scheme that can deliver high bandwidth becomes a necessity. Traditionally, bus has been the dominant structure for SoC on-chip communication as shown in the left drawing of Fig. 6.151. However, it does not scale well with the increased number of cores. This leads to the recent architecture of networked on-chip communication (Bjerregaard and Mahadevan 2006; Kolodny 2009). In the

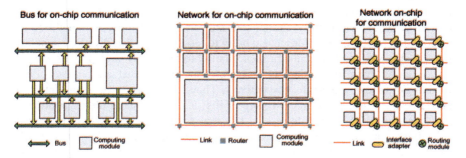

Fig. 6.151 On-chip communication architectures: bus (left), network (middle), and Network-on-Chip (NoC) (right)

middle drawing of this figure, every computing module is surrounded by communication links and routers. From any source to any destination, data is routed by using logical or physical links that follow a predefined protocol. The drawing at the right-hand side of Fig. 6.151 shows a generic Network-on-Chip (NoC) architecture where multiple cores are interconnected with sophisticated on-chip networks. Each computing module is supported by a pair of interface adapter and routing module. All of them are connected on-chip in a meshed network.

One of the key characteristics of NoC is heterogeneous clocking. For a SoC designed in NoC ideology, each computing module operates at its own optimum clock frequency. As a result, overall power consumption can be optimized since each module sets its frequency based on its own loading. More importantly, the large amount of power consumed by the high-frequency global clock distribution network is eliminated. Further, NoC architecture also alleviates the signal delay problem caused by the time-of-flight of long wires.

NoC is a SoC design strategy that separates the tasks of computation and communication in a controlled way so that each of them can be addressed efficiently. It facilitates the integration of various IPs that can come from different venders but are all eventually merged into a particular SoC. A direct consequence of this parctice is the existence of many frequencies in a chip (i.e., the heterogeneous clocking) since each individual IP has its preferred operating frequency. Therefore, in NoC architecture, one of the key design challenges is the data synchronization among the cores that may run synchronously at different frequencies, or operate at mixed mode of synchronous and asynchronous (Chelcea and Nowick 2004; Mullins 2005). Within each computing module (or any other type of IP), the implementation can be realized in synchronous fashion which is the common and efficient design method used today. The data communication among modules must be carried out in asynchronous style to accommodate the different operating frequencies. This is the so-called Globally Asynchronous and Locally Synchronous (GALS) strategy (Mullins 2005; Teehan et al. 2007).

When data transfer between modules of different frequencies takes place, there is a crucial issue of preventing metastability that must be addressed carefully. Metastability is a digital circuit design issue. It describes the condition that a signal's voltage level is at an intermediate level: it can be neither interpreted as "0" nor "1". And this condition may persist for an indeterminate amount of time. Two methods can be used to deal with metastability: timing-safe and value-safe. Timing-safe method allocates a fixed period of time for metastability to resolve (e.g., two flip-flops synchronizer). Value-safe method waits for metastability to resolve by stretching or pausing clock signal. Figure 6.152 depicts two circuit techniques that are representatives of timing-safe and value-safe, respectively. They can be included in the interface adapter, shown in the NoC architecture of Fig. 6.151, to handle the asynchronous data transfer. A FIFO module resides between a transmitter (TX) and a receiver (RX). It has ports of DIN, OUT and OK_to_PUT facing the TX, and ports of DOUT, TAKE and OK_to_TAKE facing the RX. In the left scheme, the TX clock CLKT and RX clock CLKR are all generated locally within the interface module. They can be paused (stopped) or stretched based on the FIFO status. This circuit uses the

6.18 TAF-DPS for Network-On-Chip GALS Strategy

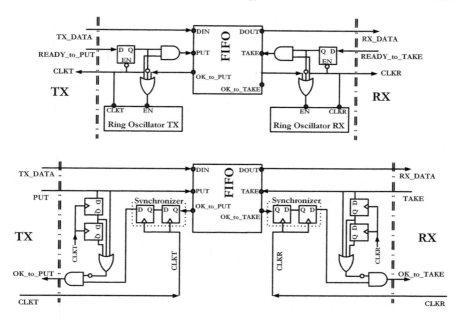

Fig. 6.152 Interface for GALS: stretching or pausing clock (top) and using synchronizer (bottom)

value-safe approach. In the right-hand side, CLKT and CLKR are provided from TX and RX, respectively. They are not locally controllable (e.g., they cannot be paused or stretched locally within the interface module). But the OK_to_PUT and OK_to_TAKE signals from the FIFO are synchronized by the synchronizers to the CLKT and CLKR domains, respectively. This circuit utilizes timing-safe method. As shown, two clock cycles are used to resolve metastability.

In GALS design strategy, local modules are synchronous blocks that can be designed and implemented using conventional CAD tools that most designers are familiar with. In the design of interfacing modules, however, special care is required so that metastability can be avoided under various interfacing scenarios of same-frequency-but-different-phase (mesochronous), averagely-same-but-not-exactly-same-frequency (plesiochronous) and averagely-not-same-frequency (heterochronous). The circuit techniques shown in Fig. 6.152 are invented to address this challenge. The design of such circuits is nontrivial. It requires skill of transistor level circuit design, which is usually not the focus of digital circuit engineer. Further, from pure circuit perspective, they are not without their own problems. The value-safe approach (in the left) relies on the ring oscillator that must be stopped and started frequently. This can degrade the clock quality significantly (i.e., large jitter). The timing-save method (in the right) uses only two (or three) cycles to allow the metastability to settle down, which might not be enough if high degree of reliability is required. Moreover, this method has the problem of large latency due to the multi-cycle synchronizers used.

Using the TAF-DPS as circuit level enabler for system level innovation, it is possible to build a more efficient NoC structure since TAF-DPS is powerful in generating ample supply of frequency and quick in switching its output frequency. In the generic NoC architecture depicted in the right-hand side of Fig. 6.151, all the computation modules are allowed to run at their optimal frequencies. The interface adapters, using the frequency-savvy TAF-DPS, can handle all the rate-related communication issues, such as the clock generation for each computation module, the data launching and capturing between each computation module and the link. The structure of this TAF-DPS based interface adapter is illustrated in Fig. 6.153 (Xiu 2017e).

Each interface adapter sits between a computation module and the link network. It handles the data communication between this particular module and the rest of the chip. When this module needs to send data to the link, it functions as TX. When receiving data from the link, it becomes RX. Similarly, the link functions as RX and TX correspondingly. A FIFO having n storage cells resides between the computation module and the link. They are operating at f_C and f_L, respectively. Each storage cell has separate write and read ports that can be independently controlled by their respective clocks (i.e., a write and a read clock). Moreover, the write and read operations are enabled by their respective enable signals. All the cells' data ports are connected to the buses DATA_C and DATA_L. At any given moment, only one cell is selected for read or write. Data in this FIFO is immobile. In other words, once a bit of data is latched into the FIFO, it will not be moved between the cells. All the storage cells form a circular array. Two tokens are created to control the input and output behavior of the FIFO. Put-token is used to enqueue data items and take-token is for dequene data items. Put-token is at the tail and take-token is at the head. Once a token is used by a cell for a data operation, it is passed to the next cell. This FIFO design is similar to the synchronous-synchronous FIFO described in Chelcea and Nowick (2004).

Two clocks, CLK_C and CLK_L, are produced by two TAF-DPS circuits that are configured to generate f_C and f_L, respectively. The base units used by the TAF-DPS

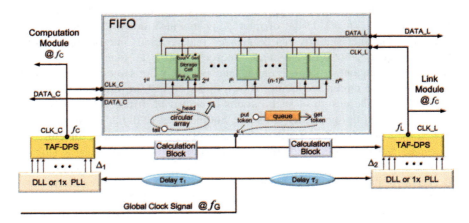

Fig. 6.153 The structure of TAF-DPS based interface adapter for GALS

6.18 TAF-DPS for Network-On-Chip GALS Strategy

circuits, Δ_L and Δ_C, are generated from the multiphase generators in the sender and receiver. The multiphase generator can be a DLL or 1x PLL using a MORO as its frequency oscillator (please refer to Sect. 6.16 and Fig. 6.143). The reference for the two multiphase generators is the global clock signal at frequency f_G. For the purposes of low power consumption and high-quality distribution for long distance, f_G is preferably at a low value (such as below 100 MHz).

Figure 6.154 includes some transistor level simulation results from this TAF-DPS cricuit, showing its capability in handling rate difference. The TAF-DPS can generate higher frequencies than f_G for CLK_C f_C and CLK_L f_L. In operation, both the signals of CLK_C and CLK_L can be stopped and stretched by the TAF-DPS circuit, as shown in the left and middle graphics of Fig. 6.154. Inside the FIFO, based on the put and get token fed to the TAF-DPS circuits, the head and tail of the queue can be properly adjusted in real time through the mechanisms of clock-stop and clock-stretch. In certain case, the clock frequency on the receiving side can be adjusted dynamically according to the sender's data rate if it is known or can be detected. This scenario is shown in the simulation in the right-hand side of Fig. 6.154.

Compared to the existing techniques used in GALS synchronization interfacing, the advantages of this TAF-DPS architecture can be listed as follows.

- Elimination of handshake signals
- Reduction in FIFO size
- Lower power consumption: data is immobile inside FIFO.
- No latency penalty: no multi-cycle synchronizer is used.
- Low-frequency global clock distribution
- One module to support all operating modes without the need of circuit modification: mesochronous, plesiochronous and heterochronous.
- Operating frequency for each core can be optimized and adjusted in real time.
- This interface module can be standardized as a universal IP for GALS.

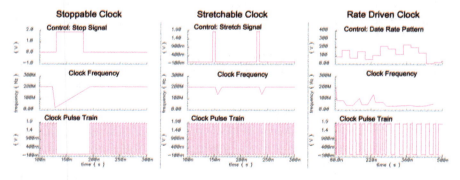

Fig. 6.154 The features of TAP-DPS interfacing circuit: stoppable clock (left), stretchable clock (right) and rate driven clock (right)

Gestalt Switch:

A powerful new tool in circuit level enables novel solution in system level. Emerging architectures for large and sophisticated chip design must be supported from lower level at circuit design. This section presents such an example.

6.19 One-Wire Communication Using Edge-For-Clock-Duty-Cycle-For-Data

Digital transmission is a critical task in modern electronic system design. During the process of transmission, information is first encoded in a line code that uses a pattern of voltage or current to represent digital data. The most commonly used methods for line code are non-return-to-zero (NRZ), return-to-zero (RZ) and Manchester code, as shown in the top row of Fig. 6.155. For NRZ, logic "1" is represented by one significant condition, usually a positive voltage, while "0" is represented by some other significant condition, usually a negative voltage, with no other neutral or rest condition. As a matter of fact, NRZ format is universally employed in digital circuit design (inside a chip) where VDD is used for logic "1" and VSS for "0". In RZ, after reaching the appropriate level, the signal returns to a neutral level within every pulse. In Manchester code, signal transition is used to represent logic "1" and "0", namely, the encoding of each data bit is expressed as the transition of either low-to-high, or high-to-low. Each of those line codes has some variants, all having its advantages and disadvantages. In application, a particular code style is chosen to meet one or more of the following criteria: minimizing transmission hardware, facilitating synchronization, easing error detection and correction, achieving a target spectral density, or eliminating DC component.

In all the line codes, the bits are bounded by clock cycles. For a successful transmission, besides the intended data, the information regarding clock has to be passed

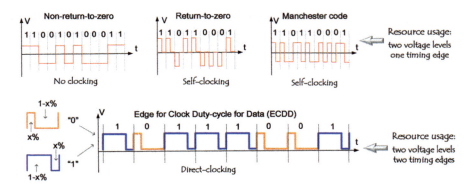

Fig. 6.155 Three most common line codes (top row) and the ECDD code (bottom)

6.19 One-Wire Communication Using Edge-For-Clock-Duty-Cycle-For-Data

from sender to receiver as well. For NRZ, additional synchronization technique must be used to convey clock message since the data itself does not contain much clock-related information. For RZ and Manchester, the neutral level (in RZ) or the level-transition (in Manchester) occurs in every clock cycle. Thus, clock information is inherently embedded in the data. However, even though clock-related information is presented in the data, it cannot be used directly as clock. The clock signal itself has to be recreated by using some additional circuitries. It is therefore natural to ask for a code style that can directly use an electrical pulse to convey both data and clock messages. This aspiration can only be fulfilled through the help from using additional resource. This extra resource can come from the dimension of time.

In the bottom row of Fig. 6.155, the principle of a coding style of "Edge-for-Clock-Duty-Cycle-for-Data" is illustrated. As shown, an electrical pulse is used to convey two kinds of message: its edge is for clock and its duty-cycle for logic value. For this reason, this coding style is shortened as ECDD. In each ECDD pulse (bit), the waveform is made in such a way that its duty cycle is significantly different from 50%. The low and high voltage portions are either "x% & (1 − x)%" or "(1 − x)% & x%", depending on the logic value that it intends to represent.

The most significant difference between ECDD and other line codes is on the ideology of how the resources of voltage and time are treated. In conventional approaches, time is used for indexing and voltage for differentiating logic states. In ECDD, besides indexing, time is further used as a differentiator for logic value. In contrast to conventional wisdom, voltage is used here just as an index tool to discern two conditions. Its value actually does not matter as long as the two conditions can be differentiated and recognized for indexing. On the other hand, time is used as the key differentiator to make the decision of logic "1" or "0" by inspecting the time durations that the pulse stays at these two conditions: long-high–short-low or long-low–short-high, as depicted in the left of bottom row of Fig. 6.155. During the process of doing this, a particular transition (e.g., the low-to-high rising edge) still functions as the data boundary. Put it in another way, for an electrical pulse, only one of its two edges is used in conventional codes (the rising edge for separating bits) while both edges are utilized in ECDD (the rising edge for separating bits and the falling edge for judging the two conditions). Therefore, resource is more efficiently exploited in ECDD (practically speaking, the resource of voltage is used in the similar way for both ideologies).

As demonstrated in Sects. 6.9 and 6.10, TAF-DPS is a powerful tool for manipulating electrical pulse. For instance, as shown in Fig. 6.112, its duty cycle can be precisely controlled. In operation, the duty-cycle of a TAF-DPS output pulse can be set to any value in the range of $\{1/I, (I - 1)/I\}$ where I is the integer part of frequency control word F. Therefore, theoretically, a multi-value logic system can be developed using duty-cycle as the differentiator. In practice, for easing the constraint on bandwidth, it is suggested to design a scheme that uses only two distinct states. In other word, the circuit is required to only discriminate two kinds of duty-cycle: long-high–short-low or long-low–short-high. This relaxation, besides lowering the bandwidth requirement on transmission medium, can significantly simplify the design of decoding circuit.

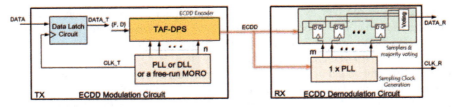

Fig. 6.156 The architecture of ECDD one-wire data communication system

As discussed in Chap. 4 of Xiu (2012a), both the TAF-DPS output pulse's length and its duty cycle can be synthesized by fixed numbers of Δ. The key design consideration in this ECDD architecture thus is the determination of the x value (i.e., the low to high ratio as illustrated in Fig. 6.155). The smaller the x is, the larger the ratio will be, and the easier for decoding circuit to differentiate the two states. However, smaller x value leads to narrower pulse which bears higher risk of being swallowed during transmission. In other words, it requires higher bandwidth on transmission medium. On the receiving side, the decoding for ECDD signal can be accomplished by a circuit made of multiple samplers. Each sampler is driven by a clock from a plurality of clock signals which are evenly spaced in one cycle. The resultant bit stream can be fed into a voting circuit. Based on the number of "0" and "1" received, the circuit can recognize by majority voting whether a digital zero or one is sent from the transmitter.

The architecture of ECDD one-wire data communication system is depicted in Fig. 6.156 (Xiu 2015c). Inside TX, the ECDD modulation circuit is a TAF-DPS circuit. A PLL or DLL can be used to generate the multiple outputs required by the TAF-DPS. In some case, a free-run MORO may be simply used. The data latch circuit captures the incoming NRZ data. The resultant DATA_T is fed to the TAF-DPS as the instruction for duty-cycle (the frequency control F can be chosen by user for his/her convenience). On the RX side, the ECDD signal is used as a reference for a PLL (or DLL) which generates a group of clock signals that have same frequency. The phases of those clock signals are evenly spaced in one cycle (e.g., a multi-stage VCO). At the same time, the ECDD signal is fed to the demodulation circuit which could be a majority voting circuit as described above. The resulting DATA_R is the received data, which is accompanied by its clock CLK_R (one of the outputs from the PLL).

Figure 6.157 shows a transistor level simulation result of ECDD one-wire communication system using circuit constructed according to Fig. 6.156, using a 0.18 um CMOS process. The ECDD system is designed at frequency of 300 MHz. A 4-differential-stage VCO (i.e., the MORO) is used to generate 8 outputs at 300 MHz for supporting the TAF-DPS ($\Delta = 0.417$ ns). The pulse length of ECDD signal is 8Δ (setting of F = 8). The duty cycle value x is chosen as x = 25% (i.e., the ratio of high to low is 6Δ to 2Δ). On the left-hand side of the figure, the bottom trace is a sequence of to-be-transmitted digital data in NRZ format. The second trace from bottom is its corresponding ECDD signal generated by the ECDD encoder. The third

6.19 One-Wire Communication Using Edge-For-Clock-Duty-Cycle-For-Data

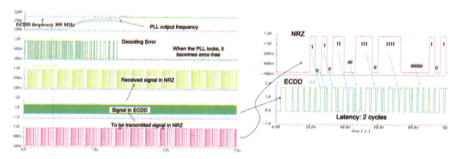

Fig. 6.157 A simulation result of ECDD one-wire data communication

trace from bottom is the output data from the ECDD demodulation circuit in NRZ format (the DATA_R in Fig. 6.156). The 2nd trace from top is the output from an error-checking circuit. When the received data has error, the error-checking circuit outputs a logic high. Otherwise, it is in logic low. The plot at the very top has two traces. The trace of "PLL output frequency" is the frequency measurement on one of the sampling clocks generated by the PLL in the receiver (inside RX). The trace of "ECDD frequency 300 MHz" is the frequency measurement of the ECDD signal sent from the TX.

In this simulation, the original NRZ data is generated by a 9 stages PRBS (Pseudorandom Binary Sequence) encoder using a clock of 300 MHz. The error-checking circuit is made of its corresponding 9 stages PRBS decoder. Thus, the inherent structures of the PRBS encoder and decoder match. As a result, the error-checking circuit will report no error if the data from the transmitter is received correctly by the receiver. In this simulation, the error-checking circuit is clocked by one of the PLL outputs (inside the RX). As shown in the top traces, when PLL reaches lock, the sampling clock's frequency matches that of ECDD signal (300 MHz, same as the PRBS encoder frequency). As seen, before the PLL reaches lock, the decoding circuit continuously reports error. After lock, the decoding error signal becomes logic low and stays at low all the time, indicating that the sampling circuit and the majority voting circuit correctly receives and decodes the ECDD signal. On the right-hand side, the zoom-in versions of the original NRZ data waveform and its corresponding ECDD waveform are shown. The ECDD signal is delayed by 2 cycles (latency = 2 cycles) since it takes 2 clock cycles for the system of Fig. 6.156 to do its work.

The biggest advantage that ECDD offers is the minimization of hardware cost since only one wire is required for transmission. There is absolutely no synchronization issue since the clock signal is directly transmitted along the data. It is therefore rightfully called direct clocking one-wire communication.

Gestalt Switch:

In the new paradigm, the issue of using electrical pulse to encode digital data can be motivated and facilitated by utilizing time resource more efficiently. This can lead to new circuit and system for conveying message over various transmission mediums.

6.20 Spectrum Pattern as Message for Communication

TAF-DPS is a circuit level enabler for system level innovation. In previous sections of this chapter, we have discussed its innovation power through a variety of topics that have already been studied in the past by researcher and engineer in the old space-dominated paradigm. One subject that has not been touched before, to our best knowledge, is the utilization of spectrum pattern for some meaningful purposes. TAF-DPS is a powerful tool for generating various spectrum patterns. Naturally, we want to use this feature for creating something useful.

$$\rho \equiv \frac{2/\pi - \|V_{fa}\|}{2/\pi} = \left(1 - \frac{1}{L} \cdot \frac{\sin\left(\frac{2^{n-m}\pi}{2w}\right)}{\sin\left(\frac{g\pi}{2w}\right)}\right) \quad (6.20.1)$$

TAF waveform is different from that of conventional frequency which is a pulse train made of a single type of cycle with 50% duty-cycle. TAF waveform consists of multiple types of cycles. The degree of the dissimilarity between the two waveforms depends on the values of I and r, which are integer and fractional parts of the frequency control word F, respectively (please refer to Sect. 6.1 and Eq. (6.1.2)). It is however difficult to visualize this dissimilarity by using I and r directly. In Xiu (2012a), a parameter of irregularity ρ is introduced to gauge this dissimilarity. The formula for calculating ρ is expressed in (6.20.1), where n, m, g, w and L are all related to I and r.

Figure 6.158 can be used to illustrate the difference between those two types of frequency. On the left, the waveform and the spectrum of conventional frequency

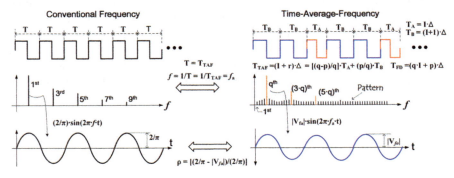

Fig. 6.158 The difference between conventional frequency and TAF

6.20 Spectrum Pattern as Message for Communication

f are shown in the top two drawings. The bottom drawing is its 1st harmonic with amplitude of $2/\pi$. On the right, the corresponding drawings for TAF are displayed with the setting of $f_a = 1/T_{TAF} = 1/T = f$. For TAF signal, it can be decomposed into harmonics by Fourier analysis. The fundamental period (frequency) is $T_{FD} = 1/f_{FD} = (q \cdot I + p) \cdot \Delta$, where $r = p/q$ and $\gcd(p, q) = 1$. Its qth harmonic is the Time-Average-Frequency $f_a = 1/T_{TAF}$. ρ is used to measure the difference between f_{TAF}'s amplitude $\|V_{fa}\|$ and $2/\pi$, which is the amplitude of conventional frequency at same frequency value. Since 1st harmonic plays dominant role, it is appropriate and convenient to use ρ for gauging the waveform. The smaller the ρ is, the closer the TAF waveform is to that of conventional waveform. Table 6.15 lists some exemplary F values and the corresponding ρ values. It is seen that I plays the dominant role in determining the value of ρ, and thus determining the shape of the waveform and the pattern of the spectrum.

As already understood, from a given control word F, the TAF-DPS can produce a series of pulses, each with desired instant frequency (or period, or length-in-time, of each pulse) and duty cycle. This pulse train can be used as clock signal, as medium for signal processing and as tool for sensing and control. Those activities take place

Table 6.15 Irregularity ρ for gauging TAF waveform

F = I + r	p and q	T_A Size	T_B Size	# of T_A	# of T_B	Irregularity ρ (%)
$40 + 2^{-16}$	p = 1, q = 2^{16} = 65,536	40Δ	41Δ	65,535	1	0.1027767
$40 + 2^{-14}$	p = 1, q = 2^{14} = 16,384	40Δ	41Δ	16,383	1	0.1027764
$40 + 2^{-12}$	p = 1, q = 2^{12} = 4096	40Δ	41Δ	4095	1	0.1027754
$40 + (1-2^{-12})$	p = 4095, q = 4096	40Δ	41Δ	1	4095	0.0978269
$40 + 2^{-2}$	p = 1, q = 2^2 = 4	40Δ	41Δ	3	1	0.1015042
$8 + 2^{-14}$	p = 1, q = 2^{14} = 16,384	8Δ	9Δ	16,383	1	2.5504255
$8 + 2^{-12}$	p = 1, q = 2^{12} = 4096	8Δ	9Δ	4095	1	2.5503096
$8 + (1-2^{-12})$	p = 4095, q = 4096	8Δ	9Δ	1	4095	2.0185552
$8 + 2^{-2}$	p = 1, q = 2^2 = 4	8Δ	9Δ	3	1	2.3624707
$120 + 2^{-14}$	p = 1, q = 2^{14} = 16,384	120Δ	121Δ	16,383	1	0.01142275
$120 + 2^{-12}$	p = 1, q = 2^{12} = 4096	120Δ	121Δ	4095	1	0.01142271
$120 + (1-2^{-12})$	p = 4095, q = 4096	120Δ	121Δ	1	4095	0.01123479
$120 + 2^{-2}$	p = 1, q = 2^2 = 4	120Δ	121Δ	3	1	0.01119759

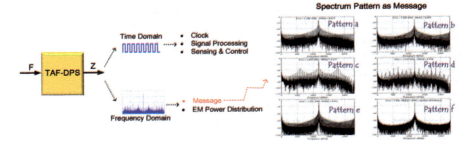

Fig. 6.159 The scheme of using spectrum pattern as message for communication

straightforwardly in time domain and they can be studied directly in that domain accordingly.

On the other hand, each value of ρ leads to a unique pattern in spectrum. The unique pattern in TAF signal's spectrum is information. If properly structured, it can be useful for conveying message. Figure 6.159 illustrates one of such ideas of utilizing spectrum pattern. When the TAF pulse train is examined in frequency domain, its characteristic is manifested through patterns created in its frequency spectrum. As can be appreciated from discussion in Sect. 6.4, TAF-DPS's output is very rich in its spectrum content. This feature can potentially be used as a vehicle for communication. As illustrated in the right-hand side of Fig. 6.159, several significantly different spectrum patterns are created by simply setting control word F to certain values. This process is uncomplicated, almost effortless.

This scheme of using patterns in spectrum as message is feasible because of two reasons: (a) there is one-to-one correspondence between the control word and the spectrum, (b) the spectrum is mathematically deterministic and is precisely calculable (please refer to Chap. 5 of Xiu (2012a)). At current stage, using patterns in frequency spectrum as a means for communication is an open research field. There are many issues to be explored.

Paradigm Shift:

In the emerging Time-Oriented paradigm, the subject of using electrical pulses to convey message can be investigated from an unconventional approach. This might lead to some unexpected and pleasant outcomes.

6.21 Conversion Between Time Span and Digital Value for Sensing and Actuating

In the practice of electronic engineering, like voltage and current, time can be used as medium for signal processing. The well-known example is the use of PWM (pulse width modulation) signal for various applications (please refer to Sect. 6.9). PWM is

6.21 Conversion Between Time Span and Digital Value for Sensing and Actuating

a circuit design technique for more or less analog-like applications. For digital signal processing, the time-domain stochastic computing discussed in Sect. 6.10 is a case of using time as medium for logic and athematic operations. Broadly speaking, in a general sense, there is a strong motivation for conversion between time-span and digital value. The circuit, in one way, is called Digital-to-Time Converter (DTC). In the other direction, it is called Time-to-Digital Converter (TDC). These two types of conversions are illustrated in Fig. 6.160.

For DTC, the input is a digital value $D(n)$ represented by n bits. The output can be one signal Z or two signals Z_1 and Z_2. For one signal type DTC, the time duration of logic high is $\delta T = D(n) \cdot R$, where R is the time resolution. The region of logic high is bounded by two events E_1 and E_2. In two-signal version, the time duration $\delta T = D(n) \cdot R$ is the time span that separates two events E_1 and E_2. Those two events take place in two signals. For the case of TDC, the conversion occurs in the opposite direction as shown in the bottom part of the figure: from two time-separated events to a digital value $D(n) = \delta T/R$.

In both cases of DTC and TDC, the digital value $D(n)$ is used in processing and the event E is used for the actions of sensing and actuating. Typically, DTC is used for applications that demand actuation. It is a control-oriented electronic device. On the other hand, TDC is for sensor applications. It is a sensing-oriented device. When DTC and TDC are used in practice, one crucial issue is the time resolution R. The finer the resolution is, the better the quality of the control and sensing will be. In electronic engineering, time resolution is materialized through frequency resolution associated with electrical signals. Finer frequency resolution leads to finer time resolution. TAF-DPS, thanks to its strong capability in manipulating the frequency of electrical pulse train, is squarely suited for being used in DTC and TDC designs.

$$R = T_2 - T_1 = [(I_2 + r_2) - (I_1 + r_1)] \cdot \Delta = [(I_2 - I_1) + (r_2 - r_1)] \cdot \Delta \quad (6.21.1)$$

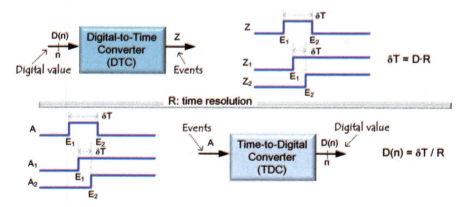

Fig. 6.160 The conversion between time-span and digital value: DTC (top) and TDC (bottom)

Figure 6.161 depicts the TAF-DPS based DTC architecture. As seen, two TAF-DPS circuits, based on the same base-time-unit generator, are used to produce two pulse trains of frequencies f_1 and f_2, $f_1 > f_2$. From a starting point in the dimension of time called Point-of-Coincidence, their elapsed times grow in different scales. After $D(n) = N$ cycles, the difference in between can be derived as $\delta T = N \cdot (T_2 - T_1)$. Therefore, this is a DTC function since a time duration δT is controllable by a digital value $D(n)$. The programmability on time resolution R depends on the value of $T_2 - T_1$, which is expressed in (6.21.1). For high resolution, a small value of R is desired. In operation, we often set $I_2 = I_1$ and $r_1 = 0$. Under this condition, it becomes $R = r_2 \cdot \Delta$. As an example, if $\Delta = 100$ ps and $r_2 = 2^{-10}$ (10 bits for fraction r), the time resolution can reach to ~ 0.1 ps. The frequency resolution can also be derived easily: $df/f = (f_1 - f_2)/f_2 \approx r_2/I$ (under the condition of $I_2 = I_1 = I$ and $r_1 = 0$). Operation-wise, after the EN signal is asserted, the circuit starts to search for the Point-of-Coincidence. When found, the counter is activated. After passing N cycles, the counter is stopped and the designated time-span δT is reached between the signals Z_1 and Z_2 as $\delta T = N \cdot (T_2 - T_1) = N \cdot r_2 \cdot \Delta$.

The TAF-DPS based TDC architecture is shown in Fig. 6.162 (Xiu 2016). It follows the principle of Vernier method: using two scales of main scale and Vernier scale to make fine measurement on the length of a physical object. In our case, the two scales are two frequency sources with values of f_1 and f_2, respectively. Those two frequencies are very close to each other so that fine measurement on time can be made. As shown in the figure, two frequencies can be created as $f_1 = 1/T_1 = 1/[(I_1 + r_1) \cdot \Delta]$, and $f_2 = 1/T_2 = 1/[(I_2 + r_2) \cdot \Delta]$. We can make signal Z_1 slightly faster than Z_2, or $f_1 > f_2$. The resulting slow signal Z_2 and fast one Z_1 are used as clocks to drive the counter #2 and #1, respectively. In operation, the slow clock Z_2 is activated first by the signal START. Then, after some time duration δT, the STOP signal activates the faster clock Z_1. When the Point-of-Coincidence between Z_1 and Z_2 is reached, the two counters are stopped and their contents are read out as N_1 and

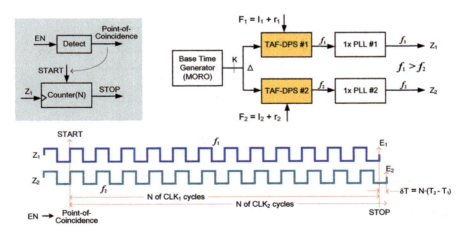

Fig. 6.161 TAF-DPS based DTC architecture

6.21 Conversion Between Time Span and Digital Value for Sensing and Actuating

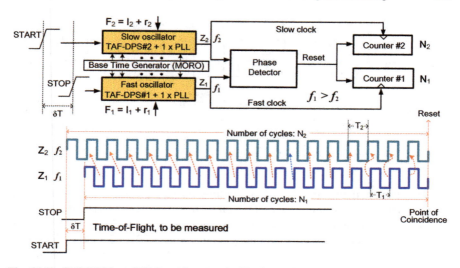

Fig. 6.162 TAF-DPS based TDC architecture, the Vernier method

N_2. The δT can then be derived as $\delta T = N_2 \cdot T_2 - N_1 \cdot T_1 = N_2 \cdot (I_2 + r_2) \cdot \Delta - N_1 \cdot (I_1 + r_1) \cdot \Delta$. If we choose $I_2 = I_1 = I$ and $r_1 = 0$, then $\delta T = (N_2 - N_1) \cdot I \cdot \Delta + N_2 \cdot r_2 \cdot \Delta = (N_2 - N_1) \cdot I \cdot \Delta + N_2 \cdot R$, where $R = r_2 \cdot \Delta$ is the time resolution.

There are three technical issues in the architectures of Figs. 6.161 and 6.162 that need to be addressed in more details. The first one is the detection of Point-of-Coincidence, which is a common issue for both structures. In the field of circuit design, there are several ways to detect the phase difference between two signals having closely-matched frequencies. The most known one is the phase frequency detector, or PFD, often used in PLL design. Its output is a narrow pulse which is suitable for analog signal processing but not friendly to be a flag for commanding action. A design that is suitable for this purpose can be found in Xiu (2012b). In here, we describe another method that is simpler. This method is based on the assumption that the two frequencies are very close to each other, or $(T_2-T_1)/T_1 \ll 1$. When we use the fast signal f_1 to sample the slow signal f_2, the output will be consecutive "1" or "0" in most of the times. This is due to the reason that, most likely, two consecutive sampling edges of f_1 will see logic-levels from two consecutive cycles of f_2. This scenario is illustrated in Fig. 6.163. However, two consecutive sampling edges will see the logic-levels from the SAME cycle of f_2 when the streams of "1" becomes streams of "0" for the first time, or vice versa. In other words, it is the first moment of time that one cycle of f_1 is completely enclosed within one cycle of f_2 (remember $f_1 > f_2$). This is the Point-of-Coincidence and it can be detected by the circuit shown in Fig. 6.163.

The second issue is related to the difference between Time-Average-Frequency and conventional frequency. Although TAF-DPS can produce frequency in a very fine resolution by adjusting fraction r in a very small step, its cycles are however not equal-length (i.e., it uses two types of cycles T_A and T_B). This abrupt jump in time

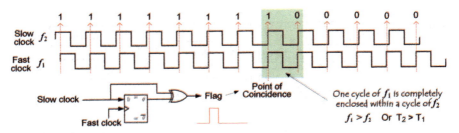

Fig. 6.163 The detection of Point-of-Coincidence

can cause trouble in time measurement. For this reason, there is a PLL attached to the output of each TAF-DPS in both architectures. It is used to convert Time-Average-Frequency to conventional frequency (by the way, the PLL after TAF-DPS #1 can be eliminated if we set $r_1 = 0$ in the frequency control word of TAF-DPS #1). The design of this PLL is in the style of jitter cleaner, or a $N = 1$ integer PLL for cleaning the input jitter. The only requirement for this simple PLL is low bandwidth (preferable in the range of below KHz range). Its function is to filter out the spurious contents in the TAF-DPS output. Its purpose is to convert the Time-Average-Frequency f_{TAF} to conventional frequency $f = f_{TAF}$.

In Chap. 5 of Xiu (2012a), there is an intensive discussion on the spectrum of TAF-DPS output. The interleaved use of T_A and T_B introduces modulation that has impact on the signal's spectrum. One repeatable cycle of pattern-in-spectrum is the fundamental period $T_{FD} = p \cdot T_B + (q-p) \cdot T_A = q \cdot T_{TAF}$. The fundamental frequency is $f_{FD} = 1/T_{FD}$. The TAF frequency $f_{TAF} = 1/T_{TAF} = q \cdot f_{FD}$ is the qth harmonic of the f_{FD} (Xiu 2012a). The magnitude of the ith harmonic going away from f_{TAF} can be calculated using (6.21.2) where J is the spectrum magnitude (Talwalkar 2012).

$$\left|\frac{J^{ith}}{J_{f_{TAF}}}\right| \approx \frac{\pi}{(q \cdot I + p)\left|\left[1 - \exp\left(-j\frac{i\pi}{q}\right)\right]\right|} \text{ dBc} \quad (6.21.2)$$

The top row of Fig. 6.164 provides several simulation examples that demonstrate the TAF-DPS fine adjustment capability in frequency. In the left, there are two simulations overlapping each other: $f_1 = 500$ MHz (setting: $I = 40$, $p = 0$, $q = 65{,}536$, brown) and $f_2 = 499.9998$ MHz (setting: $I = 40$, $r = 2^{-16}$, $p = 1$, $q = 65{,}536$, blue). The spectrum of f_1 is clean without any spurious tones. In the case of f_2, there are spurious tones about 2 db higher than the floor set by f_1. Those spurs are spaced at $f_{FD} = f_{TAF}/q = 499.9998/65536 = 7.6$ kHz. They are so close to each other that individual tones can only be visible by deep zoom-in. In the right, two more cases are provided: $f_3 = 499.9992$ MHz (setting: $I = 40$, $r = 2^{-14}$, $p = 1$, $q = 16{,}384$, blue) and $f_4 = 499.9969$ MHz (setting: $I = 40$, $r = 2^{-12}$, $p = 1$, $q = 4096$, green). The spacings separating spur tones are 30.5 kHz and 122 kHz, respectively. At about 10 db and 20 db, their strengths are higher than that of f_1 (sparser-and-stronger and denser-and-weaker, respectively). The spurs are induced by the regular

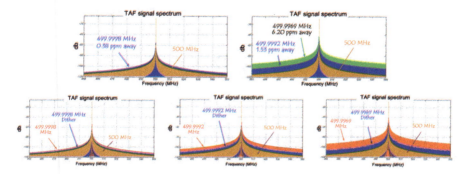

Fig. 6.164 Spectrums due to the use of small values in r: original (top) and dithered (bottom)

carry-overflow of the accumulator used in the TAF-DPS circuit, which is a 1st order delta-sigma modulator. Their strengths can be significantly alleviated by randomizing the carries. This operation is often called dithering. In the bottom row of Fig. 6.164, the results of dithering are shown. In each of the plots, the brown trace is the case of $f_1 = 500$ MHz (used as a reference frame). The red trace represents the original f_2, f_3 and f_4 (no randomizing). The blue ones are results after the dithering. As seen, the spur magnitudes are reduced. The reason is that the fundamental period is prolonged (q is enlarged $\rightarrow f_{FD}$ is reduced). At the meantime, the frequency $f_{TAF} = q \cdot f_{FD}$ is kept intact.

As an example, the frequency $f_1 = 500$ MHz can be used in architectures of Fig. 6.161 or 6.162 as the fast clock. It can be outputted directly from the TAF-DPS #1, or from the PLL #1. The other three frequencies f_2, or f_3 or f_4, can be the output from the "TAF-DPS #2 plus PLL#2". Their frequency deviations from f_1 are 0.38, 1.53 and 6.20 ppm, respectively. The time resolution $R = r_2 \cdot \Delta$ are $2^{-16} \cdot \Delta$, $2^{-14} \cdot \Delta$, $2^{-12} \cdot \Delta$, respectively. If $\Delta = 1$ ns (can be easily achieved with modern technology), the time resolution R becomes 0.15 ps, 0.6 ps and 2.4 ps, respectively. Those sets of design parameters are feasible. As can be appreciated from Fig. 6.164, the spur tones from TAF-DPS #2 can be readily filtered out by the low bandwidth (below KHz) PLL#2.

The third issue concerns only the TDC architecture of Fig. 6.162. As seen, the START and STOP signals are used to begin the Z_1 and Z_2 pulse train. In other words, they are the flags for the transition-of-state from no-oscillation to oscillation. In circuit design, the START or STOP signal can be designated as the enable signal for the oscillator (i.e., the VCO) in PLL. Only when the flag signal is at logic high, the VCO can oscillate. Figure 6.165 provides a transistor level simulation to illustrate this point. In operation, the TAF-DPS output can be always-on to provide a stable reference frequency for the PLL. The oscillation of the PLL VCO can be controlled by the START or STOP signal.

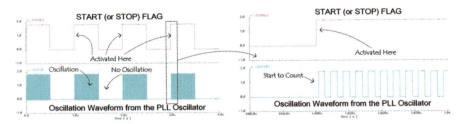

Fig. 6.165 The START (STOP) signal to enable oscillation

Gestalt Switch:

A powerful new tool in circuit level enables novel solution in system level. The classical problem of conversion between time-span and digital value is facilitated by the help of a new frequency concept and tool.

6.22 Connection-Adaptive Dynamic-Frequency Clocking for Sparse CNN

Artificial Neural Network (ANN) is computational model that is inspired by the biological neural networks in the human brain for processing information. ANN is crucial for Machine Learning (ML), a technology widely used in applications such as natural language processing, image and video recognition (self-driving cars, robotics, drones) and etc. Structurally, an ANN is organized as a collection of connected nodes called artificial neurons that approximately imitate the neurons in a biological brain. Each connection, like the synapses in a biological brain, functions as a transmission line that can be used to transmit a signal to other neurons. In operation, an artificial neuron first receives a signal then processes it. Based on the processing result, this neuron can signal neurons connected to it when necessary. The "signal" transmitted in a connection is a real number. The output of each neuron is computed by some non-linear function of the sum of its inputs.

Typically, neurons are organized into layers. Different layers may perform different transformations on their inputs. Signals travel from the first layer (the input layer) toward the last layer (the output layer). Along the paths, they might traverse multiple layers that are located in between. Operationally, the connection between any two neurons is assigned a weight to it, which is a real number. This weight is dynamically adjusted as the network gradually becomes mature through some kind of learning process. The weight increases or decreases the strength of the signal at a connection. Neurons may have a threshold assigned to it so that a signal can be sent out if the aggregated signal exceeds that threshold.

From a functional perspective, neural networks learn (or are trained) by processing examples. An example contains a set of known input and result, forming a group of probability-weighted associations between the input and the result. Each association is stored within the data structure of the connection itself. The training of a neural network from a given example is usually conducted by determining the difference between the processed output of the network (often a prediction) and a target output. This difference is the error. From this error value, the network can adjust its weighted associations according to a learning rule. Successive adjustments will cause the neural network to produce output which is increasingly similar to the target output. After a sufficient number of these adjustments the training can be terminated based upon certain criteria.

One of the most successful types of ANNs is the convolutional neural network (CNN), which is particularly useful in processing two-dimensional visual data. CNN is a type of feed forward ANN. It is inspired by the organization of animal visual cortex. Mathematically, the response is approximated by a convolution operation. From a functional perspective, a CNN has two missions: feature-extraction and classification. From implementation standpoint, CNN executes two tasks: computation and communication. The works involved in computation include the operation of convolution and the floating-point matrix multiplication (GEMM). The task of communication is for passing information between layers. Computation mostly occurs in the convolution layer, which is computational centric. Meanwhile, communication takes place in the fully connected layer that is memory centric.

Nowadays, CNN has been found useful in many applications. Two popular platforms for its implementation are GPU and FPGA. GPU is suitable for regular parallelism due to its regular structure. It is good at GEMM (floating point matrix multiplication). It can achieve performance of over TFLOP/s (trillion floating-point calculations every second). FPGA is reconfigurable and thus is appropriate for handling irregular parallelism. For all CNN implementations, the two most important goals in its execution are speed and power: performing task as fast as possible and using power as less as possible. There is a strong motivation on the searching of methods for CNN acceleration. The directions of exploration include exploiting parallelism, ameliorating data reuse, reducing data bandwidth requirement (minimizing memory footprint), among others. In addition, one particularly promising research area is to exploit sparsity in CNN. This can significantly cut down the cost of computation (Liu et al. 2015).

In the layered structure of CNNs, in principle, each node in one layer is connected to every node in the next layer (the densely connected network). In recent studies, however, it is recognized that this does not have to be the case. This observation leads to the sparsely-connected-layer architecture, the sparse CNN. Figure 6.166 illustrates the idea of sparsely-connected two layers. As shown, there are two layers of input and output, each having three nodes. The three nodes in the output layer are connected to 3, 2, and 1 nodes of the input layer, respectively. For example, output nodes #0 receives inputs from all the input nodes #0, #1 and #2 while output nodes #2 only needs input from input node #2. Therefore, as can be understood, there are three "multiply + add" operations required for computing in output node #0. A weighting factor is hence assigned to this node as $W_0 = 3$. The weighting factors for other two nodes are $W_1 = 2$ and $W_2 = 1$, respectively. The processing circuit is depicted in the

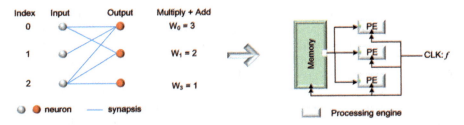

Fig. 6.166 An illustration of sparsely-connected network (left) and its processing circuit (right)

drawing on the right. The memory unit is used to store the information associated with the three input nodes. The three PE (processing engine) units are responsible for performing the "multiply + add" operations for the three nodes.

In the structure of Fig. 6.166, all the circuit units are driven by a clock of frequency f. In other words, they all work at the same pace. It is immediately clear that this scheme is not efficient since the PE units for node #1 and #2 have less computation loading (less connections) than that of node #0. If all the three PE units work at same frequency, some portion of power will be wasted. A better scheme is illustrated in Fig. 6.167 where the three PE units are all driven by their respective clocks with proper frequencies. Power consumption can be reduced since PE #1 and PE #2 are running at lower frequencies now. Or, from another perspective, the overall throughout is improved since PE #0 and PE #1 works at higher frequencies than that of PE #0. It utilizes time more efficiently.

The scheme depicted in Fig. 6.167 works perfectly if the data associated with input nodes (i.e., the operands for the operations of "multiply + add") are available to the PEs whenever they are needed. This, however, is not always the case since those data are stored in a memory unit and they can be read out only one-by-one sequentially. This problem can be explained by using Fig. 6.168. In the left of bottom row, the data associated with the three input nodes are read out in the order of $0 \rightarrow 1 \rightarrow 2$. As a result, the "multiply + add" operations of output node #2 and node #1, or

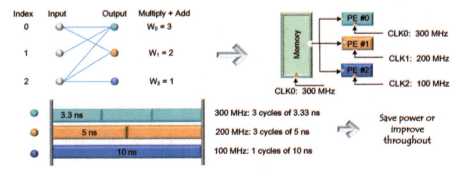

Fig. 6.167 The idea of connection-adapted dynamic clock frequency adjustment

6.22 Connection-Adaptive Dynamic-Frequency Clocking for Sparse CNN

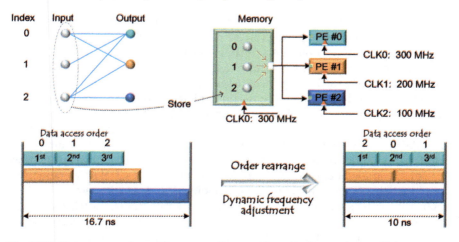

Fig. 6.168 Data access order and frequency adjustment for achieving maximum efficiency

the circuits of PE #2 and #1, can only start at the 3rd CLK0 cycle. The time required for completing the overall computation is thus 16.7 ns. If the order of read-out is changed to 2 → 0 → 1 as shown in the right-hand side of Fig. 6.168, the computation can be completed in one CLK2 cycle of 10 ns.

This example leads to the development of a general guideline for the connection-adapted dynamic clock frequency adjustment methodology, which is summarized as follow.

- For a given sparse CNN structure, splitting the total number of N neurons in a layer into M groups, assigning each group to a PE.
- For the ith group in this layer, defining $S_i = \sum_{j=1}^{n_i} W_j$ where W_j is the number of "multiply + add" operations for the jth node in this group and n_i is the number of neurons in this group, $\sum_{i=1}^{M} n_i = N$. W_j can also be considered as the number of connections between this jth node and the nodes in its previous layer.
- For the ith group, driving the corresponding PE circuit with a CLK_i of frequency f_i.
- The goal of optimization is to minimize the maximum value of S_i/f_i, or $\min(\max(S_i/f_i))$.
- For data access order, fetching the data from the node (in previous layer) that has the largest number of connections first. Then, the data of second most connected node, and so on so forth.

The example discussed and the guideline presented point to a need of clock generator for producing the frequency f_i for each PE. This frequency has to be dynamically adjustable since its value will be dependent on number-of-connections. Only after the structure of the sparse CNN is finalized, the information regarding number-of-connections for each node can become available. TAF-DPS is ideally suitable for being such a clock generator due to its feature of arbitrary frequency generation and instantaneous frequency switching (please refer to Fig. 4.10).

As discussed, in CNN operation, there are two major tasks of computation and communication. Currently, CNN acceleration mainly takes place on two dimensions: computation resource and IO bandwidth (memory access). Bringing in more resource for computing (such as parallel processors) can improve operation speed. Reducing memory footprint can also enhance performance. However, looking from another angle, clock rate (frequency) can be a 3rd dimension for exploration. For computation, the speed of computation is controlled by clock rate. For communication, the flow of data is also controlled by clock rate. Therefore, there is great potential to explore when this 3rd dimension of clock rate is introduced to the field of CNN acceleration. We want to provide CNN architect with a new tool: a flexible clock source. The clock frequency can be set at any rate desired; the clock rate can be switched instantly (adaptive to data flow, buffer status, etc.). Moreover, this flexible clock source can be inserted anywhere in the system you wanted (especially for the case of FPGA). Furthermore, this flexible clock can be employed as many as you want since it can be made low cost.

This methodology of connection-adapted dynamic clock frequency adjustment for driving sparse CNN is only briefly elucidated here. Its purpose is to inspire further exploration.

Paradigm Shift:

In the Time-Oriented paradigm, the issue of CNN acceleration can be investigated for higher efficiency from a new dimension.

6.23 Latency Awareness Computer: Frequency Scalable ISA and Microarchitecture

In general, the term of computer architecture is referred to a set of rules and methods that describe the functionality, organization, and implementation of computer system. It represents the means of interconnectivity for a computer's hardware components as well as the mode of data transfer and processing executed. It is the organization of the components making up a computer system. This organization determines the composite of functional units of which the system is composed and the structure of their interconnectivity. It is the semantics or meaning of the operations that guide its function. The semantics defines the meaning of what the systems do under user input and how the functional units are controlled to work together.

To have a better understanding on computer architecture, it is helpful to study the issue in a bigger perspective. In Fig. 6.169, the computer system is investigated in an abstraction fashion of multiple layers from user to physics. Starting from a user of computer, the abstraction level descends from application (problem to-be-solved), algorithm, program (computer language), to system software. Those layers can be

6.23 Latency Awareness Computer: Frequency Scalable ISA and Microarchitecture

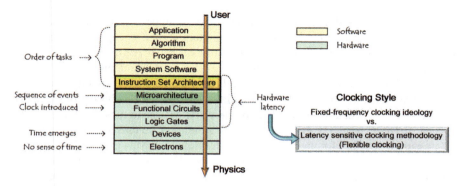

Fig. 6.169 Abstraction layers, time, clock signal and clocking styles

grouped together and be called software as a whole. From the fundamental physics at the very bottom, the first layer is electron. Up one level is electronic devices (resistor, capacitor, inductor, memristor → transistor, diode, and etc.). Still up are layers of logic gates and functional circuits. This group of layers is categorized as hardware. In between the software and hardware, there are two interfacing layers of instruction set architecture (ISA) and microarchitecture. It is those two layers that are often collectively referred to as computer architecture.

The ISA is the embodiment of semantics of a computer system. From one hand (the hardware side), it is a logical representation encoding the basic set of distinct operations that a computer architecture can perform. On the other hand (the software side), it is the vehicle by which application program uses to specify the work to be done. The most known examples of ISA are ARM, x86, MIPS, SPARC, and PowerPC. An ISA does not define the underlying hardware implementation. Oftentimes, many different hardware implementations exist for a particular ISA. Those implementations use different underlying hardware and therefore offer trade-offs in performance, price, and power. Some are optimized for high-performance whereas others are optimized for long battery life. The specific arrangement of hardware pieces (such as registers, memories, ALUs, and other building blocks) to materialize an ISA is called the microarchitecture. Microarchitecture is the hardware blueprint that a given ISA is implemented in a particular processor.

Between the twos, a given ISA may be implemented in different microarchitectures. Moreover, even a particular implementation may vary due to a different design goal or due to a change in technology. Commercially, it is quite common that many different microarchitectures exist for a single ISA. For example, both AMD and Intel have a variety of microarchitectures developed around x86 ISA. ARM's ISA has been used in many mobile and embedded systems built by various companies, all with their own microarchitectures. PowerPC ISA is used by IBM and others, each with its unique microarchitecture, for building servers.

Over the years, many different computer architectures (i.e., ISA and microarchitecture) have been developed to speed up the movement and handling of data, allowing for increased data processing speed or for improved power efficiency. It is a never-ending game for building better system (faster, cheaper, smaller, more reliable) and for enabling emerging applications (virtual reality, AI, self-driving, personalized genomics, and etc.). Currently, there are several research directions for advancing this science and art into the future: architecture for security, energy-efficient architecture, architecture for low-latency (high predictability), architecture for data-driven and data-aware applications (e.g., AI/ML, genomics, medicine, health and etc.). There are many new demands from the top (the user side in Fig. 6.169) to pressure researcher for new ideas in ISA development. Meanwhile, thanks to the advance of semiconductor technology (the Moore's Law), there are many new issues and opportunities at the bottom (the physics side in Fig. 6.169) for new schemes of implementation in the domain of microarchitecture. Among the countless proposals of new ideas, how to use time more efficiently (or more efficient clocking) in the levels of ISA and microarchitecture is an area that has not drawn much attention. It is a direction that has great potential since, besides silicon area, time is the only other resource that we can use (please refer to Sect. 3.3).

Referring back to the discussion around Fig. 3.3 in Sect. 3.1, it is recognized that the concept of time emerges only after matter is formed from atoms and electrons. The mechanism of clocking arises when we cross the boundary from device to circuit. This comprehension is reflected in Fig. 6.169 as well. In electronic engineering, except the phenomenon of component aging, we do not care much about the thing called time below the level of logic gates. The clock signal, which is the mechanism reflecting the flow of time, is introduced only when we start to make circuit and system. In microarchitecture level, the concept of event starts to be used and time is encapsulated in the sequence that events happen. In ISA and above levels, relevant events are organized into task. Time manifests its significance through the order that those tasks have to be completed.

Starting from the level of logic gates, the concept of hardware latency emerges. It is used to communicate a crucial fact that it needs certain amount of time for an electronic device to complete a designated piece of work. The length-in-time of a clock period is the allocated time slot for a circuit (a collection of electronic devices) to finish a given task (a group of relevant events). In all the currently known forms of ISA and microarchitecture, they all follow the fixed-frequency clocking ideology. In other words, the time slots for doing all kinds of different works are devised as same size. Is this approach efficient?

Following this principle of fixed-frequency clocking, the highest clock frequency (or the minimum clock period, the minimum cycle time) is determined by the longest running instruction, or the most complex instruction, or the one having largest latency. All other instructions, no matter how fast their associated works can be finished, have to accommodate it. This is clearly a waste of time resource. Naturally, an alternative is the latency sensitive clocking methodology as shown in the right-hand side of Fig. 6.169. This is not an ideologically sophisticated idea. It has not been proposed before only because of the reason that there is no circuit level support for providing such a clock signal generator that can dynamically adjust its output frequency. Now,

6.23 Latency Awareness Computer: Frequency Scalable ISA and Microarchitecture

with the emergence of the TAF-DPS technology, it becomes feasible to investigate this idea seriously.

When the length-in-time of clock cycle is fixed, one approach of improving time-utilization efficiency is to miniature the individual task (reduce the amount of work load) assigned to each clock cycle and then carry out many those small tasks in parallel. This is the so-called pipeline technique. By doing so, the clock cycle can be shortened (clock frequency risen) and, hopefully, time utilization can be improved and data processing throughput can be boosted. Nowadays, the most sophisticated pipeline structures in some advanced microprocessors can have over a hundred stages. It certainly improves the time utilization efficiency. It however has some obvious drawbacks, such as data dependence among instructions, loss in the memory hierarchy, the large area overhead for handling the much-complicated pipeline circuit, or the extra circuit for dealing with the issue of jump and branch. As a matter of fact, the distinction between two important ISA design ideologies, CISC (Complex Instruction Set Computer) and RISC (Reduced Instruction Set Computer), mostly lies in the way of how friendly their instructions are tailored for pipeline optimization. Pipelining in RISC spends a great deal of effort on reducing the clock cycle time of its simple single-cycle datapath while multiple-instruction-issue focuses on reducing clock cycles per instruction (CPI). On the other hand, CISC has a much complex instruction structure that is not very friendly to pipelining. Its main focus is to improve code density. Overall, deep down in the core, the distinction between CISC and RISC can be traced to the issue of how to use time resource more efficiently.

Pipeline is certainly a powerful technique for pushing the performance (the data processing speed) to the cutting edge. But, on the other hand, it is also one of the major factors in sucking power from the supply. In the pursuit of new architectures for such applications of high-security, low-latency, data-driven, and especially energy-efficient, the latency sensitive clocking methodology provides another perspective and is certainly worth exploring. Several decades ago, the great John von Neumann envisioned that computer should consist of three major parts: the central processing part (CPU), the memory and the input/output unit. He further envisaged that all computer instructions would be executed in four steps: fetch, decode, execute and store. Amazingly, after so many years of dramatic developments, this visionary guideline still hold its value. In light of this fact, the proposed latency sensitive clocking methodology has to be investigated from this legacy framework.

From Neumann's four-step course of action, the key idea of latency sensitive clocking methodology is illustrated in Fig. 6.170. As seen, it is recognized that the four steps of fetch, decode, execute and store all have their own unique latencies. This is based on the very fact that they all do different things and thus require different time durations to finish them. Therefore, it is natural to use different cycle-times to accommodate their particular needs. Fundamentally, this approach is good for both energy consumption and time resource utilization. To materialize the idea, the feature of latency sensitive has to be built into the instructions. As an action initializer, each instruction knows its task-to-be-accomplished and the associated work load. It is therefore aware of the approximate time required (proportionally to a reference) to finish the work. This information can be embedded inside the instruction itself. At

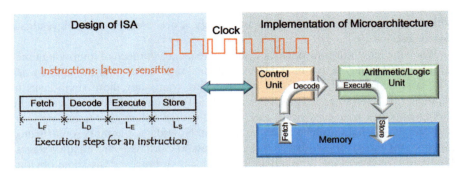

Fig. 6.170 Latency sensitive ISA and cycle-time scalable implementation in microarchitecture

the microarchitecture side, from this latency information, the implementation can be planned accordingly and appropriate hardware can be built for serving various purposes.

This latency sensitive clocking methodology is different from the dynamic frequency scaling technique used in many modern processors. Dynamic frequency scaling is used to save energy or temporarily boost performance by varying the clock rate (please refer to Sect. 6.13). For example, the Intel Core i7 has a so-called Turbo mode that will temporarily increase clock rate by about 10% until the chip gets too warm. This technique however is not instruction sensitive. All instructions, regardless their complexities, are executed in a higher or slower pace as a whole. In contrast, the latency sensitive clocking is instruction dependent. The clock rate is tuned for each instruction. Theoretically, it is compatible to the idea of so-called energy-proportional computing. As mentioned, today's microprocess design leans towards less aggressive and more efficient style. This trend has caused computer architect to reassess the energy-performance implications of some of the inventions since the mid-1990s. It results in a simplification of pipeline structure in the more recent developments of microarchitectures. Latency sensitive clocking can be considered as a continuation of this trend and is definitely a welcomed tool in the development of tomorrow's ISA and microarchitecture.

From operational perspective, the TAF-DPS circuit can be made as a standard component inside the timing & control unit, as shown in Fig. 6.171. The frequency control signal for operating the TAF-DPS can come from the to-be-executed instruction. It can also come from the compiler if that is more convenient. As shown in the left of Fig. 6.171, latency awareness computer architecture involves at least three levels: ISA, microarchitecture and compiler. It has great potential and also has many issue-to-be-addressed. It is a wildly open field.

Paradigm Shift:

In the emerging Time-Oriented paradigm, the design of computer architecture can be investigated from a broader and deeper view.

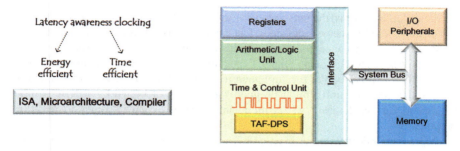

Fig. 6.171 Latency awareness computer architecture

6.24 The Missing Piece in FPGA and Reconfigurable SoC: Programmable Frequency Source

The field programmable gate array (FPGA) is a type of integrated circuit that is intended to be configured by its user after manufacture, hence the term field programmable. Structurally, a basic FPGA must contain an array of programmable logic blocks and a hierarchy of reconfigurable interconnects allowing blocks to be wired together. Inside FPGA, logic blocks can be configured to perform complex combinational functions, or act as simple logic gates such as AND and XOR. The first FPGA was introduced more than three decades ago. Since then, it has seen a rapid growth and has become a popular implementation media for digital circuits. The advancement in process technology has greatly enhanced the logic capacity of FPGA and has in turn made it a viable implementation alternative for larger and complex designs. Performance-wise, the programmable nature of their logic and routing resources has a dramatic impact on the quality of final device's area, speed, and power consumption.

FPGA can be used to solve any problem which is computable. An entire microprocessor can be implemented using FPGA, leading to the so-called soft processor which is implemented via different semiconductor devices containing programmable logic including both high-end and commodity variations. In its early years, FPGA has been reserved for specific vertical applications where the volume of production is small. For these low-volume applications, the premium that companies pay in hardware cost per unit for a programmable chip is more affordable than the development resources spent on creating an ASIC (application-specific integrated circuit). Another trend in the early use of FPGAs is hardware acceleration, where one can use the FPGA to accelerate certain parts of an algorithm and share the computation between the FPGA and a generic processor. As their size, capabilities, and speed increased, modern FPGAs have equipped increasingly more functions to the degree that some FPGAs are now marketed as full systems on chips (SoCs). For example, with the introduction of dedicated multipliers into FPGA architectures in the late 1990s, applications which had traditionally only employed digital signal processor (DSP) hardware began to incorporate FPGA. Nowadays, new cost-and-performance dynamic has broadened the range of viable applications to almost the entire electronic spectrum, including aerospace and defense, automotive, consumer electronics, data center, high performance computing, industrial, financial, medical, security, video and image processing, among others. Especially, owing to its high-parallelism structure, FPGA is very popular for implementing artificial neural networks. While GPU (graphic processing unit) might be good for training of networks, FPGA is more adaptable for accelerating AI (artificial intelligence) when it comes to real-time applications.

Moreover, industry is now driving FPGAs towards playing a big part in the heterogeneous computing paradigm. Heterogeneous computing refers to such systems that use more than one type of processor or core to perform specialized processing duties. Those systems gain performance or energy efficiency not just by adding the same type of processors but by adding dissimilar coprocessors to handle particular tasks. Furthermore, FPGA has enabled another emerging trend of reconfigurable computing. Reconfigurable computing is a computer architecture combining some of the flexibility of software with the high performance of hardware. Data processing is done with very flexible high speed computing fabrics found in FPGAs. The principal dissimilarity, when compared to ordinary microprocessors, lies in its ability to make substantial changes to the datapath itself in addition to the control flow. On the other hand, its difference from custom hardware, the ASIC, is the possibility of adapting the hardware during runtime by inscribing a new circuit into the reconfigurable fabric. In both the heterogeneous computing and reconfigurable computing regimes, FPGA is an indispensable part.

Fundamentally, FPGA consists of three main parts: configurable logic block for implementing logic functions, programmable interconnect for implementing signal routing and programmable I/O blocks for connecting with external components. The modern FPGAs have also acted on putting adders, multipliers and DSP logic inside the configurable logic block to reduce latency, boost computation speed, shorten signal routing, and improve throughput. Furthermore, modern FPGAs not only contain logic element but are further equipped with many types of embedded hardware like memories, DSP core, ARM core, high-speed serial transceiver and etc., as shown in the left-hand side of Fig. 6.172, to form a complete "system on a programmable chip". The aim is to solve real-time problem in a much efficient and timely manner.

As illustrated in the left of Fig. 6.172, modern FPGAs have become so complicated that they contain a lot more functional components than the traditional three parts of configurable logic block, programmable interconnect and programable IO. However, no matter how sophisticated those components are, they are created to manipulate the voltage (the magnitude of electric charge), through either analog or digital means. At a more fundamental level, there is a need of such tool that can manipulate the variable of time, preferably at the same degree of sophistication as we have achieved on voltage. As discussed in Sects. 3.3 and 4.5, magnitude and time are the two variables required to describe any signal, either analog signal or digital one. At current stage, the class of FPGA needs to equip a powerful tool for manipulating time, which is reflected as the frequency of the clock signal. TAF-DPS, owing to its two features of arbitrary

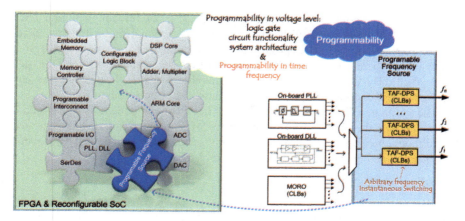

Fig. 6.172 Programable frequency source in FPGA and reconfigurable SoC

frequency generation and instantaneous frequency switching, is squarely suited for serving this goal. It is the type of component called programmable frequency source, as illustrated in Fig. 6.172, the missing piece in the FPGA which is characteristically known for its programmability.

Most of the components built inside of an FPGA are synchronous circuits, each requiring a clock signal to drive its operation. For this reason, FPGAs generally have a significant number of on-board phase-locked loop (PLL) components to synthesize clock frequencies needed for operations. Also, FPGAs have to contain dedicated global and regional routing networks for distributing the clock signals so they can be delivered to the clock sinks with minimal skew. For compensating the skew problem, FPGAs might also equip with components called delay-locked loop (DLL). PLL and DLL are however not qualified as a programable frequency source, simply due to the reason that they do not possess the features of arbitrary frequency generation and instantaneous frequency switching (please refer to Sect. 4.5).

As shown in the right-hand side of Fig. 6.172, the group of phase-evenly-distributed signals needed for TAF-DPS operation can be generated by three means: from the on-board PLL whose VCO is most likely a MORO (multiple output ring oscillator), from the on-board DLL which contains a VCDL (voltage-controlled delay line) made of multiple delay stages, or a MORO directly made from the configurable logic blocks (the CLB). The TAF-DPS circuit itself can be created from the CLBs since TAF-DPS is digitally implementable (can be made 100% from digital standard cells) (Xiu 2017d). Further, the TAF-DPS circuit can be created readily at whatever location that a programmable frequency source is needed. Multiple TAF-DPS copies can be made since the size of TAF-DPS circuit is reasonably small. There are already many real examples of TAF-DPS on FPGA and readers are refer to references (Xiu and Chen 2017; Xiu 2017b; Xiu et al. 2019; Xiu and Wei 2019; Ma et al. 2020; Wei and Xiu 2020; Li et al. 2021).

An interesting point worth mentioning is that the task of creating a digital TAF-DPS on FPGA can be accomplished by using the conventional digital design method. It does not require any sophisticated skill such as analog transistor-level design knowledge. Therefore, digital TAF-DPS on FPGA is an ideal tool for DIY (Do It Yourself). By giving this tool to a vast number of researchers, engineers and hobbyists, it is expected that countless innovations may come out of it.

The addition of TAF-DPS into the FPGA's family of components is significant. It mends a hole in the puzzle. From the perspective of paradigm change that is advocated in this book, it equips FPGA with the power of manipulating the time, the piece that has been missed in the puzzle for a long time.

Paradigm Shift:

In the emerging Time-Oriented paradigm, the creative power enabled by FPGA can be expanded to a higher level by equipping FPGA with a frequency-manipulation tool.

6.25 TAF-DPS Clock Source as a Tool in Design for Manufacture

The manufacturing of semiconductor chips requires many process steps. In every of those steps, there can be imperfections, there might be variability and alignment issues. Some of those problems can cause the complete failure of a die or even a whole wafer. In reality, there are several types of variability. Some of them are global, such as those mask or optics related, and they affect all the transistors in a design in a similar manner. Also, temperature variation can also translate into changes across all transistors. However, more problematic is the local variability that affects just one or a few transistors. This could be caused by crystal variation, contaminant or a number of other mechanisms.

In the past, variation in semiconductor is merely considered as a foundry issue and is largely ignored by most chip design companies. With the continuous technology advancement, new process comes online every couple of years. Progressively, variation is becoming more problematic as chips become increasingly heterogeneous and as chips are used in more emerging applications and more divergent locations, sparking concerns about what the full impact will be and how to solve those issues to an acceptable level. Now, foundries have to figure out how to correlate what is manufactured with what is originally specified, which is mostly dealt with in the process rule deck. At current stage, this problem of process variation is no longer "someone else's problem". It concerns all the parties involved in the entire chain of a product creation, for example, especially for automotive and medical-AI related markets where problems can be life-threatening.

In the development of an end-product, the financial cost due to reliability and manufacturing issues can increase by orders of magnitude if they are addressed at

6.25 TAF-DPS Clock Source as a Tool in Design for Manufacture

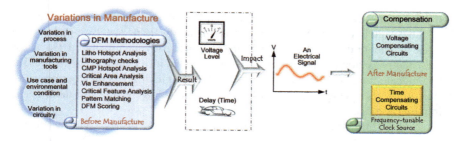

Fig. 6.173 Variations in IC manufacture and the techniques for its compensation

the later stages of development cycle. The general rule of thumb is to address those issues early and proactively. This can pay big dividends since spending money on the front-end to catch to-be-occurring-issues on the back-end can lead to substantial cost-savings down the road. This leads to the practice of Design for manufacturing (DFM). DFM refers to the actions taken during the physical design stage of IC development, to ensure that a design-in-paper can be accurately manufactured. Design techniques and analysis tools can be used to minimize the range of variability that can cause failures. During the years, experience accumulated in developing countless products has provided evidences to support the claim that using those DFM techniques can improve yield. The importance of DFM will only increase as we move along the path of CMOS scaling since, at each manufacturing node, the dimensions get smaller and thus a constant variability actually means greater percentage change. Considering that some dimensions can now be measured in terms of the number of atoms, a difference of 1 atom may represent a significant discrepancy in deep-submicron processes whereas at large geometries that may not have been a problem at all.

In Fig. 6.173, the impact of variation in semiconductor manufacture is illustrated in a big perspective. As shown, there are several types of variation that can occur in the manufacture of a chip. The most noticeable variations occur during the various steps in the manufacture process, including critical dimensions and overlay variation, pattern line edge roughness, localized pattern-dependent lithography variation, film thickness variation, surface roughness, dopant variation (both in dose and location), etch bottom roughness, pattern loading, CMP dishing variation and loading, material stoichiometric and intermixing, among others. Variation in manufacturing tools can also contribute to the uncertainty of final product. Moreover, different use-case and environmental condition can induce variation as well. Finally, some circuitries, especially analog circuits, can initiate variation due to their sensitive to noises (voltage level).

The various DFM methodologies, as shown in the figure, are used to predict those variations in the physical design stage of IC development and, hopefully, prevent them to become real problems in the actual manufacture process later. If some of the variations do happen, they can actually occur even after applying all the DFM techniques, their impact will be manifested as variations in either voltage level or time delay associated with signal traveling between locations. These variations in voltage and delay (time) can change the behavior of an electrical signal, making it deviate from the design intention. At this stage, from the hardware perspective, nothing can be done to ameliorate the situation since the chip is already manufactured. However, certain compensation mechanisms can be incorporated inside the chip beforehand to accommodate the circumstance. One approach is employing voltage compensating circuits (circuits having large margin for tolerating voltage error) and the other is using time compensating circuits.

In term of time compensating circuits, a feasible solution is the frequency-tunable clock source whose frequency can be adjusted in fine step. TAF-DPS, owing to its powerful capability in manipulating pulse length (please refer to Fig. 6.111), is squarely suited for this purpose. By slightly increasing or decreasing the length of clock pulses (i.e., clock frequency), delay deviation induced by the variations in manufacturing or caused by environmental changes can be compensated. In other words, those chips that would be thrown away in conventional standard can be requalified as useful chips, perhaps running at a slightly slower frequency. Or, in the opposite case, nominal-speed grade chips can be upgraded to run at a slightly higher frequency. In short, TAF-DPS is a weapon in the battle of Design for Manufacture, planned in the design stage and reserved for the remedy after the chip is manufacture.

Gestalt Switch:

In the new paradigm, the problems caused by variation in manufacturing can sometimes be remedied by a time-manipulation tool after all existing means fail.

6.26 Digital-To-Frequency Converter and Frequency as a Variable in Programming

TAF-DPS's capability of arbitrary frequency generation (small frequency granularity) and instantaneous frequency switching (two cycles) makes it a new type of signal converter: digital-to-frequency converter (DFC). The process of digital-to-frequency conversion is similar to the process of digital-to-analog conversion (DAC) while the difference lies at their outputs. DAC's output is an analog signal where the voltage value on the waveform is the focus point. In contrast, DFC's output is a pulse train where the frequency of each individual pulse (the pulse length, the period, the instant frequency) is the intended information. The input to DFC is digital value, which can be the product of any digital signal processing task. DFC's output can be used for serving many purposes. For example, the resultant pulse train can be used to control a variety of pumps such as charge pump, fluid pump, flux pump, light pump, among others (e.g., Sect. 6.9). The characteristics embedded in the waveform of the pulse train can be reflected on its spectrum, which can be useful to certain applications (e.g., Sect. 6.20). Moreover, the rate-of-switch associated with this pulse train can be interpreted as meaningful message and be used for serving some purposes. Of course, the most typical use of this pulse train is for functioning as a clock to drive circuits (as trigger for actuating events). All of those actions can be controlled by the task executed in microprocessor through the DFC. This concept of DFC is illustrated in Fig. 6.174.

The introduction of DFC completes the family of converters as illustrated in Fig. 6.175. In voltage domain, the analog-to-digital and digital-to-analog converters have been long established. In time domain, the operation of frequency-to-digital conversion (i.e., the FDC) can be easily accomplished by using a frequency counter. The construction of DFC however is not trivial. It requires a mechanism that can create pulse train of arbitrary frequency. Further, the pulse length has to be individually controllable (i.e., instantaneous frequency switching). Only until the emergence of TAF-DPS technology, the idea of DFC becomes feasible.

By the addition of DFC, as illustrated in the Fig. 6.175, the interface between physical world and electronic world can be established not only through voltage but also time. An information processing flow based on "rate-of-switching" becomes feasible. In this approach, the sensor outputs a pulse train which could be in the form of voltage, current, sound, light and etc. Instead of voltage level as in the case of ADC, the intended information in this case is the number of zero-crossing points occurred within a given time window. Those points represent the "rate-of-switching" of some physical phenomenon that is monitored by the sensor. This information can then be converted into a digital value. The digital signal processing unit further processes the information and produces desired outcome in digital format. On the

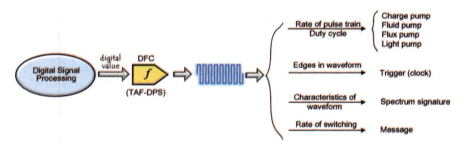

Fig. 6.174 The concept of digital-to-frequency converter

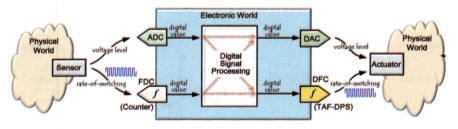

Fig. 6.175 A suite of converters in voltage domain and time domain for interfacing electronic world and physical world

actuating side, the DFC can be used to carry out the desirable action by controlling the "rate-of-switching" type of pump (e.g., digital pump and valve). For certain physical phenomenon, this "rate-of-switching" flow could be more cost-effective or more precise in processing the information that we are interested in. This is due to the fact that many real-world phenomena are vibration. In those cases, the speed of vibration is the information that we care.

In the picture painted in Fig. 6.175, there are four types of converters ADC, FDC, DAC and DFC, and two types of information formats: level and "rate-of-switching". Depending on the problem investigated, those converters and formats can be used in various combinations for best efficiency. An example can be helpful to elucidate the Fig. 6.175. The majority of applications of DAC are in the audio and video markets. In these applications, audio/video information is originally processed in digital domain where high precision (high fidelity) can be preserved. Only at the last stage, the audio/video information is converted back into analog voltage by DAC so that it can be heard/viewed by human. In this chain of signal processing, everything is voltage-based, either in analog or digital formats. In the case of PLL design, as another example, the output of PFD (phase and frequency detector) is pulse trains of varying rate and varying duty-cycle. Those pulse trains are used to drive a charge pump, which is used to control the VCO. In this chain of signal processing, the information carrier is pulse trains where only the duration of time (i.e., the time span of pulse in high state) is meaningful. For this reason, the PFD can be viewed as a rate-of-switching type actuator, although a very rudimentary one.

Although the introduced DFC is created from TAF-DPS, its purpose is not just for frequency synthesis (clock generation). It can be used as an on–off switch to control the flow of certain physical materials such as fluid, gas, light, electrical charge, magnetic flux, etc. In those cases, the term DFC can also be interrupted as Digital-to-Flow converter. This type of converter uses the high/low percentage (please refer to Sect. 6.9 for an example) to control the intended operation. In the case of DAC, its output voltage value can be programmed through manipulating its input digital value. Thus, the voltage is considered as a programmable variable. Similarly, for DFC, the frequency, or rate-of-switching, can be considered as a programmable variable. This argument is depicted in Fig. 6.176. The debut of DFC provides software programmer with one more tool, namely the frequency or rate-of-switching, to control the world. The LED drone light show is an excellent demonstration of the power of using software programming to control the way human perceive the world. By making

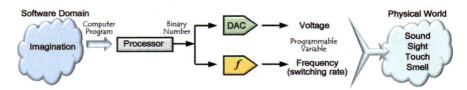

Fig. 6.176 Frequency as a variable in software programming

frequency (or rate-of-switching) programmable, it is believed that software engineer could become even more creative.

Paradigm Shift:

In the emerging Time-Oriented paradigm, a new type of signal converter, Digital-to-Frequency converter, becomes a valuable tool for interfacing the electronic world and physical world.

References

T. Addabbo, M. Alioto, A. Fort, S. Rocchi and V. Vignoli, "Efficient Post-Processing Module for a Chaos-based Random Bit Generator," 2006 *13th IEEE International Conference on Electronics, Circuits and Systems*, Nice, 2006.

A. Alaghi, C. Li, and J. P. Hayes, "Stochastic circuits for real-time image-processing applications," in *Proc. 50th ACM/EDAC/IEEE Design Autom. Conf. (DAC)*, Jun. 2013, pp. 1–6.

A. Alaghi and J. P. Hayes, "Fast and accurate computation using stochastic circuits," *2014 Design, Automation & Test in Europe Conference & Exhibition (DATE)*, 2014, pp. 1-4.

A. Alaghi, W. Qian, and J. P. Hayes, "The promise and challenge of stochastic computing," *IEEE Trans. Comput.-Aided Design Integr. Circuits Syst.*, vol. 37, no. 8, pp. 1515–1531, Aug. 2018.

J. D. H. Alexander, "Clock Recovery from Random Binary Signals," *IEE J. Electronics Letters*, vol. 11, pp. 541-542, Oct. 1975.

M. Alioto, "Trends in Hardware Security: From Basics to ASICs," *IEEE Solid-State Circuits Magazine*, vol. 11, no. 3, pp. 56–74, Summer 2019.

T. Amaki, M. Hashimoto, and T. Onoye, "A process and temperature tolerant oscillator-based true random number generator with dynamic 0/1 bias correction," in *Proc. IEEE Asian Solid-State Circuits Conf. (A-SSCC'13)*, Nov. 2013, pp. 133–136.

Analog Devices, "A Technical Tutorial on Digital Signal Synthesis", available: https://www.ieee.li/pdf/essay/dds.pdf, 1999.

A. Ardakani, F. Leduc-Primeau, N. Onizawa, T. Hanyu, and W. J. Gross, "VLSI implementation of deep neural network using integral stochastic computing," *IEEE Trans. Very Large Scale Integr. (VLSI) Syst.*, vol. 25, no. 10, pp. 2688–2699, Oct. 2017.

Aweya, J., "Implementing synchronous ethernet in telecommunication systems," *IEEE Commun. Surv. Tutorials*, vol. 16, no. 2, pp. 1080–1113, 2014.

M. J. Azhar, F. Amsaad and S. Köse, "Duty-Cycle-Based Controlled Physical Unclonable Function," in *IEEE Transactions on Very Large Scale Integration (VLSI) Systems*, vol. 26, no. 9, pp. 1647–1658, Sept. 2018.

Azpúrua, M. A., Pous, M., and Silva, F., "On the Statistical Properties of the Peak Detection for Time-Domain EMI Measurements," *IEEE Trans. Electromagn. Compat.*, vol. 57, no. 6, pp. 1374–1381, 2015.

M. Barbareschi, G. Di Natale, L. Torres and A. Mazzeo, "A Ring Oscillator-Based Identification Mechanism Immune to Aging and External Working Conditions," in *IEEE Transactions on Circuits and Systems I: Regular Papers*, vol. 65, no. 2, pp. 700-711, Feb. 2018.

Bendicks, A., Frei, S., Hees, N., and Wiegand, M., "Systematic Reduction of Peak and Average Emissions of Power Electronic Converters by the Application of Spread Spectrum," *IEEE Trans. Electromagn. Compat.*, vol. 60, no. 5, pp. 1571–1580, 2018.

S. Ben Dodo, R. Bishnoi, S. Mohanachandran Nair and M. B. Tahoori, "A Spintronics Memory PUF for Resilience Against Cloning Counterfeit," in *IEEE Transactions on Very Large Scale Integration (VLSI) Systems*, vol. 27, no. 11, pp. 2511–2522, Nov. 2019.

T. Bjerregaard and S. Mahadevan, "A Survey of Research and Practices of Network-on-Chip," *ACM computing surveys*, vol. 38, March 2006, article 1.

R. Brederlow, R. Prakash, C. Paulus, and R. Thewes, "A low-power true random number generator using random telegraph noise of single oxidetraps," in *IEEE Int. Solid-State Circuits Conf. (ISSCC) Dig. Tech. Papers*, Feb. 2006, pp. 1666–1675.

B. D. Brown and H. C. Card, "Stochastic neural computation. I. Computational elements," *IEEE Trans. Comput.*, vol. 50, no. 9, pp. 891–905, Sep. 2001.

B. D. Brown and H. C. Card, "Stochastic neural computation. II. Soft competitive learning," *IEEE Trans. Comput.*, vol. 50, no. 9, pp. 906–920, Sep. 2001.

M. Bucci, L. Germani, R. Luzzi, A. Trifiletti, and M. Varanonuovo, "A high-speed oscillator-based truly random number source for cryptographic applications on a smart card IC," *IEEE Trans. Comput.*, vol. 52, no. 4, pp. 403–409, Apr. 2003.

Buevich, M.; Rajagopal, N., and Rowe, A., "Hardware assisted clock synchronization for real-time sensor networks," *Proc. - Real-Time Syst. Symp.*, pp. 268–277, 2013.

Y. Cao, L. Zhang, C. Chang and S. Chen, "A Low-Power Hybrid RO PUF With Improved Thermal Stability for Lightweight Applications," *IEEE Trans. Computer-Aided Design of Integrated Circuits and Syst.*, vol. 34, no. 7, pp. 1143-1147, July 2015.

Chang, H. H., Hua, I. H., and Liu, S. I., "A spread-spectrum clock generator with triangular modulation," *IEEE J. Solid-State Circuits*, vol. 38, no. 4, pp. 673–676, 2003.

S. C. Chan, K.L. Shepard, and P.J. Restle., "Design of resonant global clock distributions", *Proceedings of International Conference on Computer Design*, pages 248–253, Oct. 2003.

S. C. Chan, P. Restle, K. Shepard, N. James, and R. Franch. "A 4.6GHz resonant global clock distribution network", *In Proceedings of IEEE International Solid-State Circuits Conference*, volume 1, Feb. 2004.

S. C. Chan, et.al, "A Resonant Global Clock Distribution for the Cell Broadband Engine Processor", IEEE Journal of Solid-State Circuits, vol. 44, no. 1, pp. 64 –72, Jan. 2009.

Chang, Y., et al, "A Differential Digitally Controlled Crystal Oscillator With a 14-Bit Tuning Resolution and Sine Wave Outputs for Cellular Applications," *IEEE J. Solid-State Circuits*, vol. 47, pp.421-434, 2012.

G. Chance, et al., "Integrated MEMS oscillator for cellular transceivers," *IEEE Int. Freq. Cntl. Symp., (FCS)*, pp.1–3, 2014.

C. Chang, Y. Zheng and L. Zhang, "A Retrospective and a Look Forward: Fifteen Years of Physical Unclonable Function Advancement," *IEEE Circuits and Systems Magazine*, vol. 17, no. 3, pp. 32–62, third quarter 2017.

T. Chelcea and S. M. Nowick, "Robust Interfaces for Mixed-Timing Systems," *IEEE Trans. On VLSI system*, vol. 12, no.8, pp. 857-873, Aug. 2004.

S. Chellappa and L. T. Clark, "SRAM-Based Unique Chip Identifier Techniques," *in IEEE Transactions on Very Large Scale Integration (VLSI) Systems*, vol. 24, no. 4, pp. 1213–1222, April 2016.

Chung, C. C., Su W. S., and Lo, C. K., "A 0.52/1 V Fast Lock-in ADPLL for Supporting Dynamic Voltage and Frequency Scaling," *IEEE Trans. on VLSI*, vol.24, pp.408-412, Jan. 2016.

M. Danesh et al., "Unified Analog PUF and TRNG Based on Current-Steering DAC and VCO," in *IEEE Transactions on Very Large Scale Integration (VLSI) Systems*, vol. 28, no. 11, pp. 2280–2289, Nov. 2020.

Dean, R. N., and Rane, A. K., "A Digital Frequency-Locked Loop System for Capacitance Measurement," *IEEE Trans. on Instrumentation and Measurement*, vol. 62, pp. 777-784, April 2013

De Caro, D., "Optimal discontinuous frequency modulation for spread-spectrum clocking," *IEEE Trans. Electromagn. Compat.*, vol. 55, no. 5, pp. 891–900, 2013.

De Caro, D., et al., "A 3.3 GHz Spread-Spectrum Clock Generator Supporting Discontinuous Frequency Modulations in 28 nm CMOS," *IEEE J. Solid-State Circuits*, vol. 50, no. 9, pp. 2074–2089, 2015.

M. Ding, Z. Zhou, Y. H. Liu, S. Traferro, C. Bachmann, K. Philips, and F. Sebastiano, "A 0.7-V 0.43-pJ/cycle Wakeup Timer Based on a Bang-Bang Digital-Intensive Frequency-Locked-Loop for IoT Applications," *IEEE J. Solid State Circuits Letters*, vol.1, pp. 30–33, 2018.

Djemonai, A., Swan, M. A., and Slamani, M., "New Frequency-Locked Loop Based on Frequency-to-Voltage Converter: Design and Implementation" *IEEE Trans. on Circuit And System II*, vol. 56, pp.441–449, May, 2001

Donegan, P. "IEEE 1588 timing for mobile backhaul: The road to 'plug & play,'" Heavy Reading, New York, NY, USA, Tech. Rep., 2011, http://www.portals.aviatnetworks.com/exLink.asp?115 273260E13W14I75165336.

Ebuchi, T., et al., "A 125-1250 MHz process-independent adaptive bandwidth spread spectrum clock generator with digital controlled self-calibration," *IEEE J. Solid-State Circuits*, vol. 44, no. 3, pp. 763–774, 2009

Eidson, J. C., "Measurement, Control, and Communication Using IEEE 1588," New York, NY, USA: *Springer*, 2006.

Estrela, P. V., Neusüß, S., and Owczarek, W., "Using a multi-source NTP watchdog to increase the robustness of PTPv2 in financial industry networks," in *Proc. IEEE Symp. Precis. Clock Synchronization Meas. Control Commun. (ISPCS)*, Austin, TX, USA, 2014, pp. 87–92.

Ferrant, J. L., et al., "Synchronous Ethernet and IEEE 1588 in Telecoms: Next Generation Synchronization Networks," *ISTE and John Wiley & Sons*, 2013.

FIPS 140–2, "Security Requirements for Cryptographic Modules," *National Institute of Standards and Technology*, GPO, Washington DC, 1999.

Al-Fuqaha, A., Guizani. M, Mohammadi, M., Aledhari, M., and Ayyash, M., "Internet of Things: A Survey on Enabling Technologies, Protocols, and Applications," *IEEE Commun. Surv. Tutorials*, vol. 17, no. 4, pp. 2347–2376, 2015.

Gaderer, G., Nagy, A., Loschmidt, P., and Sauter, T., "Achieving a realistic notion of time in discrete event simulation," *Int. J. Distrib. Sensor Netw.*, vol. 2011, pp. 1–11, Jul. 2011.

B. R. Gaines, *Stochastic Computing Systems*, Boston, MA, USA, Springer, 1967.

B. R. Gaines, "Stochastic computing," *In Proceedings of the AFIPS Spring Joint Computer Conference*, pp. 149–156, 1967b.

B. R. Gaines, "Stochastic computing systems," *Adv. Inform. Syst. Sci.* 2, pp. 37–172, 1969.

Glass, M. E., and O'Donoghue, K. F., "Navy shipboard time synchronization service options and analysis," in *Proc. IEEE Symp. Precis. Clock Synchronization Meas. Control Commun. (ISPCS)*, Ann Arbor, MI, USA, 2008, pp. 105–109.

B. G. Goldberg, "Digital Frequency Synthesis Demystified", LLH Technology Publishing, 1999.

G. Gonzalez, "Foundations of Oscillator Circuit," Ch. IV, Artech House, Inc., London, 2007.

R. Govindaraj, S. Ghosh and S. Katkoori, "Design, Analysis and Application of Embedded Resistive RAM Based Strong Arbiter PUF," in *IEEE Transactions on Dependable and Secure Computing*, vol. 17, no. 6, pp. 1232–1242, 1 Nov.-Dec. 2020.

Griffith, D, Dülger, F., Feygin, G., Mohieldin, A. N., Vallur, P., "A 65nm CMOS DCXO system for generating 38.4MHz and a real time clock from a single crystal in 0.09mm," *IEEE Radio Frequency Integrated Circuits Symposium*, pp.321–324, 2010.

V. Gutnik and A. P. Chandrakasan. "Active GHz clock network using distributed PLLs." IEEE JSCC, vol. 35, no. 11 (2000): 1553-1560.

V. Gutnik and A. P. Chandrakasan. "Active GHz clock network using distributed PLLs." IEEE JSCC, vol. 35, no. 11 (2000b): 1553–1560.

Hardin, K. B., Fessler, J. T., and Bush, D. R., "Spread spectrum clock generation for the reduction of radiated emissions," *Electromagn. Compat. 1994. Symp. Rec. Compat. Loop., IEEE Int. Symp.*, pp. 227–231, 1994.

Hardin, K. B., Oglesbee, R. A., and Fisher, F., "Investigation into the interference potential of spread-spectrum clock generation to broadband digital communications," *IEEE Trans. Electromagn. Compat.*, vol. 45, no. 1, pp. 10–21, 2003.

Harris, K., "An application of IEEE 1588 to industrial automation," in *Proc. IEEE Symp. Precis. Clock Synchronization Meas. Control Commun. (ISPCS)*, Ann Arbor, MI, USA, 2008, pp. 71–76.

J. Hazarika and P. Sumathi, "Moving Window Filter Based Frequency-Locked Loop for Capacitance Measurement," *IEEE Trans. on Industrial Electronics*, vol. 62, pp.7821-7823, Dec. 2015.

Z. He, M. Wan, J. Deng, C. Bai and K. Dai, "A Reliable Strong PUF Based on Switched-Capacitor Circuit," in *IEEE Transactions on Very Large Scale Integration (VLSI) Systems*, vol. 26, no. 6, pp. 1073–1083, June 2018.

J. Helle, "VCXO Theory and Practice," *29th Annual Symp. on Freq. Cntl.*, pp.300–307, 1975.

C. Herder, M. Yu, F. Koushanfar and S. Devadas, "Physical Unclonable Functions and Applications: A Tutorial," in *Proceedings of the IEEE*, vol. 102, no. 8, pp. 1126-1141, Aug. 2014.

D. E. Holcomb, W. P. Burleson and K. Fu, "Power-Up SRAM State as an Identifying Fingerprint and Source of True Random Numbers," in *IEEE Transactions on Computers*, vol. 58, no. 9, pp. 1198-1210, Sept. 2009.

Ho, Y. H. and Yao, C. Y., "A Low-Jitter Fast-Locked All-Digital Phase-Locked Loop With Phase–Frequency-Error Compensation," *IEEE Trans. on VLSI*, vol.24, pp.1984-1992, May 2016.

Horauer, M., "Clock synchronization in distributed systems," University of Technology Vienna, Austria, 2004.

Hsieh, Y. Bin and Kao, Y. H., "A fully integrated spread-spectrum clock generator by using direct VCO modulation," *IEEE Trans. Circuits Syst. I Regul. Pap.*, vol. 55, no. 7, pp. 1845–1853, 2008.

H. Ichihara, S. Ishii, D. Sunamori, T. Iwagaki, and T. Inoue, "Compact and accurate stochastic circuits with shared random number sources," in *Proc. IEEE 32nd Int. Conf. Comput. Design (ICCD)*, Oct. 2014, pp. 361–366.

IEC Technical Committee 65 and IEEE Standards Association (IEEE-SA) Standards Board, IEEE Standard for a Precision Clock Synchronization Protocol for Networked Measurement and ControlSystems, IEEE Standard 1588(E)-2008, 2008.

IEEE, 802.1AS-2011 - IEEE Standard for Local and Metropolitan Area Networks - Timing and Synchronization for Time-Sensitive Applications in Bridged Local Area Networks, IEEE Std., 2011.

ISCRI Part 1–1, International Special Committee on Radio Interference, "Specification for Radio Disturbance and Immunity Measuring Apparatus and Methods—Part 1–1: Radio Disturbance and Immunity Measuring Apparatus, Document CISPR 16–1–1," 2015a.

ISCRI Part 2–1, International Special Committee on Radio Interference, "Specification for Radio Disturbance and Immunity Measuring Apparatus and Methods—Part 2–1: Methods of Measurement of Disturbances and Immunity—Conducted Disturbance Measurements, Document CISPR 16–2–1," 2015b.

O. Ishii, T. Shibata, T. Ohshima, "High frequency fundamental VCXO for SDH system," *IEEE Int. Freq. Cntl. Symp.*, pp.714–721, 1996.

"Definitions and terminology for synchronization networks", ITU-T G.810, 08/1996, ITU-T, https://www.itu.int/rec/T-REC-G.810-199608-I/en

"Timing characteristics of telecom boundary clocks and telecom time slave clocks for use with full timing support from the network," ITU-T G.8273.2/Y.1368.2, 10/2020, ITU-T, https://www.itu.int/rec/T-REC-G.8273.2-202010-I/en

B. Jun and P. Kocher, "The Intel RNG," White Paper, 1999 [Online]. Available: http://www.cryptography.com/intelRNG.pdf

Karaca, T., Deutschmann, B., and Winkler, G., "EMI-receiver simulation model with quasi-peak detector," *IEEE Int. Symp. Electromagn. Compat.*, vol. 2015–Septm, pp. 891–896, 2015.

R. L. Kent, "The Voltage-Controlled Crystal Oscillator, Its Capabilities and Limitations," *19th Annual Symp. on Freq. Cntl.*, pp.642–654, 1965.

W. Khalil, S. Shashidharan, T. Copani, S. Chakraborty, S. Kiaei and B. Bakkaloglu, "A 700-uA A 405-MHz All-Digital Fractional-N Frequency-Locked Loop for ISM Band Applications," *IEEE Trans. on Microwave theory and Techniques*, vol. 59, pp.1319–1326, May, 2011

J. Kim et al., "A Physical Unclonable Function With Redox-Based Nanoionic Resistive Memory," *IEEE Trans. Inf. Forensics Security*, vol. 13, no. 2, pp. 437-448, Feb. 2018.

K. Kobayashi, et al., "High-performance DSP-TCXO using twin-crystal oscillator," *IEEE Int. Freq. Cntl. Symp., (FCS)*, pp.1–4, 2014.

References

Avinoam Kolodny, "Networks on Chips: A New Paradigm," tutorial 2009.

Koskiahde, T., Kujala, J., and Norolampi, T., "A sensor network architecture for military and crisis management," in *Proc. IEEE Symp. Precis. Clock Synchronization Meas. Control Commun. (ISPCS)*, Ann Arbor, MI, USA, 2008, pp. 110–114.

V. F. Kroupa, "Direct Digital Frequency Synthesis", *IEEE Press*, 1998.

M. Kumm, H. Klingbeil, and P. Zipf, "An FPGA-Based Linear All-Digital Phase-Locked Loop," *IEEE Trans. on Circuit And System I*, vol. 57, pp.2487–2497, Sep., 2010.

Lamport, L., "Time, clocks, and the ordering of events in a distributed system," *Commun. ACM*, vol. 21, no. 7, pp. 558–565, 1978.

Lamport, L. and Melliar-Smith, P. M., "Synchronizing clocks in the presence of faults," *J. ACM*, vol. 32, no. 1, pp. 52–78, 1985.

Latremouille, D., Harper, K., and Subrahmanyan, R., "An architecture for embedded IEEE 1588 support," in *Proc. IEEE Symp. Precis. Clock Synchronization Meas. Control Commun. (ISPCS)*, Vienna, Austria, 2007, pp. 128–133.

Lee, W. Y. and Kim, L. S., "A spread spectrum clock generator for displayPort main link," *IEEE Trans. Circuits Syst. II Express Briefs*, vol. 58, no. 6, pp. 361–365, 2011.

Lévesque, M. and Tipper, D., "A Survey of Clock Synchronization Over Packet-Switched Networks," *IEEE Commun. Surv. Tutorials*, vol. 18, no. 4, pp. 2926–2947, 2016.

P. Li, D. J. Lilja, W. Qian, K. Bazargan, and M. D. Riedel, "Computation on stochastic bit streams digital image processing case studies," *IEEE Trans. Very Large Scale Integr. (VLSI) Syst.*, vol. 22, no. 3, pp. 449–462, Mar. 2014.

C. Li, S. Lin, Y. Li, H. Zhang, X. Wei and L. Xiu, "A Method of Low-Cost Pure-Digital GPS Disciplined Clock for Improving Frequency Accuracy and Steering Frequency Through TAF-DPS Frequency Synthesizer," in IEEE Transactions on Instrumentation and Measurement, vol. 70, pp. 1-10, 2021.

Lin, F., and Chen, D. Y., "Reduction of Power Supply EMI Emission by Switching Frequency Modulation," *IEEE Trans. Power Electron.*, vol. 9, no. 1, pp. 132–137, 1994.

L. Lin, S. Srivathsa, D. K. Krishnappa, P. Shabadi, and W. Burleson, "Design and validation of arbiter-based PUFs for sub-45-nm low-power security applications," *IEEE Trans. Inf. Forensics Security*, vol. 7, no. 4, pp. 1394–1403, Aug. 2012.

Baoyuan Liu, Min Wang, H. Foroosh, M. Tappen and M. Penksy, "Sparse Convolutional Neural Networks," 2015 IEEE Conference on Computer Vision and Pattern Recognition (CVPR), 2015, pp. 806–814.

Daihyun Lim, J. W. Lee, B. Gassend, G. E. Suh, M. van Dijk and S. Devadas, "Extracting secret keys from integrated circuits," in *IEEE Transactions on Very Large Scale Integration (VLSI) Systems*, vol. 13, no. 10, pp. 1200–1205, Oct. 2005.

S. Lim, B. Song and S. Jung, "Highly Independent MTJ-Based PUF System Using Diode-Connected Transistor and Two-Step Postprocessing for Improved Response Stability," *IEEE Trans. Inf. Forensics Security*, vol. 15, pp. 2798-2807, 2020.

Y. Liu and K. K. Parhi, "Architectures for recursive digital filters using stochastic computing," *IEEE Trans. Signal Process.*, vol. 64, no. 14, pp. 3705–3718, Jul. 2016.

Liu, X., Kong C., Gao Y. and Tang Z., "A 0.07-ppm/step differential digitally controlled crystal oscillator with guaranteed monotonicity," *IEEE 12th International Conference on ASIC (ASICON)*, pp. 406–409, 2017.

C. Q. Liu, Y. Cao and C. H. Chang, "ACRO-PUF: A Low-power, Reliable and Aging-Resilient Current Starved Inverter-Based Ring Oscillator Physical Unclonable Function," *in IEEE Transactions on Circuits and Systems I: Regular Papers*, vol. 64, no. 12, pp. 3138-3149, Dec. 2017.

Y. Liu, L. Liu, F. Lombardi, and J. Han, "An energy-efficient and noise-tolerant recurrent neural network using stochastic computing," *IEEE Trans. Very Large Scale Integr. (VLSI) Syst.*, vol. 27, no. 9, pp. 2213–2221, Sep. 2019.

Y. Liu, S. Liu, Y. Wang, F. Lombardi, and J. Han, "A survey of stochastic computing neural networks for machine learning applications," *IEEE Trans. Neural Netw. Learn. Syst.*, vol. 32, no. 7, pp. 2809–2824, Aug. 2020.

Lombradi, M. A., Ch. 17, "Fundamentals of Time and Frequency", The Mechatronics Handbook, CRC Presss, 2002.

Loy, D., "GPS-Linked High Accuracy NTP Time Processor for Distributed Fault-Tolerant Real-Time Systems," Vienna University of Technology, Austria, 1996.

Ma Y., Wei X. and Xiu L., "A Novel Spread Spectrum Clock Generation Technique: Always-On Boundary Spread SSCG," in *IEEE Transactions on Electromagnetic Compatibility*, vol. 62, no. 2, pp. 364-376, April 2020.

Mach, H., Grim, E., Holmeide, O., and Calley, C., "PTP enabled network for flight test data acquisition and recording," in *Proc. IEEE Symp. Precis. Clock Synchronization Meas. Control Commun. (ISPCS)*, Vienna, Austria, 2007, pp. 110–115.

Mahmood, A., Exel, R., Trsek, H., and Sauter, T., "Clock Synchronization Over IEEE 802.11—A Survey of Methodologies and Protocols," *IEEE Trans. Ind. Informatics.*, vol. 13, no. 2, pp. 907–920, April 2017.

Mair, H. and Xiu, L., "Architecture of high-performance frequency and phase synthesis," *IEEE J. Solid-State Circuits*, vol. 35, no. 6, pp. 835–846, 2000.

Matsumoto, Y., Fujii, K., and Sugiura, A., "An analytical method for determining the optimal modulating waveform for dithered clock generation," *IEEE Trans. Electromagn. Compat.*, vol. 47, no. 3, pp. 577–584, 2005.

Matsumoto, Y., Fujii, K., and Sugiura, A., "Estimating the amplitude reduction of clock harmonics due to frequency modulation," *IEEE Trans. Electromagn. Compat.*, vol. 48, no. 4, pp. 734–741, 2006.

M. Matsumoto, S. Yasuda, R. Ohba, K. Ikegami, T. Tanamoto, and S. Fujita, "1200 μm2 physical random-number generators based on SiN MOSFET for secure smart-card application," in *IEEE Int. Solid-State Circuits Conf. (ISSCC) Dig. Tech. Papers*, Feb. 2008, pp. 414–624.

S. K. Mathew et al., "2.4 Gbps, 7 mW All-digital PVT-variation tolerant true random number generator for 45 nm CMOS high-performance microprocessors," *IEEE J. Solid-State Circuits*, vol. 47, no. 11, pp. 2807–2821, Nov. 2012.

Mills, D. L., "Internet Time Synchronization: The Network Time Protocol," *IEEE Trans. Commun.*, vol. 39, no. 10, pp. 1482–1493, 1991.

Mirabella, O., Brischetto, M., Raucea, A., Banno, F., and Caruso, N., "Improving the Dynamic Continuous Clock Synchronization for WSNs," *IECON 2010–36th Annu. Conf. IEEE Ind. Electron. Soc.*, pp. 2126–2133, 2010.

Mock, M., Frings, R., Nett, E., and Trikaliotis, S., "Continuous clock synchronization in wireless real-time applications," *Proc. 19th IEEE Symp. Reliab. Distrib. Syst.* SRDS-2000, pp. 125–132, 2000.

S. Mohajer, Z. Wang, K. Bazargan, M. Riedel, D. Lilja, and S. Faraji, "Parallel computing using stochastic circuits and deterministic shuffling networks," U.S. Patent Appl. 16/165 713, Apr. 25, 2019.

A. Morro, et al., "A stochastic spiking neural network for virtual screening," *IEEE Trans. Neural Netw. Learn. Syst.*, vol. 29, no. 4, pp. 1371–1375, Apr. 2018.

R. Mullins, "Asynchronous versus Synchronous Design Techniques for NoCs," tutorial, *international symposium on System-on-Chip*, Tampere, Finland, Nov., 2005.

M. H. Najafi, S. Jamali-Zavareh, D. J. Lilja, M. D. Riedel, K. Bazargan and R. Harjani, "Time-Encoded Values for Highly Efficient Stochastic Circuits," in *IEEE Transactions on Very Large Scale Integration (VLSI) Systems*, vol. 25, no. 5, pp. 1644–1657, May 2017.

M. H. Najafi, D. Jenson, D. J. Lilja, and M. D. Riedel, "Performing stochastic computation deterministically," *IEEE Trans. Very Large Scale Integr. (VLSI) Syst.*, vol. 27, no. 12, pp. 2925–2938, Dec. 2019.

Namgoong, W., "Observer-Controller Digital PLL," *IEEE Trans. on Circuit And System I*, vol. 57, pp.631–641, Sep., 2010.

Neagoe, T., Cristea, V., and Banica, L., "NTP versus PTP in computer networks clock synchronization," in *Proc. IEEE Int. Symp. Ind. Electron.*, Montreal, QC, Canada, 2006, pp. 317–362.

J. Neumann, *Lectures on probabilistic logics and the synthesis of reliable organisms from unreliable components*, California Institute of Technology, 1st edition, 1952.

"A Statistical Test Suite for the validation of Random Number Generators and Pseudo Random Number Generators for Cryptographic Applications," National Inst. Standards and Technology, Pub 800–22, 2010.

"Framework for Cyber-Physical Systems, Release 1.0," National Inst. Standards and Technology, May, 2016.

"Recommendation for the Entropy Sources Used for Random Bit Generation," National Inst. Standards and Technology, Pub 800–90B, Jan. 2018.

O'Mahony, F. ; Yue, C.P. ; Horowitz, M.A. ; Wong, S.S., "A 10-GHz global clock distribution using coupled standing-wave oscillators," *IEEE Journal of Solid-State Circuits*, vol. 38, no. 11, pp. 1813–1820, Nov. 2003.

Pareschi, F., Setti, G., Rovatti, R., and Frattini, G., "Short-term optimized spread spectrum clock generator for EMI reduction in switching DC/DC converters," *IEEE Trans. Circuits Syst. I Regul. Pap.*, vol. 61, no. 10, pp. 3044–3053, 2014.

Pareschi, F., Setti, G., Rovatti, R., and Frattini, G., "Practical optimization of EMI reduction in spread spectrum clock generators with application to switching DC/DC converters," *IEEE Trans. Power Electron.*, vol. 29, no. 9, pp. 4646–4657, 2014.

Pareschi, F., Rovatti, R., and Setti, G., "EMI reduction via spread spectrum in DC/DC converters: State of the art, optimization, and tradeoffs," *IEEE Access*, vol. 3, pp. 2857–2874, 2015.

A. Partridge, H. C. Lee, P. Hagelin, V. Menon, , "We know that MEMS is replacing quartz. But why? And why now?" *Joint European Frequency and Time Forum & International Frequency Control Symposium (EFTF/IFC)*, pp.411–416, 2013.

G. A. Pratt et al., "Distributed Synchronous clocking", IEEE transaction on parallel and distributed systems, vol. 6, n. 3, march 1995, pp. 314-328.

C. S. Petrie and J. A. Connelly, "A noise-based IC random number generator for applications in cryptography," *IEEE Trans. Circuits Syst. I, Fundam. Theory Appl.*, vol. 47, no. 5, pp. 615–621, May 2000.

M. Petrowski, R. Clark, "DSP-based oscillator technology greatly simplifies timing architectures in multi-service platforms," *Optical Fiber Communication Conference*, 2006.

Pinchas, M., and Cohen, R., "A combined PTP and circuit-emulation system," in *Proc. IEEE Symp. Precis. Clock Synchronization Meas. Control Commun. (ISPCS)*, Vienna, Austria, 2007, pp. 143–147.

Poore, R., "Overview on phase noise and jitter," Agilent Technologies, 2001, available: http://cp.literature.agilent.com/litweb/pdf/5990-3108EN.pdf.

W. J. Poppelbaum, "Statistical processors," *Adv. Computers* 14, pp. 187–230, 1976.

W. Qian, X. Li, M. D. Riedel, K. Bazargan, and D. J. Lilja, "An architecture for fault-tolerant computation with stochastic logic," *IEEE Trans. Comput.*, vol. 60, no. 1, pp. 93–105, Jan. 2011.

F. Rahman, B. Shakya, X. Xu, D. Forte and M. Tehranipoor, "Security Beyond CMOS: Fundamentals, Applications, and Roadmap," in *IEEE Transactions on Very Large Scale Integration (VLSI) Systems*, vol. 25, no. 12, pp. 3420–3433, Dec. 2017.

S. T. Ribeiro, "Random-Pulse Machines," in *IEEE Transactions on Electronic Computers*, vol. EC-16, no. 3, pp. 261–276, June 1967.

S. A. Salehi, "Low-Cost Stochastic Number Generators for Stochastic Computing," *in IEEE Transactions on Very Large Scale Integration (VLSI) Systems*, vol. 28, no. 4, pp. 992–1001, April 2020.

N. Saraf, K. Bazargan, D. J. Lilja and M. D. Riedel, "IIR filters using stochastic arithmetic," *2014 Design, Automation & Test in Europe Conference & Exhibition (DATE)*, 2014, pp. 1-6.

Sasaki, M., "A High-Frequency Clock Distribution Network Using Inductively Loaded Standing-Wave Oscillators," *IEEE Journal of Solid-State Circuits*, vol. 44, no. 10, pp. 2800–2807, Oct. 2009.

Sathe, V.S. ; Kao, J.C. ; Papaefthymiou, M.C., "Resonant-Clock Latch-Based Design," *IEEE Journal of Solid-State Circuits*, vol. 43, no. 4, pp. 864 –873, Apr. 2008.

Sauter, T., "The three generations of field-level networks–evolution and compatibility issues," *IEEE Trans. Ind. Electron.*, vol. 57, no. 11, pp. 3585–3595, Nov. 2010.

Schmuck, F., and Cristian, F., "Continuous clock amortization need not affect the precision of a clock synchronization algorithm," in *Proceedings of the ninth annual ACM symposium on Principles of distributed computing*, (New York, NY, USA), pp. 133–143, ACM, 1990.

Silicon Labs Inc., Si570/Si571 datasheet, rev. 1.5 4/14, 2014.

Silicon Labs Inc., Voltage-Controlled Crystal Oscillator (VCXO) 10 MHz To 1.4 GHz, Si550, Rev. 1.2 6/18, 2018.

Song, M., Ahn, S., Jung, I., Kim, Y., and Kim, C., "Piecewise linear modulation technique for spread spectrum clock generation," *IEEE Trans. Very Large Scale Integr. Syst.*, vol. 21, no. 7, pp. 1234–1245, 2013.

Staszewski, R. B. and et al., "All-Digital PLL and Transmitter for Mobile Phones," *IEEE J. Solid State Circuits*, vol.40, pp. 2469–2481, 2005.

Subrahmanyan, R., "Implementation considerations for IEEE 1588v2 applications in telecommunications," in *Proc. IEEE Symp. Precis. Clock Synchronization Meas. Control Commun. (ISPCS)*, Vienna, Austria, 2007, pp. 148–154.

Sullivan, D., Allan, D., Howe, D. and Walls, F., "Characterization of clocks and oscillators," National Institute of Standards and Technology, Technical Note 1337, 1990.

B. Sunar, W. J. Martin and D. R. Stinson, "A Provably Secure True Random Number Generator with Built-In Tolerance to Active Attacks," *in IEEE Transactions on Computers*, vol. 56, no. 1, pp. 109-119, Jan. 2007

Q. Tang, B. Kim, Y. Lao, K. K. Parhi, and C. H. Kim, "True random number generator circuits based on single- and multi-phase beat frequency detection," *in Proc. IEEE Custom Integr. Circuits Conf. (CICC'14)*, 2014, pp. 1–4.

S. Talwalkar, "Quantization error spectra structure of a DTC synthesizer via the DFT axis scaling property", *IEEE Trans. Circuits Syst. I*, vol. 59, no. 6, pp.1242 – 1250, June 2012.

Tatsukawa, J., "Spread-Spectrum Clocking Reception for Displays," vol. XAPP469. Xilinx Application Note, pp. 1–8, 2008.

Paul Teehan, et al., "A Survey and Taxonomy of GALS Design Styles," *IEEE Design & Test of Computers*, Sep-Oct., 2007.

S. S. Tehrani, W. J. Gross, and S. Mannor, "Stochastic decoding of LDPC codes," *IEEE Commun. Lett.*, vol. 10, no. 10, pp. 716–718, Oct. 2006.

F. Tehranipoor, N. Karimian, W. Yan and J. A. Chandy, "DRAM-Based Intrinsic Physically Unclonable Functions for System-Level Security and Authentication," in *IEEE Transactions on Very Large Scale Integration (VLSI) Systems*, vol. 25, no. 3, pp. 1085–1097, March 2017.

Tektronix, "Understanding and Characterizing Timing Jitter," application note 55W-16146–6, April 2017, available: https://www.tek.com/primer/understanding-and-characterizing-timing-jitter-primer.

N. Temenos and P. P. Sotiriadis, "Nonscaling Adders and Subtracters for Stochastic Computing Using Markov Chains," in IEEE Transactions on Very Large Scale Integration (VLSI) Systems, vol. 29, no. 9, pp. 1612–1623, Sept. 2021.

Ying Teng; Jianchao Lu; Taskin, B., "ROA-brick topology for rotary resonant clocks," *Computer Design (ICCD), 2011 IEEE 29th International Conference on*, 2011, pp. 273–278.

Texas Instruments Inc., VCXO Application Guideline for CDCE(L)9xx Family, SCAA085A, June 2007.

S. L. Toral, J. M. Quero and L. G. Franquelo, "Stochastic pulse coded arithmetic," *2000 IEEE International Symposium on Circuits and Systems (ISCAS)*, 2000, pp. 599–602 vol. 1.

C. Tokunaga, D. Blaauw, and T. Mudge, "True random number generator with a metastability-based quality control," *IEEE J. Solid-State Circuits*, vol. 43, no. 1, pp. 78–85, Jan. 2008.

T. H. Tran, et al., "A Low-ppm Digitally Controlled Crystal Oscillator Compensated by a New 0.19-mm2 Time-Domain Temperature Sensor," *IEEE Sensors Journal*, vol.17, pp. 51-62, 2017.

M. D. Tsai, et al., "A Temperature-Compensated Low-Noise Digitally-Controlled Crystal Oscillator for Multi-Standard Applications," *IEEE Radio Frequency Integrated Circuits Symposium*, 2008.

Ullmann, M.,, and Vögeler, M., "Delay attacks—Implication on NTP and PTP time synchronization," in *Proc. IEEE Symp. Precis. Clock Synchronization Meas. Control Commun. (ISPCS)*, Brescia, Italy, 2009, pp. 1–6.

John R. Vig, "Quartz Crystal Resonators and Oscillators For Frequency Control and Timing Applications: A Tutorial," January 2000. https://www.am1.us/wp-content/uploads/Documents/U11625_VIG-TUTORIAL.pdf.

M. Wan, Z. He, S. Han, K. Dai and X. Zou, "An Invasive-Attack-Resistant PUF Based On Switched-Capacitor Circuit," in *IEEE Transactions on Circuits and Systems I*: Regular Papers, vol. 62, no. 8, pp. 2024-2034, Aug. 2015.

Wang, W., Xu, Y. and Khanna, M., "A survey on the communication architectures in smart grid," *Comput. Netw.*, vol. 55, no. 15, pp. 3604–3629, Oct. 2011.

Z. Wang et al., "Current Mirror Array: A Novel Circuit Topology for Combining Physical Unclonable Function and Machine Learning," in *IEEE Transactions on Circuits and Systems I*: Regular Papers, vol. 65, no. 4, pp. 1314-1326, April 2018.

Weiss, M. A., et al., "Time-Aware Applications, Computers, and Communication Systems (TAACCS)," National Institute of Standards and Technology, Tech. Note, vol. 1867, 2015.

Wei, Xiangye and Xiu, L., "An All-Digital Highly Programable TAF-DPS Based True Random Number Generator Working on Principles of Frequency-Mixing and Frequency-Tracking", *ISCAS2020*, 2020.

Wiklundh, K., "Relation between the amplitude probability distribution of an interfering signal and its impact on digital radio receivers," *IEEE Trans. Electromagn. Compat.*, vol. 48, no. 3, pp. 537–544, 2006.

J. Wood, T. C. Edwards and S. Lipa, "Rotary traveling-wave oscillator arrays: a new clock technology," *in IEEE Journal of Solid-State Circuits*, vol. 36, no. 11, pp. 1654-1665, Nov. 2001.

Wu, X., Chan, M. C., Ananda, A. L., and Ganjihal, C., "Sync-TCP: A new approach to high speed congestion control," in *Proc. IEEE Int. Conf. Netw. Protocols, Princeton, NJ, USA, 2009*, pp. 181–192.

Wu, Y., Shahmohammadi, M., Chen, Y., Lu, P., Staszewski, R. B., "A 3.5–6.8-GHz Wide-Bandwidth DTC-Assisted Fractional-N All-Digital PLL With a MASH," *IEEE J. Solid State Circuits*, vol.52, pp. 1885–1903, 2017.

Xiu, L. and You, Z. "A 'flying-adder' architecture of frequency and phase synthesis with scalability," IEEE Trans. Very Large Scale Integr. Syst., vol. 10, no. 5, pp. 637–649, 2002.

Xiu, L. and Z. You, "A New Frequency Synthesis Method Based on Flying-Adder Architecture," *IEEE Trans. on Circuit And System II*, vol. 50, pp.130–134, March 2003.

L. Xiu, W. Li, J. Meiners and R. Padakanti, "A Novel All-Digital PLL with Software Adaptive Filter," *IEEE J. Solid-State Circuit*, vol. 39, no. 3, pp. 476-483, March 2004.

Xiu, L., Z. You, "A Flying-Adder Frequency Synthesis Architecture of Reducing VCO Stages," *IEEE Trans. on VLSI*, vol.13, pp. 201–210, Feb., 2005.

Xiu, L., "A Flying-Adder Based on-chip Frequency Generator for Complex SoC," *IEEE Trans. on Circuit And System II*, vol. 54, pp.1067–1071, Dec. 2007.

Xiu, L., "A Flying-Adder PLL Technique Enabling Novel Approaches for Video/Graphic Applications," *IEEE Trans. on Consumer Electronic*, vol. 54, pp.591-599, May, 2008.

Xiu, L., "A Novel DCXO Module for Clock Synchronization in MPEG2 Transport System," *IEEE Trans. on Circuit And System I*, vol. 55, pp.2226–2237, Sep. 2008.

Xiu, L., "The Concept of Time-Average-Frequency and Mathematical Analysis of Flying-Adder Frequency Synthesis Architecture," *IEEE Circuit And System Magazine*, 3rd quarter, pp.27–51, Sep.2008.

Xiu, L., "A Fast and Power-Area Efficient Accumulator for Flying-Adder Frequency Synthesizer," *IEEE Trans. on Circuit And System I*, vol. 56, pp.2439–2448, Nov., 2009.

Xiu, L., Nanometer Frequency Synthesis Beyond the Phase-Locked Loop. Piscataway, NJ 08854, USA: John Wiley IEEE-press, 2012a. Nanometer Frequency Synthesis Beyond the Phase-Locked Loop. Piscataway, NJ 08854, USA: *John Wiley IEEE-press*, 2012a.

Xiu, L., Kun-Ho Lin and M. Ling, "The Impact of Input-Mismatch on Flying-Adder Direct Period Synthesizer," *IEEE Trans. on Circuit And ystem I*, vol. 59, pp. 1942–1951, Sep. 2012.

Xiu, L., "Lock detector, method applicable thereto, and phase lock loop applying the same", US patent 8,258,834, Sep. 2012b.

Xiu, L., Lin, W. T. and Lee, T., "A Flying-Adder Fractional-Divider based integer-N PLL: the 2nd generation Flying-Adder PLL as clock generator for SoC", *IEEE J. Solid-State Circuits*, vol. 48, pp.441-455, Feb. 2013.

Xiu, L., "Circuits and methods for clock generation using a flying-adder divider inside and optionally outside a phase locked loop", US patent 8,664,988, March, 2014a.

Xiu, L., "Circuit And Method Of Using Time-Average-Frequency Direct Period Synthesizer For Improving Crystal-Less Frequency Generator's Frequency Stability", US 8,890,591, Nov. 2014b.

Xiu, L., "From Frequency to Time-Average-Frequency: A Paradigm Shift in the Design of Electronic system," May 2015a, John Wiley IEEE press (IEEE Press Series on Microelectronic Systems).

Xiu, L., "Direct Period Synthesis for Achieving Sub-PPM Frequency Resolution through Time Average Frequency: The Principle, The Experimental Demonstration, and Its Application in Digital Communication," *IEEE Trans. on VLSI*, vol.23, no.7, pp.1335-1344, 2015.

Xiu, L., "Circuits and methods for one-wire communication bus of using pulse-edge for clock and pulse-duty-cycle for data", US patent 8,929,467, Jan. 2015c.

Xiu, L., "Circuits And Methods For Using A Flying-Adder Synthesizer As A Fractional Frequency Divider", US 9,008,261, April 14, 2015d.

Xiu, L., "Circuits and methods for time-average frequency based clock data recovery", US patent 9,036,755, May 2015e.

Xiu, L., "Circuit And Method For Adaptive Clock Generation Using Dynamic-Time-Average-Frequency", US 9,118,275, August 25, 2015f.

Xiu, L., "Microelectronic System Using Time-Average-Frequency Clock Signal As Its Timekeeper", US 9,143,139, Sep. 22, 2015g.

Xiu, L., "Circuits and methods of TAF-DPS vernier caliper for time-of-flight measurement", US patent 9,379,714, June 2016.

Xiu, L., and Chen, P. L., "A Reconfigurable TAF-DPS Frequency Synthesizer on FPGA achieving 2 ppb Frequency Granularity and Two-Cycle Switching Speed," *IEEE Trans. on Industrial Electronics*, vol. 64, pp. 1233-1240, Feb. 2017.

Xiu, L., "Clock Technology: The Next Frontier", *IEEE Circuit And System Magazine*, vol. 17, no.2, pp. 27–46, 2017a.

Xiu, L., "All digital FPGA-implementable time-average-frequency direct period synthesis for IoT applications", ISCAS2017, 2017b.

Xiu, L., "Circuits and methods of TAF-DPS based chip level global clock signal distribution", US patent 9,582,028, Feb. 2017c.

Xiu, L., "Circuits and methods of implementing time-average-frequency direct period synthesizer on programmable logic chip and driving applications using the same", US patent 9,621,173, April 2017d.

Xiu, L., "Circuits and methods of TAF-DPS based interface adapter for heterogeneously clocked Network-on-Chip system", US patent 9,740,235, Aug. 2017e.

L. Xiu, Xiangye Wei, Yuhai Ma, "A Full Digital Fractional-N TAF-FLL for Digital Applications: Demonstration of the Principle of a Frequency-Locked Loop Built on Time-Average-Frequency", *IEEE Trans. on VLSI*, vol. 27, no.3, pp. 524-534, March 2019.

Xiu, L., and Wei, X., "A 0.02 ppb/step Wide Range DCXO Based on Time-Average-Frequency: Demonstration on FPGA," *2019 IEEE International Symposium on Circuits and Systems (ISCAS2019)*, 26–29 May 2019, Sapporo, Japan.

Xiu, L., "Time Moore: Exploiting Moore's Law from Time Perspective", *IEEE Solid-State Circuits Magazine*, vol. 11, no.1, pp. 39-55, 2019.

References

Xiu, L., chapter 5 "A Cost-Effective TAF-DPS Syntonization Scheme of Improving Clock Frequency Accuracy and Long-Term Frequency Stability for Universal Applications" in book of "Low-Power Circuits for Emerging Applications", 2019b, CRC press.

Xiu, L., "Method and apparatus for improving frequency source frequency accuracy and frequency stability", US patent, 10,686,458, June 2020.

L. Xiu and X. Wei, "A New Perspective of Using Integrated On-Chip Syntonistor for Time Synchronization in Network: Meeting the TAACCS Challenge," *IEEE Circuit And System Magazine*, 2nd quarter, No.20, issue 2, pp. 8–29, June 2020.

Yan, Y., Qian, Y., Sharif, H., and Tipper, D., "A survey on smart grid communication infrastructures: Motivations, requirements and challenges," *IEEE Commun. Surveys Tuts.*, vol. 15, no. 1, pp. 5–20, 1st Quart. 2013.

Yang, C. Y., Chang, C. H., and Wong, W. G., "A Δ-Σ PLL-based spread-spectrum clock generator with a ditherless fractional topology," *IEEE Trans. Circuits Syst. I Regul. Pap.*, vol. 56, no. 1, pp. 51–59, 2009.

Yang, Y., and Haider, T., "MPEG-2 transport stream packet synchronizer", US patent 8249171, 2012.

K. Yang, D. Fick, M. B. Henry, Y. Lee, D. Blaauw, and D. Sylvester, "A 23Mb/s 23pJ/b fully synthesized true-random-number generator in 28 nm and 65 nm CMOS," *in IEEE Int. Solid-State Circuits Conf. (ISSCC) Dig. Tech. Papers*, Feb. 2014, pp. 280–281.

K. Yang, D. Blaauw, and D. Sylvester, "An All-Digital Edge Racing True Random Number Generator Robust Against PVT Variations," *IEEE J. Solid-State Circuits*, vol. 51, no. 4, pp. 1022–1031, March 2016.

H. Yu, P. H. W. Leong and Q. Xu, "An FPGA Chip Identification Generator Using Configurable Ring Oscillators," in *IEEE Transactions on Very Large Scale Integration (VLSI) Systems*, vol. 20, no. 12, pp. 2198–2207, Dec. 2012.

Zeidler, E., and Hunt, B., *Oxford Users' Guide to Mathematics(in Chinese)*. 2012.

Yulei Zhang ; Buckwalter, J.F. ; Chung-Kuan Cheng, "On-chip global clock distribution using directional rotary traveling-wave oscillator," *Electrical Performance of Electronic Packaging and Systems, 2009. EPEPS '09.* IEEE 18th Conference on, pp. 251–254.

J. Zhang and C. Shen, "Set-Based Obfuscation for Strong PUFs Against Machine Learning Attacks," in *IEEE Transactions on Circuits and Systems I*: Regular Papers, vol. 68, no. 1, pp. 288-300, Jan. 2021.

E. Zianbetov, D. Galayko, F. Anceau, M. Javidan, C. Shan, O. Billoint, A. Korniienko, E. Colinet, G. Scorletti and JM. Akrea "Distributed clock generator for synchronous SoC using ADPLL network", Custom Integrated Circuits Conference (CICC), 2013 IEEE, pp. 1–4.

H. Zhuang, X. Xi, N. Sun and M. Orshansky, "A Strong Subthreshold Current Array PUF Resilient to Machine Learning Attacks," in *IEEE Transactions on Circuits and Systems I*: Regular Papers, vol. 67, no. 1, pp. 135-144, Jan. 2020.

Chapter 7
Forefront of the New Paradigm: Using Time to Encode Information

7.1 The State of Extraordinary Research and Development Activities

Figure 7.1 shows the trend of 40 years' microprocessor development (https://arch2030.cs.washington.edu/slides/arch2030_tom_conte.pdf). From 1970s to current, the number of on-board transistors grows at exponential rate. In the middle of 2000s, unfortunately, Dennard Scaling stopped working. As a result, clock rate ceased to increase. The top speed stabilized at around 3–4 GHz and power consumption peaked at a few hundred watts range. In the good old time of Moore's Law (especially when Dennard Scaling held), the prowess of process-shrinking overwhelmed innovations at all other technological areas. The unfortunate side effect of this single-technology-dominance is that it forces the market to adapt to a fixed technology rather than, in a healthier fashion, to have a variety of technologies compete for supporting the market needs.

After several decades of rapid growth under this single-technology-dominance, the semiconductor industry landscape starts to change. The focus of research and development now shifts from the miniaturization of long-established CMOS technology to the coordinated introduction of new devices, new integration technologies, and new architectures for computing. We are now entering into a new period that is considerably different from the normal engineering of puzzle-solving that the golden time of Moore's Law has enjoyed for three or four decades. As revealed in Fig. 2.1, after a long period of normal engineering characterized by activities of creation, invention and design, we are gearing more towards such type of work of scientific discovery and theorization, which also occupied the majority of activities in the early stage of semiconductor industry. We are returning to science, from engineering, for seeking new ideas to advance.

When the semiconductor industry is inspected under the lens of evolution as previously discussed around Figss. 2.2 and 4.1, the possible paths for future evolution are along the three directions of "devices and materials", "computation models" and "architectures and packaging", as illustrated in Fig. 7.2. Following those paths, a

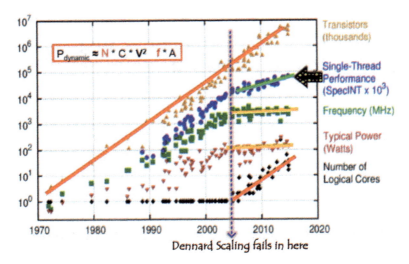

Fig. 7.1 Dennard scaling failed circa middle of 2000s

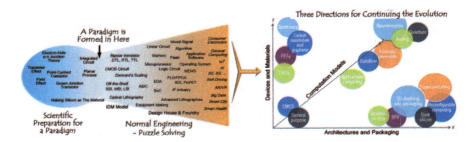

Fig. 7.2 The three directions for continuing the evolution in microelectronics

myriad of activities have already occurred and they can be roughly classified into the following categories: *architecture-and-software-advance* (advanced energy management, advanced circuit design, system-on-chip specialization, logic specialization, dark silicon, near-threshold voltage operation), *3D-integration-and-packaging* (chip stacking in 3D using through-silicon vias, metal layers, active layers), *resistance-reduction* (superconductors, crystalline metals), *millivolt-switches-or-better-transistors* (tunnel field-effect transistors, heterogeneous semiconductors, strained silicon, carbon nanotubes and graphene, piezo-electric transistors) and *beyond-transistor-new-logic-paradigm* (spintonics, topological insulators, nanophotonics, biological and chemical computing) (Cavin et al. 2012; Dreslinski et al. 2010; Nikonov and Young 2013; Bernstein et al. 2010; Theis and Solomon 2010; Shalf and Leland 2015; Wolf et al. 2001; Zutic et al. 2004).

The picture painted in Fig. 7.2 intelligibly reflects the certainty that the semiconductor industry will relentlessly march forward regardless of any obstacle that may stand in its way. This is driven by the fact that people have a never-satisfied appetite for computation power. In the pursuit of ever-greater computing power, people are willing to try anything by any price, just to gain several percentiles or so more processing power. The illustrations presented in Figs. 2.1, 2.2, 4.1, 7.1 and 7.2 manifests a sign that we have reached a state of extraordinary research and development activities. This is a symptom of crisis, a sign of cry for a big change. This somehow chaotic circumstance however can be the breeding ground for big ideas, ideas of wild imagination, ideas that can lead to next era. One of such ideas, which focuses on the foremost medium used for embodying information, will be the subject of this chapter.

7.2 Computation Variable and Data Representation in Electronic System

In the past practices of microelectronics, the dominant method of encoding information is through the help of electric charge. The measure of electrical voltage, or electrical current, can be made meaningful to people, can be utilized to act or speak on behalf of our thought, can be used in conveying message for communication over long distance, and etc. All those traits are various facets of something that we call information. Electrical voltage is the collective effect of a group of electrons, each of them has a designated amount of electric charge. Following this reasoning, electric charge is functioning as a state variable at the very bottom level to materialize the information. This state variable lies at the core of Moore's Law, and the microelectronics at large. Sooner or later, however, fundamental physics limitations will destine CMOS scaling to a conclusion. In light of this, extraordinary research and development has taken place on seeking alternatives. In the quest of new possibilities, an option is to identify new types of state variable for embodying information. Those new found state variables, whether you like it or not, could be fundamentally different from what we are familiar with. Their operation could be based on some entirely different mechanisms.

The International Technology Roadmap for Semiconductors (ITRS) is a set of documents produced by a group of semiconductor industry experts. The documents generated by this group represent the most valuable opinion on the directions of research and the time-line up to about 15 years into the future for all the areas of technology related to semiconductor industry. As of 2017, however, ITRS is no longer being updated due to the difficulty of CMOS scaling. Figure 7.3 is one of the works created by ITRS. It shows, in a style of high-level abstraction, the five building layers for electronic system.

At the very top is the architecture of computing. In this category, the renowned Von Neumann computer architecture is a classical example. Multicore is a recent example of 2000s. The method of analog based signal processing also has a long history. Quantum computing, holographic computation, morphic computing, soft computing,

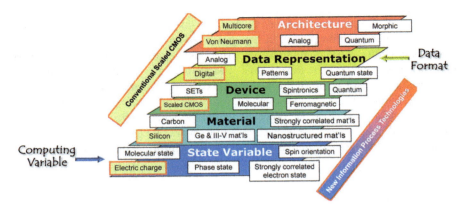

Fig. 7.3 Five abstract layers for describing electronic system (www.itrs.net)

and DNA computing are some types of non-mainstream computing architectures, representing the emerging trends. Under the architecture layer is the layer of data representation. It defines how information can be organized as data in various ways for different applications and then be processed by specific devices created in their respective methods for best performance. The two most widely used approaches are the binary digital signal processing method and the multiple-values analog method. Quantum state has been used to encode message in quantum computing, which has already shown some encouraging signs of great potential in recent years. Moreover, in principle, pattern can be used for expressing useful meaning as well since different patterns can be discerned and be used for different significations.

Beneath the layer of data representation lies the device layer. Device is the workbee that literally carries out the task of signal processing. The most popular device is the celebrated CMOS transistor. There are some other types of non-mainstream devices that are emerging in recent years, such as molecular, spintronics, quantum, ferromagnetic and etc. Those devices support the various data representations and architectures listed in the upper layers. Under the device layer lies the layer of material. Materials are used in the manufacture process of fabricating those devices. For the past several decades, the king of all materials used in semiconductor industry is the silicon. As the juice in silicon is almost fully squeezed out, people have turned their attention to new materials such as carbon and Ge & III-V materials.

At the very bottom lies something that is purely subjective in its nature. It is the issue of how we extract and purify certain attribute, out of the material world, that can be used as a means to distinguish different states, such as differentiating true from false, discerning one from zero, separating good from bad, and etc. With anticipation, we can use those states to create information, to express our thought. The most widely used carriage nowadays for this purpose is electric charge. The amount of electric charge can be handily corresponded to information. This method is straightforward and thus is easy to be comprehended. More significantly, it is convenient or friendly to implementation. However, it might not bear a preeminent efficiency in terms of material used and energy consumed for each unit of information processed.

People therefore are actively pursuing other ventures at this level of state variable, in hope to improve computing performance and reduce consumed energy. This can be significant since this level plays the fundamentally dominant role. The options that have already been explored, in one way or another, include electron spin, magnetic dipoles, electric dipole, phase state, orbital state, photons and etc. They target at creating devices of small feature on the order of a few nanometers and beyond. Those nano-scale structures pass tokens in the spin, excitonic, photonic, magnetic, quantum, or even heat domains. Further, alternative tokens also require new transport mechanisms to replace the conventional chip wire interconnect schemes of charge-based computing. With intensified anticipation, emergent physical behaviors and idiosyncrasies of these small-feature switches can execute specific algorithms by enabling unique architectures. As an example, quantum computing is a computing method using quantum–mechanical phenomena of superposition and entanglement (Nielsen and Chuang 2000; Gomes 2018). It is completely different from the traditional digital electronic computing that is based on binary logic (which is in turn based on electric charge). Instead of using bit which is always in one of the two definite states of 0 or 1 (i.e., less- or more-amount of electric charge), quantum computing uses qubit that can be in superposition of states. Hence, quantum computing does not fall in the category of binary logic but uses multiple states simultaneously. It therefore can provide much higher processing efficiency for computing.

In this organization of five abstract layers, it is worth to provide some supplementary explanation to distinguish the two layers of state variable and data representation. State variable is the fundamental manner of extracting attributes from materials found in the physical world to carry abstracted information. Data representation deals with the methods of utilizing the extracted attributes. Taking electrical charge for example, it is an attribute extracted from the physical entity of electron for conveying message. By this action of extraction, this attribute is made useful to us. This attribute can be used in a variety of ways when we reach the level of data representation. For digital approach, having only two conditions of "more or less, have or null, zero or one" is sufficient to do the trick. For analog, however, every amount in the aggregation of electric charge is meaningful. Those two different doctrines reflect different philosophies of viewing the world. They subsequently lead to quite dissimilar ways of conducting engineering.

7.3 Multi-value Rate-of-Switching as a Method for Signal Processing

In the territory of mainstream computation, electrical charge (electrons' movement, collectively represented as voltage or current) is used for encoding information. This practice has underpinned the growth of semiconductor industry in the past several decades. Using electrical charge involves matter and subsequently requires space: the amount of charge is essentially proportional to the volume of space. As Moore's Law running out of steam, is it feasible to use the other piece of real estate, time, to encode information? Before answering this question, it has to be explained that this pursuit is different from what we are chasing in Time Moore Strategy. In the story of Time Moore, information is still embodied in the traditional way of gauging the amount of electrical charge. The hallmark in Time Moore Strategy is the nonuniform-flow-of-time. Its distinguished practice is to use time wisely through the adoption of flexible clocking ideology. In current subject-under-study, however, we are trying to change the old game at a more fundamental level, namely, the method of expressing information (the state variable in Fig. 7.3).

It is helpful to first examine how we have used the electrical charge to express our thought in the old world. Figure 7.4 illustrates the two doctrines adopted by circuit professional to design electrical circuit. On the left is the analog design approach where signal processing is based on voltage value. Every value has its divine meaning. And, further, every value is meaningful only when tagged by a particular time stamp. In other words, each pair of (V_A, t_A), (V_B, t_B), (V_C, t_C), (V_D, t_D) uniquely represents something. They are the so-called information if we define proper rules for using them. On the left is the digital design method. In this doctrine, voltage value is only differentiated between two levels: high and low. All other values in between have no meaning.

Typically, although not an absolute necessity, a clock signal is employed for serving the operation of digital-based circuits. It is this clock signal that is used to mark the locations, in the dimension of time, of those high and low voltage states. For analog-based signal processing circuit, naturally, clock signal is meaningless since every point in the voltage curve has its divine meaning and there are infinite number of such points. This argument is based on the reality that there is no way to construct a clock signal capable of indexing those infinite number of points. When stand-alone, analog signal processing unit can be self-timing. Putting it in another way, the timing associated with each voltage point is implicitly identified by its relationship with other points. However, when an analog signal needs to interact with a digital signal processing unit, clock signal will become necessary. For example, in their operations, Analog-to-Digital converter (ADC) and Digital-to-Analog converter (DAC) all work

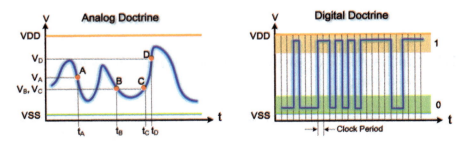

Fig. 7.4 The two different doctrines of utilizing electrical charge

7.3 Multi-value Rate-of-Switching as a Method for Signal Processing

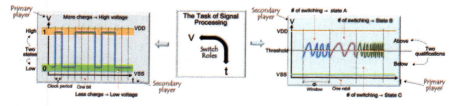

Fig. 7.5 The voltage-based (left) and the switching-based signal processing methods (right)

on the pace defined by the clock signal driving them. Electrical oscillator is another excellent example in this regard. Being an analog circuit, an electrical oscillator can output a signal as clock for driving other circuits. But this only happens when it is in the situation of interacting with external world (such as a digital circuit or an ADC or a DAC). When external world is not a concern, there is no dedicated time mechanism operating inside the circuit. All the timings are implicitly expressed by relating signals among each other.

Both the analog and digital methods follow the principle of "using the quantity of electrical charge for encoding information", under the guidance of proportionality. From Fig. 7.4, it is clear that this principle is materialized through the fact that the vertical V-axis is used predominantly. For the purpose of encoding message, the horizontal t-axis just plays a secondary role of indexing. By the beauty of symmetry, it is natural to visualize a different scenario of using t-axis as the primary player, or switching the roles of those two players. This is the idea of using t-axis as the principal factor in the game of signal processing. Figure 7.5 depicts this idea conceptually. This is a dramatic change on how the game of signal processing is played since the roles of voltage and time are switched.

On the left of Fig. 7.5, for the purpose of reference, the voltage-based doctrine is incarnated by a digital signal. On the right-hand side, the idea of using *rate-of-switching* for signaling is illustrated. In this method, the designated information is embodied in the switching activities. In particular, the rate-of-switching within a given window is encoded for message. Instead of being embodied as clock period in digital method, the dimension of time (i.e., the t-axis) is divided into a plurality of identical windows. Within each window, the switching activities are counted and encoded. In other words, the number of "zero-crossing" is calculated and is used as information. The task of signal processing is carried out on the patterns of zero-crossing. Apparently, the number of activities (i.e., the zero-crossings) within any given window can be larger than two. Therefore, this method is coined as Multi-Value-Rate-of-Switching, or MVRoS for short.

Although the window in MVRoS is structurally similar to the traditional clock period, the size of window is in principle larger than clock period since this window is used to count switching activities occurred within. The window is not used for just indexing a value (a zero or one in the case of digital circuit), as clock signal does. When clock signal is inspected, the group of switching activities (logical transitions) occurred inside a clock period is treated as a whole, internal details are of no concern.

Those switching activities have been absorbed by the logic gates and they are not observable to the outside world. On the other hand, for window in MVRoS, all the switching activities occurred within a window are meaningful.

Similar to conventional clock signal where each clock pulse is number-marked for being used as index, each window in the MVRoS approach is number-marked for the same purpose. For conventional digital logic, the information within each clock period is "high/low voltage state plus clock index number". In MVRoS, the information within each window is "number of zero-crossing plus window index number".

Table 7.1 compares the three signaling doctrines from several aspects. For MVRoS, instead of electrical charge, state variable is materialized by the pattern of switching (i.e., the number of "zero-crossing"). In its implementation, however, MVRoS may still need help from electrical charge to realize the pattern. The illustration in Fig. 7.5 clearly shows the spirit of how the electrical charge is used. For digital signaling, the electrical charge is directly used to discern two states, which is straightforwardly treated as information. In MVRoS, the electrical charge is only used for differentiating two conditions: above or below a threshold. Information comes from an extra step of work, namely, the pattern formed by the "above and below", or the number of "zero-crossing". The underpinning device for realizing MVRoS can still be the popular CMOS transistor with silicon as the key material. Other types of devices and materials however are feasible as well. A piece of good news is that transistor is switching faster with the continued shrinking of feature size. This fact could be beneficial to MVRoS.

From the discussion around Fig. 7.3, the five abstract layers used for describing all electronic systems are elucidated. In that discussion, the two layers of data representation and state variable are of special interest to us since the MVRoS method

Table 7.1 Comparison of signaling doctrines: digital, analog and MVRoS

	Digital	Analog	MVRoS
State variable	Electrical charge	Electrical charge	Pattern of switching
Data representation	Two conditions: less and more	Many conditions: all the points in a specific range	Many conditions: all the zero-crossings in a window
V-t waveform	V is dominant; t for indexing	V is dominant; t for indexing	t is dominant; V for assistance
Material	Primarily silicon; others have been explored	Primarily silicon; others have been explored	Can use silicon; also, others can be used
Device	Mainly CMOS transistor	CMOS transistor and others	Preferably CMOS transistor, can use others
Architecture	Von Neumann, Multicore, SoC and many others	Myriad techniques	Open (to be developed)

7.3 Multi-value Rate-of-Switching as a Method for Signal Processing

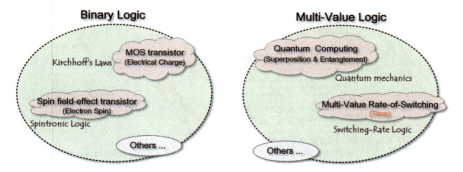

Fig. 7.6 Binary logic and multi-value logic: two different data representations and their corresponding state variables

distinguishes itself from conventional wisdom through eccentric implementations on those two layers. This unconventional approach is not without precedent. This point can be illustrated through examples presented in Fig. 7.6. When the level of data representation is inspected, one viewpoint is the number of logic values that can be supported by the particular method of data representation. As shown, it can be distinguished as two families of binary logic and multi-value logic. For each of these two families, there are a number of state variables that can support the corresponding logic system.

In the category of binary logic, the most popular state variable is electrical charge with MOS transistor as its key implementer. A promising state variable that researchers are working on recently is electron spin. It is also a binary system with two directions of electron spin. In the multi-value regime, the most celebrated one is the quantum computing, which uses superposition and entanglement as the base for state variable. MVRoS is a newcomer that uses time as the base for state variable. Among those examples, spintronic logic and quantum mechanisms are radically different from the conventional electrical-charge-based approach. They have been proven effective, at least in some special application areas. In future, the same could be said for MVRoS.

As a multi-value logic system, MVRoS naturally bears the potential of higher processing efficiency. From various aspects, it is a widely open field with many challenges that can lead to countless new possibilities. The overall big picture is painted in Fig. 7.7. A new state variable is introduced at the very bottom level and a novel pattern recognition scheme is suggested in data representation level. All the subjects in between and above are open. This openness can lead to countless new possibilities.

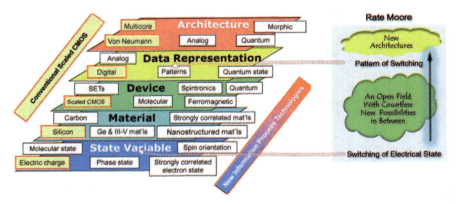

Fig. 7.7 Countless new possibilities enabled by MVRoS

7.4 The First Three Challenges

There are at least five merits in MVRoS that are worthy of attention striking. First of all, unlike the "full voltage swing between VDD and VSS" in conventional digital processing when the switch (i.e., the transistor) is opened and closed, the voltage swing in MVRoS can be much smaller than the difference-between-VDD-and-VSS since only "zero-crossing" is of interest. This can lead to the possibility of a very low-voltage design style. It can potentially reduce power consumption significantly. In modern designs, power consumption is becoming an issue of higher concern than transistor speed. MVRoS therefore is a promising option in this regard. The second advantage is that a larger number of states can be achieved (larger than the two possibilities of "0" and "1"). The use of multi-value logic can improve compute efficiency in a great stride. For example, one can imagine the scenario of using decimal system directly in computer. Thirdly, the efficiency on the usage of resource-of-time is improved by two-fold since the window is used not only for indexing but also for information processing. Fourthly, although the speed gain in each new generation of CMOS technology is not as large as in previous generations, the transistor's absolute switching speed is still increasing in every new generation. This fact favors the rate-of-switching approach. In other words, it is cheaper to take advantage of, or easier to squeeze juice from, transistor's speed gain when using rate-of-switching than in the case of traditional level-based processing. Lastly, as Dennard Scaling was, MVRoS can be scalable with technology advancement (smaller feature size on transistor, new device, new material and etc.).

Out of the four currently already explored computational variables (electrical charge, electric dipole, magnetic dipole and orbital state), all of them share one common characteristic. They all rely on some kind of material property to come into being. Those mechanisms directly involve space (since material occupies space). MVRoS introduces a new computational variable from another dimension, namely, the time. In this domain, the computation power is not necessarily proportional

7.4 The First Three Challenges

to the amount of material, or the volume of space. For the most straightforward implementation, MVRoS can rely on electrical charge for its execution. This seems like "game as usual" since, in principle, it does not introduce any new kind of device. It however provides a new perspective to use electrical charge. When viewing from the surface, its control (to control transistor's on and off) is still voltage or current that are the embodiments of electrical charge. Similarly, its output (the transition or change-of-state occurred at transistor's output terminal) is still voltage or current. But its state variable is not the quantity of the electrical charge but rate-of-switching. Putting it in another way, only the pattern of switching is of interest to us, not the voltage level. This is fundamentally different from the conventional ways of using electrical charge.

Currently, however, there is a long way to go before MVRoS can be any useful to mainstream computation. At this stage, MVRoS is just an idea at its infant stage. Figure 7.8 presents a projected roadmap for developing the MVRoS. After a consensus is established around the idea, there are at least three challenges that must be surmounted before MVRoS can be adopted for mainstream applications. The first one is to develop a mechanism of using switching activity for logic operation. It will be a many-valued logic system. The field of many-valued logic is started around 1920s (Rosser and Turquette 1952; Ackermann 1967; Rine 1984; Epstein 1993; Cignoli et al. 2000). It has enjoyed great progress since then. However, for its specific application on MVRoS, more concrete work is required. The second challenge is the creation of standard circuits for fundamental logic operations. A library consisting of those circuits needs to have the capability of handling key basic operations defined in this multi-valued logic. This is similar to the creation of binary logic gates of NOT, AND, OR, XOR and etc. This is definitely a new research field. The third challenge is to, using the standard circuits created as the underpinning support, explore architectures at higher level for various computation tasks. The aim is low power consumption and high computation efficiency.

With the benefit of hindsight, modern computing models can be roughly classified into four categories (Joneckis et al. 2014): classical digital computing (CDC), analog computing (AC), neuro-inspired computing (NC) and quantum computing (QC).

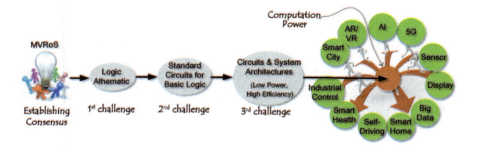

Fig. 7.8 The roadmap for developing MVRoS

CDC includes all the binary digital electronics that form the basis of the microelectronics. AC includes nonbinary devices that implement computation through direct physical principles. NC includes devices based on the principles of brain operation and general neuronal computation. QC is designed to solve some problems with combinatorial complexity through the selection of a desired state from a superposition of all possible answers to a problem. There are strength and weakness for each of these computation paradigms. AC could be simpler than some of its digital approximations, but it does not lend itself to general purpose computing because the devices are specialized for their dedicated computation. The computational precision is problematic to maintain and can be sensitive to its environment. CDC is good at deterministic and algorithmic calculation, but poor at simple reasoning and recognition. NC devices is proven inherently resilient and is very good at problems that CDC is not. Many unexplored opportunities exist for NC, but much is still not understood about how the brain actually computes. QC theoretically could enable the efficient solution of some combinatorial and NP-hard problems or could be used to simulate the electronic state of complex molecules. It is however surely not a suitable replacement for CDC in domains where CDC excels.

Those options discussed above provide the possible approaches that could go beyond what traditional digital electronics has performed effectively. It is however not believed that they are suitable as replacements for digital electronics in tasks that digital computing (i.e., the CDC) already performs well. On the other hand, MVRoS can be regarded as an extension of CDC, or a new technological implementation of the CDC model. It could be considered as the potential candidate to continue the phenomenon of Dennard Scaling. As transistor feature size shrinks, it switches faster. But conventional level-based transistor operation does not fully enjoy the benefit of supply voltage decrease after the Dennard Scaling breakdown. MVRoS can take advantage of the supply voltage decrease and thus facilitate the power consumption reduction further since it uses rate-of-switching instead of the full-open and full-close states of a switch. It is believed that it could be the most immediately relevant option to boost computation efficiency after the traditional digital computation.

References

Ackermann, R., "An Introduction to Many-Valued Logics," London: Routledge and Kegan Paul, 1967.

K. Bernstein, R. K. Cavin, W. Porod, A. Seabaugh, and J. Welser, B, "Device and Architecture Outlook for Beyond CMOS Switches," *Proc. IEEE*, vol. 98, no. 12, pp. 2169–2184, Dec. 2010.

R. K. Cavin III, P. Lugli, V. V. Zhirnov, "Science and Engineering Beyond Moore's Law," *Proc. IEEE*, vol. 100, pp. 1720–1749, 2012.

Cignoli, R., d'Ottaviano, I. and Mundici, D., "Algebraic Foundations of Many-Valued Reasoning," *Dordrecht: Kluwer*, 2000.

T. Conte, "IEEE Rebooting Computing Initiative & International Roadmap of Devices and Systems," IEEE Rebooting Computing - Architecture 2030 Workshop, Available: https://arch2030.cs.washington.edu/slides/arch2030_tom_conte.pdf.

References

R. G. Dreslinski, M. Wieckowski, D. Blaauw, D. Sylvester, T. Mudge, "Near-Threshold Computing: Reclaiming Moore's Law Through Energy Efficient Integrated Circuits," *Proc. IEEE*, vol. 98, no. 2, pp. 253–266, 2010.

Epstein G., "Multiple-Valued Logic Design," Bristol: *Institute of Physics Publishing*, 1993.

L. Gomes, "Quantum computing: Both here and not here," *IEEE Spectrum*, Vol. 55, Issue: 4, pp. 42-47, April 2018.

International Technology Roadmap for Semiconductors (ITRS), [Online], available: www.itrs.net.

L. Joneckis, D. Koester, and J. Alspector, "An Initial Look at Alternative Computing Technologies for the Intelligence Community," *tech. report, Inst. for Defense Analysis*, Jan. 2014, http://oai.dtic.mil/oai/oai?verb=getRecord&metadataPrefix=html&identifier=ADA610103.

M. Nielsen, I. Chuang, "Quantum Computation and Quantum Information," Cambridge, *Cambridge University Press*, 2000.

D. E. Nikonov and I. A. Young, "Overview of Be-yond-CMOS Devices and a Uniform Methodology for Their Benchmarking," *Proc. IEEE*, vol. 101, no. 12, 2013, pp. 2498–2533.

Rine, D.C. (ed.), "Computer Science and Multiple Valued Logic," Amsterdam: North-Holland, 2nd rev. ed., 1984.

Rosser, J.B. and Turquette, A.R., "Many-Valued Logics," Amsterdam: *North-Holland*, 1952.

J. M. Shalf and R. Leland, "Computing Beyond Moore's Law," *Computer*, vol. 48, no. 12, pp. 14–23, 2015.

T. N. Theis and P. M. Solomon, B, "In Quest Of The 'Next Switch': Prospects For Greatly Reduced Power Dissipation In A Successor To The Silicon Field-Effect Transistor," *Proc. IEEE*, vol. 98, no. 12, pp. 2005–2014, Dec. 2010.

S. A. Wolf, D. D. Awschalom, R. A. Buhrman, J. M. Daughton, S. von Molnár, M. L. Roukes, A. Y. Chtchelkanova, D. M. Treger, "Spintronics: A Spin-Based Electronics Vision for the Future," *Science*, vol. 294, issue 5546, pp.1488–1495, Nov. 16, 2001.

I. Zutic, J. Fabian and S. D. Sarma, "Spintronics: Fundamentals and applications," *Rev. Mod. Phys.*, vol. 76, no. 2, pp.323-410, April 2004.

Chapter 8
Epilogue: It's Time for the Big Ship of Moore's Law to Make a Turn

8.1 The Continuation of Evolution but Along a New Path

The left graph in Fig. 8.1 shows the trend of 40 years' microprocessor development. From 1970s to the middle of 2000s, owing to the CMOS scaling symbolized by the Moore's Law (i.e., the so-called "More Moore"), the number of on-board transistors grows at exponential rate. At the same time, the top clock speed, the overall power consumption and the processor's performance all increase correspondingly in a linear fashion. It is elucidated in previous chapters that this is a period of puzzle-solving normal engineering under a well-formed space-dominant paradigm. It was a perfect world free of anomaly and crisis. When Dennard Scaling fails in the middle of 2000s, clock rate stops increase. The top speed stabilizes at around 3–4 GHz and power consumption peaks at a few hundred watts range. Although the number of transistors still grows linearly, the improvement on processor's performance does not follow as expected. Some signs of a crisis start to emerge. All those problems force people to investigate other options, as illustrated in the right-hand side of Fig. 8.1. As a result, besides the traditional "More Moore", two new paths of "More than Moore" and "Beyond Moore" were being explored (please refer to Sect. 3.5).

In Sect. 4.3, the cogitation of development-in-microelectronics as an evolution process is first presented. The similarity between biological world and electronic world in the fashion of a gradual development from simple to complex form is illustrated in Fig. 4.1. Along the same line of argument, the turn of Moore's Law from space to time, or the paradigm shift from the space-dominant paradigm to the time-oriented one, is the continuation of this evolution process. It is however happening along an undoubtedly new path. This insight is depicted in Fig. 8.2. On the left is the biological evolution where the branch of mammals is depicted, as a reference for contraposition. From a basic cell, many different evolution paths have emerged during the millions or billions year of life development. Eventually, a variety of species have come into being. After ruthless competition of natural selection, each is most efficiently adapted to its environment, and each can best serve its divine purpose.

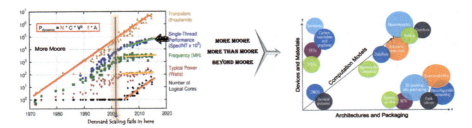

Fig. 8.1 The development of microprocessor in the last 40 years (left) and the future trends (right)

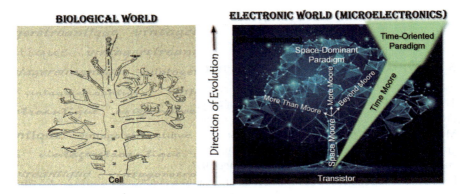

Fig. 8.2 Time Moore: the continuation of evolution along a new path

On the right-hand side of Fig. 8.2, the development of microelectronics is also recognized as an evolution process. This process started in the 1950s with transistor as the origination of life. After seven or so decades, from this seed of transistor, this field has grown into a huge tree with many branches. The More Moore, More Than Moore and Beyond Moore are the three major branches. They all can be included in the evolution package of *Space Moore*, which starts from the root of transistor. This tree probably can still sustain itself for decades to come. The growth hormone however has gradually dried up (i.e., the practice of packing more transistors into a given space for more processing power is gradually losing its magic, or dying). At this crucial stage, a turn for the future direction of Moore's Law development is suggested. Now, the growth hormone for future advance will be time, instead of space. A progenitive new branch of evolution, starting from the root, is expected to appear and it is labeled as *"Time Moore"* (to be differentiated from the "Space Moore"). Along this new path, a new paradigm will be gradually established where nonuniform-flow-of-time is the consensus and everything else will be built around it. The psyche of engineering will be changed from "packing more transistors into a given space" to "collectively using the given transistors in a more time-efficient way".

8.2 The Full Picture of Time Moore: Clock Moore and Rate Moore

It is fair to say that More Moore, More Than Moore and Beyond Moore all focus their effort on space: getting more computation capability from ever-smaller space and using ever-less matter. For this reason, they can be encapsulated in a term of Space Moore. In contrast, Time Moore is a strategy of getting more computation power from the perspective of time. It focuses on this mission: squeezing out more juice from transistors by exploiting more deeply into the dimension of time. Inside Time Moore, the straightforward approach is to play with clock signal. This intelligence of "using more of clock frequency" is symbolized by the term *Clock Moore*, which is the first part of Time Moore. The entire Chap. 6 is dedicated to the support of Clock Moore. On another front, during the decades of growth in semiconductor industry, the dogma for mainstream computation is to use electrical charge for encoding information. Using charge involves matter and subsequently requires space: the quantity of charge is directly proportional to the size of space. In Chap. 7, MVRoS is introduced as a method of using rate-of-switching for signaling, or using more of the rate associated with switching activity. This ideology is signified as *Rate Moore*, the second part of Time Moore.

Together, Clock Moore and Rate Moore make up the full picture of Time Moore. In Chap. 5 and 6, the focus is on the Clock Moore that is the straightforward and relatively mature bifurcation of Time Moore. The MVRoS presented in Chap. 7 is the forefront of Time Moore. As already discussed in Sect. 7.1, Fig. 7.2 illustrates in high level the three directions for future growth: (1) create new devices, (2) invent new architectures, and (3) develop new computational paradigms. In Fig. 8.3, the roles of Clock Moore and Rate Moore are merged into the big picture of this evolution process, and marked on this map. They are drafted in this picture as operationable methodologies in the pursuit of higher computation power. Rate Moore is one of the options along the path of computation model. Clock Moore can be influential in all the three directions.

Figure 8.4 provides another angle for viewing Clock Moore and Rate Moore. They are presented in this picture as new perspectives against the background of the whole frame of microelectronics. Figure 8.3 is pragmatic in a sense that it is an engineering-oriented roadmap while Fig. 8.4 is geared more toward fundamental understanding where advance in ideology and methodology is the goal. The comprehension coming out of Fig. 8.4 can help people easy the pain in accepting the new ideas of Time Moore, Clock Moore and Rate Moore.

Figure 8.5 further provides an insight on the role of Time Moore when microelectronics is appraised in the industry setting. From the position of industry where profiting from commercial products is the ultimate goal, the creation of a product needs to go through four main stages: transistor (circuit), geometrical pattern, chip & board, and whole system. The chain of actions associated with those stages can be organized into five major working areas: basic research and process development, circuit design, manufacture, package & PCB, and finally, productization. Basic research and

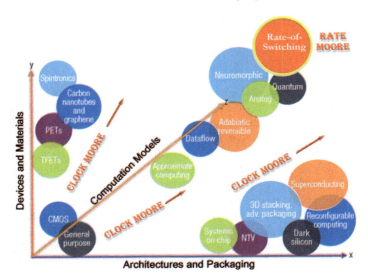

Fig. 8.3 Clock Moore and Rate Moore merged into the scene of evolution

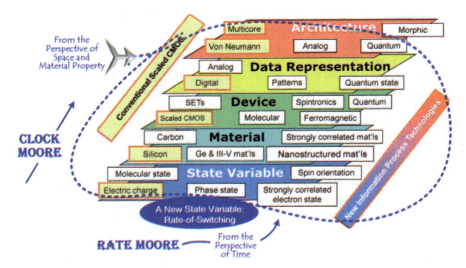

Fig. 8.4 Clock Moore and Rate Moore as perspectives against the background of the whole frame of microelectronics

process development leans more towards science than engineering. It establishes the physical foundation for creating the switch used in signal processing (e.g., CMOS transistor). Circuit design is responsible for, based on the transistor created, devising functional circuits for processing signal. Manufacture is to fabricate the designed circuits, which are presented to manufacture shop as geometrical pattern of many

8.2 The Full Picture of Time Moore: Clock Moore and Rate Moore

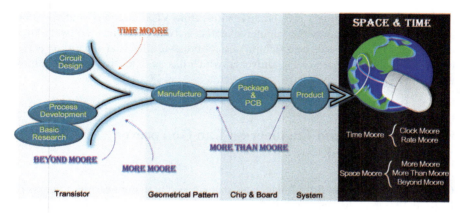

Fig. 8.5 The new addition to the Moore's Law family

layers, by implementing layers of pattern through various chemical and mechanical processes. The step of Package & PCB encapsulates the finished bare dies into sealed package for protection. The packaged die is often called chip and it can then be mounted in system board for becoming a functioning component in the final product. The step of product involves product testing and production scheduling to satisfy customer's need.

In this chain of actions, the traditional three approaches for continuing the Moore's Law all work on their respective areas where their unique strengths can be best utilized. "Beyond Moore" seeks breakthroughs in basic research for making the basic processing element (e.g., the transistor) more efficient. It rather belongs to the domain of science than engineering. "More Moore" spend its effort squarely on technology refinement, namely, how to use a large number of basic processing element more efficiently as a whole. It is mainly an engineering endeavor. "More Than Moore" focuses its attention on packaging stage, also a venture of engineering. All those three favors of Moore's Law eye on space rather than time. As said, they can be grouped as Space Moore. Space Moore works on the structure and electrical characteristics of the transistor. Its goal is to make transistor faster and smaller, and to collectively use them for high computational power through brute force. Time Moore comes from a completely different direction. It focuses on the interactions among many transistors. Its aim is to make transistors work together, in a collectively more efficient fashion.

The trendsetters in Space Moore are scientists of physicist, chemists, and other types. They define how the transistors look like, internally and externally. Meanwhile, circuit designer plays a secondary role as they mere use the transistors to create functions, more or less in a brute force fashion (i.e., by using more transistors for progressively making more complex systems). In the era of Time Moore, however, we have a whole new world to explore. It will be an endeavor beyond brute force because Time Moore focuses more on the intelligent interaction among transistors. Time Moore is at the intersection point of many disciplines. It is a challenge that requires an overwhelmingly collaborative effort from VLSI circuit designer, system architect, process researcher, computer scientist, network architect, software engineer, among others. It will be a rich process of cross-fertilization since time is one of the fundamental pieces of the Universe, which has an extensive reach to almost everything.

8.3 No Science and Engineering Can Go Forever Unchallenged

Our recorded history is full of evidences that science and engineering progress by leaps and bounds. In many occurrences, in a literal sense, advance in science and engineering is the very epitome of progress. It is observed that scientific knowledge and technological advance are cumulative, building upon previous benchmarks to scale new peaks. Although the overall progress is cumulative and most of the times even smooth, a branch of science or engineering however cannot go forever unchallenged. Occasionally, a revolution destroys the continuity and a new paradigm emerges. Table 8.1 includes the list of major scientific revolutions occurred in our civilization so far. This is by no mean a complete list but a selected group of cases which are the most representative examples of paradigm-shift-after-a-revolution.

The Turn of Moore's Law is not in the same scale, in term of scope of influence, of the others presented in Table 8.1. It is listed at the end of this table simply because of the reasons that it is the topic of this book and, more legitimately, it shares the same intrinsic nature inbuilt in those great revolutions. In the field of microelectronics, an idiosyncratic electronic world is created by human. The living cells are transistors and the sign of life is embodied in the form of signals. In the first development phase of this electronic world, a creed of uniform-flow-of-time is adopted by the world's creator (i.e., the Almighty circuit designer). This is driven by the motivation of easy implementation. Just as discussed in previous chapters, and spiritly supported by those great examples in history listed in Table 8.1, the time for a revolution might have just arrived for this technological field of microelectronics. The key in this revolution is the change from uniform- to nonuniform-flow-of-time. This change will make the electronic world more resemblance to the biological world where the circadian clock uses nonuniform-flow-of-time. The nonuniform-flow-of-time in circadian clock is developed during millions of years of evolution. It is efficient when the ratio of activity-of-life over energy-consumed is considered. This high efficiency is proven by the very fact that it survives the relentless competition of natural selection during the enduring evolution. Following the same line of reasoning, it is therefore believed that, by enforcing a new doctrine of nonuniform-flow-of-time, the field of microelectronics can advance further towards the goal of higher information processing efficiency in term of the ratio of bit-over-energy, as illustrated in Fig. 8.6.

8.3 No Science and Engineering Can Go Forever Unchallenged

Table 8.1 Major scientific revolutions in history

Emblem of revolution	Time frame	Scope of influence	The change of paradigm
Transition in mathematical practice	~Mid-4th century BC	The whole civilization	Arithmetic techniques used by Babylonia and Egypt →Proofs from postulates by Greece
The first scientific revolution	~16th–17th century	The whole civilization	Old science (geocentric astronomy, Ptolemaic system) →Copernican heliocentrism established, the emergence of modern science. Francis Bacon is its prophet, Copernicus its spark, Galileo its lighthouse, and Newton its sun
Emergence of experimental method	~16th–17th century	The whole civilization	Pure reasoning →Experiment becomes a valid method for conducting science; science laboratory emerges
The second scientific revolution	19th century	Mainly physics	Individually or independently conducted scientific activities →The whole field of science is mathematized; the trend of specialization emerges; paradigms are acquired for the specialized fields of heat, light, electricity, and magnetism →Leading to the industry revolution, the beginning of the modern technoscientific world in which we live now
Theory of evolution	Mid-19th century	Biology, the whole civilization	All things are created by God →The variety of lives might be the result of evolution process through natural selection
Relativity revolution	Early 20th century	Physics, philosophy	Space and time are separated entities →Four-dimension spacetime
Quantum revolution	Early 20th century	Physics, philosophy	Nature is deterministic →Cause and effect are mere appearance and indeterminacy is at the root of reality
The turn of Moore's Law	Now	Microelectronics (the electronic world)	Uniform-flow-of-time →Nonuniform-flow-of-time

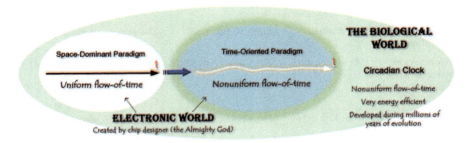

Fig. 8.6 From uniform- to nonuniform-flow-of-time is the key of this revolution

According to Kuhn, progress in science is not a simple line leading to the truth. It is more like the act of gradually progressing away from less adequate conceptions of, and interactions with, the world (Kuhn 2012). Putting differently, it is not a headway towards a preestablished goal. Rather, it is a course of gradually moving away from what once worked well but no longer can handle its new problems methodically. The change from uniform- to nonuniform-flow-of-time is a progress away from a less adequate conception of world, or away from a less efficient way of designing circuit and system. It is a challenge to the old space-dominant paradigm, which has dominated the microelectronics for seven or so decades and is leading us to the current space-related crisis. Anomaly arouse new scientific discoveries. Crisis induces the emergence of fresh scientific theories. Technological obstruction triggers novel engineering practices. No science and technology can go forever unchallenged. For microelectronics, now is the time for the big ship of Moore's law to make a turn.

8.4 Attitude Toward Paradigm Shift: Change-of-Mindset is Extremely Difficult

A 1513 quote from Italian renaissance political philosopher and statesman Niccolo Machiavelli says: "There is nothing more difficult to take in hand, more perilous to conduct, or more uncertain in its success, than to take the lead in the introduction of a new order of things". Although he probably did not have science in his mind when saying this (science is considered in its infancy, or even not existed at all, at that time), this statement is surprisingly appropriate if used in describing the adventure of science and engineering developed in the following several hundred years.

Once a paradigm is formed, a branch of science or engineering seems to act like an enterprise. It always attempts to force the subject-of-study into the preformed and relatively inflexible frame defined and provided by the paradigm. This is the characteristic of the so-called puzzle-solving normal science, first studied by Thomas S. Kuhn circa 1960s. The aim of normal science is not to call forth new sorts of phenomena. As a matter of fact, those not fitting the frame are often not seen by the regular practitioners at all. In general, normal scientific research is directed to the articulation of those phenomena and theories that the paradigm already supplies (Kuhn 2012).

Scientists and engineers normally do not aim to create new theories nor introduce new concepts, worse yet, they are often intolerant of those invented by others. A person may be attracted to science or engineering for all sorts of reasons. Among them are the desire to be useful, the excitement of exploring new territory, the hope of finding order, the drive to test established knowledge, or simply the pursuit of financial gain. Some science and engineering practitioners have acquired great reputations not from any novelty of their discoveries, not from invention of new apparatuses or new methods, but from the precision, reliability, and scope of the methods they developed for the redetermination of previously known facts, or from the improving and perfecting of previously invented apparatuses or methods. Being ordinary persons, many scientists and engineers are exceedingly difficult to rid themselves of preconceived-ideas and ingrained-habits. As physicist Max Planck put it: "A new scientific truth does not triumph by convincing its opponents and making them see the light, but rather because its opponents eventually die, and a new generation grows up that is familiar with it."

In science and engineering, novelty emerges only with difficulty; it is often manifested by resistance, against a background generated from expectation. Any new interpretation of the nature, whether a discovery or a theory, any new invention of a technology, whether a novel apparatus or an innovative method, appears first in the mind of one or few individuals. In many cases, out of crisis arises new idea, then new method, and finally a new theory or a new technology. Creative individual is critical to the success of innovative process.

Why do some ingenious persons come so close to major breakthroughs but fail to push them through? Change-of-mindset is extremely difficult. Take French mathematician and mathematical physicist Henri Poincaré and Dutch physicist Hendrik Lorentz's case on relativity for example. Prior to Einstein, Lorentz and Poincaré had done significant work on the study of space and time, which could potentially lead to the birth of relativity theory with mere one small step. They both failed on this final step due to their mindsets. Their main difference with Einstein is that they did not believe that the relativistic effect is revolutionary. They did not think that the concepts of time and space should be overhauled. Einstein's genius lays on his reinterpretation of the notion of time, a contribution that is novel to physics at the very fundamental level. Euclidean geometry is three dimensional and is considered as based on simple principles. In relativity theory, however, time is a fourth dimension on the same footing as the three spatial ones. The basic postulates of Euclid, in the view of some great mathematicians of that time which has impact on Einstein, are no longer adequate. This is one of the most powerful examples that demonstrates the importance of change-of-mindset and that is why Einstein receives and deserves the credit for revolutionizing physics and, ultimately, science as a whole.

How does a community perpetuate particular ways of carrying on from its past achievements? This is mostly accomplished through the formation of paradigm. In microelectronics, successful practices of designing electronic devices are passed among engineers from generations to generations. Effective theories and methodologies are taught in engineering schools in a daily basis. Classical examples are illustrated in the main text of science and engineering textbooks while typical cases are presented as the problems at the ends of book chapters. What is the most important thing that students, junior scientists and engineers as well, have learned from this line of training? Commitment to paradigm: follow the rules and standards honored by all the practitioners of a specialty. In the case of fostering sense-of-time inside electronic world for establishing order from chaos (i.e., the subject-of-study of this book), the most widely shared commitment is the belief that time must flow uniformly and hence every pulse in the clock pulse train has to be identical to very other pulse. This is however, as argued throughout the book, an old-school mindset.

A characteristic shared by all creative individuals is their willingness to surmount obstacles and their persistence in the pursuit. Any creative endeavor will undoubtedly face obstacles because such endeavors threaten the established or entrenched interest, or status quo. The most shocking example is the fact that one of the most revolutionary theories in our civilization, the theory of relativity, does not earn Einstein the Noble prize (He got the Noble prize in physics for his discovery of the law of the photoelectric effect in 1921, sadly not because of the theory of relativity). Maxwell's equations are as revolutionary as Einstein's, and they were resisted fiercely in their debut. The proposition of other new theories and concepts regularly evokes the same response from some of the specialists whose area of special competence they impinge on. For those people, the new theory or concept implies a change in the rules governing the prior practice of normal science or normal engineering. Inevitably, therefore, it conflicts with scientific and engineering work they have already successfully completed. That is why a new thing, being it a theory or a concept or a method, however special its range of application is, is seldom just an increment to what is already known. Its assimilation requires the reconstruction of prior theory or concept, the reevaluation of prior fact, and the modification or perhaps even the abandonment of known methods. This is an intrinsically revolutionary process that is, as history has taught us many times, never accomplished without serious fights.

As happened regularly in history, disruptive new concepts usually meet with resistance by the establishment. How are the creative individuals able to convert the entire profession or the relevant professional subgroup to their ways of seeing the world? What must they do to lead others towards the uncharted territory? This change-of-mindset requires the leadership from visionaries. It demands courage. Ultimately, it needs a strong faith. Debate on paradigm is not really about relative problem-solving ability, although for good reason it is often portrayed in this term. Instead, the real issue is which paradigm should in future guide the research and development effort on those problems that many of which neither competitor can yet claim to resolve completely. Facing this difficult situation of choosing the more appropriate paradigm for future (i.e., the paradigm shift), the decision between the alternative ways of practicing science or engineering must be based less on past achievement than on future promise. This is the attitude that we shall take when facing the serious challenge of changing paradigm. If a new candidate for paradigm had to be judged from the start by those hard-headed people examining only the relative problem-solving ability, our science and technology would experience very few major revolutions. People who embrace a new paradigm at an early stage must often do so in defiance of the evidence provided by problem-solving. They must have faith that the new paradigm will succeed in solving the many new and difficult problems that confront it. A decision of that kind can only be made on faith. This is what is needed the most in the transition from space-dominant paradigm to time-oriented paradigm.

To respond to the current crisis that we are facing, to challenge the old way of conducting business, to carry through the shift from old paradigm to new one, we must think out of the box and be prepared for change-of-mindset. This point is so vital that it has to be emphasized persistently, as being done throughout this entire book.

Reference

T. S. Kuhn, "The Structure of Scientific Revolution," the *Univ. of Chicago Press*, 2012.

Printed by Books on Demand, Germany